LDV
VISION

HOW VISUAL TECHNOLOGIES
ARE REVOLUTIONIZING BUSINESS & HUMANITY

BASED ON TALKS BY
ANDY WEISSMAN, JOANNE WILSON, LANE BECKER,
EMILY GRAY, MICHAEL COHEN, EVAN NISSELSON AND
65 OTHER EXPERTS IN VISUAL TECHNOLOGY

LDV VISION SUMMIT

LDV VISION 2015
How Visual Technologies Are
Revolutionizing Business & Humanity

ISBN 978-1-61961-370-6

All videos of these sessions can be watched at http://video.ldv.co

THE LDV VISION SUMMIT 2015 PRESENTED BY
Evan Nisselson, LDV Capital

IN COLLABORATION WITH
Rebecca Paoletti, CEO, CakeWorks
Serge Belongie, Professor, Cornell Tech

SPONSORS
AWS activate, Dropbox, JWPlayer, Olapic, Orrick,
Qualcomm, Sprout by HP, Vidcaster

MEDIA PARTNERS
Kaptur, VizWorld

GRAPHIC RECORDERS
Dean Meyers, John Bloch, Anne Gibbons,
Jill Greenbaum, Becca Wilson

CONTENTS

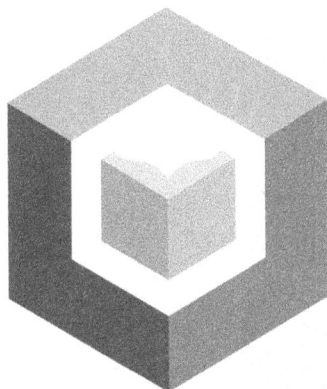

THE STORY BEHIND THIS
BOOK IS PRETTY AMAZING...

Two entrepreneurs at the LDV Entrepreneur Dinner met and decided to help each other. The result was a completely new type of publishing company: **Book in a Box, the publisher for this book.**

You can read the whole story at **http://linkd.in/1bxZnr1**

To find out more about the LDV Dinners, visit **www.ldv.co/community**

To find out more about Book in a Box, visit **bookinabox.com**

FOREWORD

Serge Belongie, Professor, Cornell Tech

THE FUTURE OF COMPUTER VISION LIES IN THE HANDS OF THE CREATORS— of the people who take this incredible technology and bring things to the next level, dream up new questions to ask, and work on new problems to solve.

Computer vision can be broken down into what's known as the "four Rs": recognition, reconstruction, registration, and reorganization. Those are age-old themes around which one can organize research and product development. In reconstruction, we produce 3D models from multiple photographs. In two of the other themes, registration and reorganization, we track objects in video and apply unsupervised learning methods to discover structure in visual data, respectively. In the last few years, however, the biggest source of excitement within computer vision is all about recognition. More than ever, recognition is looming large and this is propelled by the incredible popularity of a technique called deep learning. And that is echoed throughout the keynotes and panel discussions at this year's LDV Vision Summit. Who knows—maybe by next year's Summit, the recognition market will crash and everyone will be doing 3D reconstruction. That's what happened when Kinect came out. Microsoft Kinect created a lot of activity in 3D reconstruction. But right now recognition and deep learning have really taken over the world. That's one of the themes that was present at the Summit.

At this year's Summit, we often explored the idea of whether or not computer vision can be seen as a commodity, and the extent to which its commoditization is generating the startup excitement. And though our panelists were split down the middle on the question, this year's speaker lineup attests to the fact that computer vision is enough of a commodity to be the centerpiece of a lot of exciting technology that is coming out right now. It is at the heart of several computer vision startups and larger companies. There's Orbital Insight, which is using a network of satellites to generate a huge amount of image data to solve all kinds of problems. There is so much you can learn about the world when you tap into this resource and there are things that are apparent just by looking at satellite imagery of parking lots, of oil reserves, harbors where there are shipping containers coming in. When a major auto manufacturer ships a bunch of cars and the economy is not doing well, these lots of cars just fill up. That information is sitting there in the images if someone is capable of analyzing it. When you want to make decisions about agriculture, investing, construction progress, traffic, commerce, etc., the answers are often sitting there in plain sight. And we can rely on computer vision to give us those answers. Orbital Insight is a small startup with a lot of excitement around it. We don't know if it will succeed or fail, but it's exciting to see what it can do. Arguably, its success will hinge on whether computer vision is a commodity, and if it is, how effective it can be.

There's also the company Placemeter, which is very similar to what I've described for Orbital Insight but at street level and with a focus on retail. In the past, when a corporation wanted to know how business was in their stores, how happy their customers were, and how nice the stores looked, they would hire a professional shopper or spy to just walk around retail areas and get a sense of what is going on. Placemeter is doing that with computer vision. You can put a camera outside the store, inside the store, or next to specific promotional racks and instantly get a sense of where a person's attention goes when they walk into a store. What they're looking at, what they're drawn to, how they make their way through the store can all be determined using computer vision.

Another example is from a company called Emotient, which deals with making machines aware of emotions. Imagine a set top box with your TV or computer that can detect and analyze the facial expression you're making while watching a movie or an advertisement. 10 years ago that problem of detecting the face, classifying the expression, or simply collecting the data to be able to study it would consume all of a researcher's energy. The idea of treating emotion recognition as a commodity and thinking about what they call "the power and the promise of emotion aware machines" brings it to a completely different level. You could build a company on that and say, "We're not worried about how we're going to find the faces and classify the expressions; we're going to assume that works, and then build an ecosystem around it." Imagine how this could change the way advertisers craft commercials or how movies get scripted, if they could read our facial expressions and react to our facial cues in real time. What Emotient is doing is part of a bigger theme of crafting user experiences around how people feel.

A family affair! Serge is introducing the next Keynote speaker with his son August Flemming Belongie.

This question of whether computer vision is a commodity is on a lot of people's minds. We're not going to give you the answer, but enough people have decided that it is indeed a commodity that they are charging forward, and these are some of the ways that they are charging forward. In each case, it's some kind of business idea that may or may not work. But at its heart is a kernel of computer vision technology that provides the company an excellent chance at success.

FOREWORD

Rebecca Paoletti, CEO, CakeWorks

FOR YEARS, WE'VE BEEN SAYING, "THIS IS THE YEAR FOR VIDEO." IN FACT, I think many of us have been calling it "the year of video" going back to 1997. Yet every year, there's some technological glitch that keeps that dream from being realized. Even after we sorted the rights issues and the perception of cannibalization of television, there is always some aspect of technology that must catch up. First, we were waiting for broadband connections to actually be deployed and get faster. Then, we needed wifi to speed up and be readily available so we could actually watch video on our phones. Then, of course, publishers and creators had to commit new resources to understanding the tech to deliver content to wireless devices. iPads and phones are bigger, TVs are smarter, and streaming is lightning fast. Everyone can watch video wherever and whenever they want. Now, we are caught up.

With massive adoption by viewers, we are now at a time when moving pictures and digital video have completely transformed the way we consume content, facts, entertainment, and information. And I think we can safely say that not only have we hit the inflection point, but we have officially arrived at the era of video. Now, a whole new generation will grow up as true digital video natives, conducting all their communication, search, and Internet consumption in an immersive video environment. Powered by technology and audience expectations, the entire industry is changing, too—challenging advertisers how to monetize, creators how to innovate with more content, and consumers how to seamlessly integrate it into their daily lives.

At CakeWorks, we've kept a close watch on the surge of Instagram and Snapchat. Instagram was a big acquisition for Facebook and is a testament to how strategic mobile was to Facebook at that time. Video on these platforms is changing the game. 5 million videos were created and uploaded to Instagram the first day they made video capabilities available. Facebook has surpassed 4 billion videos. Snapchat has also taken off, becoming the playground for a younger generation that is influencing how all of us communicate and share. And for the audience of these video platforms, they're a whole new generation of consumers: kids with no media training and no media filter. They're comfortable in front of a camera, comfortable talking to a camera. They play games with their cameras on and talk to each other "face to face" in real time over the phone. This new way of talking with videos has really transformed the industry. Apple certainly did it with FaceTime, but they weren't first and they won't be last.

In just the first six months of 2015, we've seen an abundance of major media companies who are now ready to produce and upload content and make it available to a wide audience, going directly to their consumers. It's taking advertising off of the television and it's forcing companies to rethink their entire strategies and to reinvent the rules of the game. The technology has certainly made progress, but even more important is this new willingness to work in these environments. The opening of ad and subscription models, the availability of shorter clicks on phones, the new willingness to put content into YouTube and Facebook environments for the first time is changing how business is done. More than anything, the high rate of consumer adoptions is making this possible.

Watching programming on the phone has transformed everything. This was the year of live streaming video: Meerkat, Periscope, and Facebook Live all launched with their own content strategy and business model. There have been these big outputs in live video streaming and a lot of different ways that people are looking at it and monetizing it and promoting events around it and then creating personalities and content and lessons and business models.

The technology behind live video streaming is complex and evolved. Hundreds of millions of investment dollars toward platform solutions have proven impactful to programmers and consumers alike. Now, hosting a live stream (or viewing one) doesn't have to be extraordinarily complicated or expensive. There is also very little risk involved in live programming, around audiences or platforms. It used to be that when you tried to do a live event, chances are, at some point, it would crash. The frame would stutter, an ad would not be served, the user wouldn't be subscribed or fully engaged in the experience. Now, live streaming on your phone is a given. You can Tweet or post or share that stream to aggregate viewers fast. This is a huge advancement.

Looking ahead to these changes with video, we have to also look at how engagement metrics around video have become key indicators for the industry as a whole. A metric of pre-roll as a value driver and the currency for video no longer makes sense in a world where everybody's looking at longer form. Subscription-based business models are the trend—look at Netflix and Hulu Plus for proof.

Erika Trautman's keynote showcasing the growth and impact of Rapt Media gives insight into how we can begin to monetize and measure value in this new video environment. Trying to value consumer engagement around video experiences is becoming much more of a trend for advertisers and agencies. Offering consumers a video experience that's controlled by them, that's not interrupted by pre-roll or other forms of advertising while still maintaining value from an advertiser, is key to success.

The industry is becoming more creative. We can't all pay for these dynamic, awesome content creations without advertising budgets. Advertisers and agencies need to participate in that value exchange. "Ad supported" can mean a lot of things, but the better the creative, the more authentically embedded into the viewers' program, the more engaging it will be for us all. No one wants their favorite digital show to be interrupted by an "old-school" pre-roll.

The proliferation of video has social and cultural implications as well. It was a big deal for Hillary Clinton to announce her candidacy via a video. Instead of a live speech, she released a very carefully crafted and thoughtful video piece. There was a launch strategy around that video: who saw it first, where it was posted, and how it was treated. For the first time, brands, companies, and politicians really do have the power to control their own message. Clinton has obviously been through this before and knows the power of television, the power of the media and showing up places to shake hands, but she still thought that the best way to announce her candidacy, especially in a millennial-saturated environment, was to do it digitally with video. This speaks volumes.

I believe we're going to see more of that. Now when anything happens—from news to entertainment—there's an immediate availability of video clips. This is now a massive force in entertainment, news, and media and is changing how messaging gets distributed. Now we see brands shortening their message to get viral clips out, leveraging the power of social networks. The presidential campaigns this fall will continue to be a great example of this. Video simply helps everyone tell their story: brands, media personalities, politicians, you, me.

The world is changing. How we present ourselves to the world is now in our hands—just look at the proliferation of the selfie. There is a kind of visual bombardment that's happening where we are now saturated with images of things to acquire or a way of life to aspire to. It will be interesting to see how this shakes out in years to come.

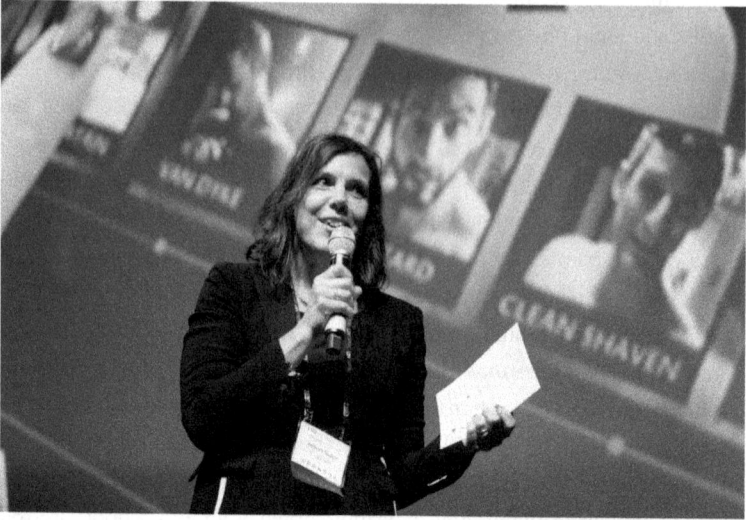

Rebecca Paoletti, CEO, CakeWorks

One of the things that I love most about video, which can get lost in all the details of the technology, is the sheer power of images. Images help us to tell amazing stories. If you look at GoPro, it's not just about a camera that's portable; the value is in awe-inspiring pictures of snowboarders on these massive cliffs or surfers blasting through huge waves. The images bridge the gap between your world and theirs. It opens up corners of the world that you might not get to see otherwise. Images help foster relationships and connections with people that you wouldn't normally get to see or meet. As the image and video revolution continues to move forward, it will be amazing to see how it can continue to make the world a little smaller, and how future generations will take the technology to the next level.

If this is the age of video, then it's only the beginning.

INTRODUCTION

Evan Nisselson, LDV Capital

VISUAL TECHNOLOGIES ARE REVOLUTIONIZING HUMANITY AND BUSINESS around the world like never before. People spend an unbelievable majority of their time every single day with technologies that help them make sense of what's going on around them, capture the moments that matter to them, and figure out how they get from point A to point B. What's unique now in the world of technology and the Internet is the evolution of a higher bandwidth, Cloud computing, real time data, and mobile devices that are allowing people to capture exponentially more visual content. Whether it be images or video—this is drastically changing how people communicate with each other. Instead of going to the dentist and texting someone that you just went to the dentist, you send a picture of yourself at the dentist. While on vacation, you don't just write an email telling your friends about your vacation. You post an image of the beach on Facebook or post what you ate that night on Instagram. You capture video of a beautiful event, like a wedding or a birthday, to preserve the moment so you can relive it later or share it with others.

Now in its second year, the annual LDV Vision Summit is all about gathering the key people in our ecosystem at one event to hear their wisdom and advice on how humanity and business will be impacted by new technologies and companies in the visual imaging world. We gather the five different categories of people—from startups, investors, computer vision and AI experts, media and brand execs, and creators—to come together to network, do deals, explore and inspire, and find co-founders to build businesses. It's all about empowering the people building the visual revolution.

The ideas and lessons from the LDV Vision Summit live on long after the event concludes via different mediums as we know everyone has different content viewing tastes. We capture videos from the event which are available online and we are also very happy to be putting together this second LDV Vision book.

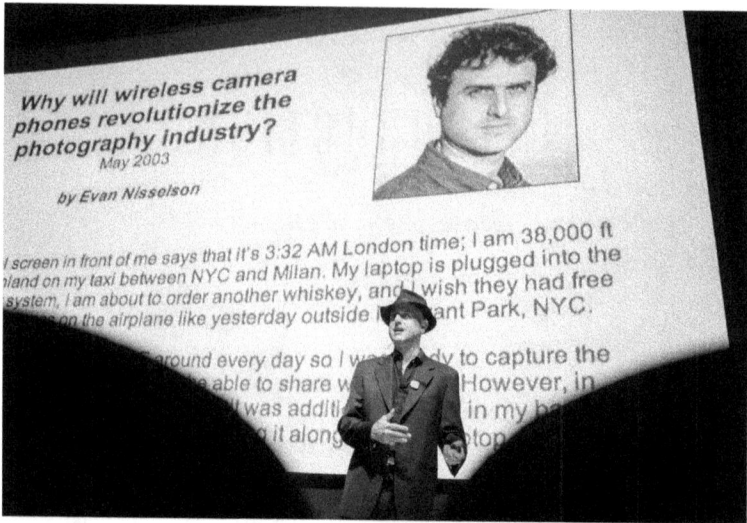

Evan Nisselson, LDV Capital

What is visual technology? Visual technology is any technology that captures, organizes, filters, learns from, or distributes visual content either for consumers or businesses. It's a horizontal focus across all business sectors and humanity. How will these technologies impact everybody's lives, in work and in play? If you look across different industries, these examples cover a wide range of sectors that visual technology impacts, from personal to business, and sub-sectors of medical, financial, sentiment analysis in advertising, in publishing, in shopping, autonomous cars, and so on. There are so many examples of visual technology affecting our world today, you may not even realize it. Facebook couldn't exist without images and video. There's been a lot of discussion recently about self-driving cars. Self-driving cars would not exist without many cameras in and around the car telling us where the pedestrians are, where other cars are, where the curb is, and how fast the car should be going. That's an unbelievable evolution in leveraging visual content, not only to communicate visually but to help us be safer in day-to-day activities. The mission of the LDV Vision Summit is to bring together—in a unique way—the men and women who work in this world and who together represent the whole ecosystem of visual imaging, but who don't typically sit in the same room.

One of the things that's extremely rewarding and unique about our Summit is that there's no other event that brings together those five categories of people to one place. It's challenging because they all have different needs, but what's unique is the serendipity of everybody coming to one place. Over the last two events we've had, there have been at least 10 to 15 companies that have raised significant funding after being highlighted at our Summit. There are several people that have been recruited by technology companies after meeting at the Summit. There are several technology service deals with media companies that have been inked because of the Summit. Some people have joined together to build new companies and also a multitude of investment opportunities. We've invested in a couple from the Summit, and we look forward to investing in more. That core of serendipity of opportunity is the foundation for why we put this together, in addition to inspirational and educational benefits.

I've been building businesses and visual technologies since the mid '90s, and prior to that I was a professional photographer and a photo agent. What's fascinating is that in the '90s, because of the bandwidth and other bottlenecks, it was harder. There were no phones with high-quality cameras. No one was walking around making photos with camera phones or iPads. It was a very slow evolution. It continued to be slow until the mid 2000s, and I think in the last five years the pace started to move faster, mainly because of the high growth of adoption of smartphones with cameras and smaller devices using hands-free cameras. Things have exploded. Take GoPro, for example.

Everybody who is doing extreme sports is using a GoPro. However, with this exponential growth of content, it's actually making it harder and harder to decipher and discover the signal through the noise. I'd say 99.9% of the content that's created and shared is not high-signal or high-quality content. One significant challenge is to figure out the single video or single video frame that's contextually relevant to me and at the contextually right time, so that I can enjoy watching it on any device. In the last five years, the industry has grown exponentially following the evolution and growth of mobile devices. I think we're just at the beginning of that growth. We cannot even imagine how many more ways there will be for us to capture, create, share, and visually communicate with content.

We talk about exponential growth of capturing and of creating but the real core that is going to make this exponential growth of visual content valuable is machine intelligence, computer vision, deep learning, and artificial intelligence, which can now contextually filter what's relevant for us at the right time. As we create more and more visual content, we're becoming overwhelmed with everything available to us. How can we make sense of it all? Because of Cloud computing and advancements in machine intelligence, we are seeing tremendous progress and opportunity. Facebook, Twitter, Google, Apple, Dropbox, Qualcomm, Intel, Microsoft, Shutterfly, and other major players are building up huge departments for machine learning by acquiring companies and teams in that space. They realize the need for filtering the signal through the noise.

All technology impacts lives for good and for bad. It's all a matter of perspective. 20 years ago I was working at an early-stage startup called @Home Network, and I was working around the clock in front of my computer buying all these devices and saw very early how it was going to impact people's lives. When a couple of us created the first broadband photo community in 1997 at @Home Network called Making Pictures as a $3 million joint venture with Intel, everybody thought I was crazy. Everybody said, "I don't want to make digital photos. That's a horrible thing. I want negatives, and I want to go to the darkroom." The point of whether or not a technology is good or bad evolves with our evolution of technology and humanity. I'd rather spend more time on creativity and helping entrepreneurs build successful businesses than doing a lot of the menial stuff that doesn't intrigue me, entertain me, or is in my mind a waste of time. One person's favorite service is another person's bane of their existence.

A big theme of this year's Summit was satellite imaging, which is something that we'll talk about a lot here. More and more satellites are being put in the air and more and more have high-definition cameras. Not only will this impact businesses and lead to the creation of new billion-dollar businesses, but satellite imaging is going to be able to move financial markets as well. Now we can actually track how many cars are in a parking lot at Walmart or Home Depot and compare the two in real time rather than months later after manual research. We can identify the abundance or lack of oil reserves from the sky. We can monitor traffic and track trends for busy roadways. We can track how popular or how active shipping ports are. We can count how many containers are on a ship, how many cars are in the port, and how fast they're moving in and out of the shipping ports.

To take it to the next level: Sure, we have camera phones and we have flying cameras attached to drones—but when will we have satellite selfies? Let's back up a second. You can take a selfie with your camera phone. You can take a selfie with a selfie-stick. You can take a drone selfie, which is called a "delfie." There's a "jelfie," which is a jumping selfie. Soon there's going to be a satellite selfie, where you'll be able to touch your smartphone, look up into the sky and a satellite will make a picture for you. Obviously there are technical hurdles that we have to get through, but it will happen in our lifetime. You'll be able to tap your smartphone, grab your friend or significant other, look up, and the satellite will take a picture of you. Maybe soon you won't even have to carry your camera phone around or take any pictures because you can say, "I want this picture" and in two seconds it will be captured by a security camera, a drone, or a satellite camera.

Obviously, there are bandwidth issues from the satellite. Obviously, there are delays from the satellite, and the quality is not yet good enough to see the freckles on your face—yet. However, the military are actually using satellite cameras that can read a paper in your hand. This technology is happening, but there are still a lot of questions and I don't have all the answers. That's why we gather the experts at the annual Vision Summit and in this book, so we can help inspire others about how visual technologies are impacting our lives for better or for worse.

Machine intelligence has great impacts within the medical world, too. Why is it that we go to the doctor for an x-ray and the radiologist only looks at three x-rays from that one visit? Imagine if they could look at the x-rays from our whole life and see trends? Even better, why don't they look at millions of anonymous x-rays to better understand whether what we have is something serious? What about skin cancer? There are several companies that are starting to create cameras that can photograph moles or different pieces of your skin so that you actually can send that to a doctor in advance of having a doctor's appointment. What about mammography? We can look at hundreds of mammogram photos to understand trends in these medical scans, helping radiologists be exponentially more efficient at seeing potentially problematic patterns. That can't be done unless we're leveraging computer vision and machine intelligence.

This potential for analysis of visual content—contextualizing, organizing, using it to predict and identify trends—is a maturation of visual technology that allows us to transition these tools from industrial applications to mass consumer use. This year, we saw tools for advertisers to analyze consumer content via facial expressions through sentiment analysis thanks to companies like Emotient and Affectiva. We learned about face recognition doorbells, exciting developments in 3D printing technology—like a custom-fitted insole—and Microsoft's hologram headsets. There's the interactive video experience from WIREWAX and Sphericam's 360 degree video camera that's taking adventure photography to the next level.

The purpose of this book is to capture the most important ideas from the Summit so they are not lost once the event is over, and to explore the latest thinking in everything from computer vision to AI to deep learning to augmented reality. During the 2015 LDV Vision Summit, we assembled 80 speakers from many countries around the world to take part in this two-day meeting of the minds in New York City. This written document of the sessions, presentations, and conversations that took place there is our attempt to preserve the knowledge and wisdom from those we brought together so that others can learn, innovate, and create. The book is broken up into themed sections, and what's really fascinating and unique about the group of people assembled here is the wide range of visual technologies that are represented across such a broad spectrum. I think that's the most exciting, most challenging, and will deliver exponential returns for everyone involved and who reads this book—hopefully for happiness, health, and financial benefit.

I have always been fascinated by and passionate about visual computing and visual communications, since getting my first Nikon FM camera at 13 years old. I've been a professional photographer, 18 years of building businesses within this visual technologies sector, and an investor for more than three years at LDV Capital. What excites me even more is that I don't have all the answers. I never have all the answers. I love surrounding myself with others who might have the answers and collaborating with them to figure things out, to be inspired, to learn.

That's the core of why several of us have worked together to bring the LDV Vision Summit to life, so that it can empower the world to learn, inspire, and create. Hopefully each and every one of you will have your own benefits, whether or not you're an expert, whether or not you love photography or video, whether or not you're thinking about being an entrepreneur or you're a computer vision expert and you want to figure out how to build a business rather than studying how to write grants for the rest of your life. There's a piece of the puzzle for everybody here, and I hope you enjoy the wisdom of all of our speakers in this book.

1

HOW VISUAL TECHNOLOGIES ARE REVOLUTIONIZING BUSINESS & HUMANITY

Evan Nisselson, LDV Capital

I'VE BEEN A PHOTOGRAPHER FOR 24 YEARS, AN ENTREPRENEUR IN SIL-icon Valley, New York, and Europe for 18 years. Along the way, there were successes and there were failures. We built the first broadband photo community in 1997 before people thought they were going to shoot digital, share digital, or make digital pictures. That was a joint venture with Intel. I wrote an article in 2003 that said camera phones were going to replace point-and-shoot cameras. A little bleeding edge, but kinda right. I'm sure a lot of the smart people here had visions of what it was going to be like and that's what excites me.

We built a platform for professional photographers and photo agencies to help them manage, market, and sell their images in 2004—another idea that people thought was crazy. But I didn't do it alone. Andy Parsons is here in the audience, our CTO, first employee, and a fantastic collaborator.

After 18 years with the successes and failures, I wanted to help other entrepreneurs avoid my mistakes and hopefully help them succeed with things that I've done well. I've mentored five accelerators and recently started LDV Capital to invest in European companies coming to the US and US companies focusing in two verticals: imaging and video tech, obviously, and B2B SaaS. What I love is when they collide, when it's a business that's a B2B SaaS that deals with visual technologies.

One of the most important things I get told all the time by entrepreneurs is, "I want to do this, I want to that, I want to do this, I want to do this." And I always ask, "What's your most important thing? What's your goal?" You can't succeed, you can't reach your goal, in my mind, unless you know your goal.

My goals are to solve problems, be creative, have fun, and make money while I sleep. If I'm not doing all of those together, I'm not happy.

Today's the deep dive day in tech and I want to start off with reiterating. So obviously a lot of these photographs are mine and they're looking pretty blown out up there but they're not. They are photographed with 3200 speed black and white film pushed to 6400 speed film. Ah, the days of film. You guys remember that? I can hear chuckles.

Visual content is exploding, it's obvious. But what's interesting is it's not just personal content. It's satellite imaging, medical imaging, mapping imaging, advertising, commerce, and more. Every one of those verticals has opportunities. Here are some stats that are unbelievable, but also just stats:

- 3.8 trillion images were photographed from 1838 to 2011
- 1 trillion were photographed this year already
- 500 million images were shared in 2013
- 1.8 billion were shared in 2014

Imagine what's going to happen this year. Everybody says, "Oh, enough of these selfies on Instagram." But selfies actually make up only 4% of the total pie. There's a lot of content out there. 5 million videos were uploaded the first day of Instagram's video platform. 65% of consumers seek out user-generated content before making a purchase. Not the horrible shots that the corporations are doing; we want to know how real people are using them. 24% of brands are using online video to market to consumers. 84% of retailers believe more images per product will increase conversions. This is fascinating.

What Is Your Goal?

©Evan Nisselson

So why do people choose which hotel? Is it the price? You'd think it's the price. Or it could be the location. It's actually the images. 63% trust consumer photos more than brands and 300 hours of video are uploaded every day to YouTube. It's probably more today and this is probably a couple of months old but that's still a perspective. And all these stats are from their sources, they're not my research. In emails, we see a 200% - 300% click-through rate for video. We're going to hear tomorrow from Vivette from Movable Ink. This is unbelievable data. Images and video are valuable.

Take Instagram user misshattan, for example. 280,000 followers. She's going to speak tomorrow.

YouTube user PewDiePie has 37 million followers.

We used to make a picture or video and share once or twice. No longer.

Think about Facebook, YouTube, Instagram, SnapChat without pictures and without video. They would not exist. We're sharing exponentially and while statistics are valuable, these are only a small part of this story. They're the surface layer. Everybody talks about statistics but we haven't gotten the real statistics yet.

Mary Meeker's annual report on the Internet and last year one of the fascinating validations for the opportunities to come, not only to build technology but to make money, is that only 7% of content was tagged last year. 1% is analyzed. We can now understand complex signals from visual content. But why now?

Many of you know this but to reiterate, for those that are getting their feet wet in computer vision and Artificial Intelligence and the opportunities of visual technology, it's all about real time access, computer vision, Artificial Intelligence, and a Cloud infrastructure that is exponentially scalable and much more affordable. Humans can't review millions of images. But computers can review billions and trillions very quickly. The focus is contextual analysis in real time. We soon will be able to know whether people are happy or sad right now in Times Square. Never could this be done before. Just by analyzing their expressions, either via satellites or by crowd-sourced images, now we can know.

Did More People Walk Into Retail Stores Today In LA or NYC What % Walked Out With Shopping Bags

Here's another perspective. How many people walked into retail stores today in LA or New York City? Which one was more? And how many walked out with shopping bags? Talk about a signal for the economy. Why do you have to wait six months to hear about research on what happened six months ago? It's ridiculous. I want real time data from analyzing images and video. Or how about: Are people healthier from their emotions? Because we can map the emotions via analyzing the faces but we can also map it to some fantastic medical technologies that we're going to hear more about tomorrow.

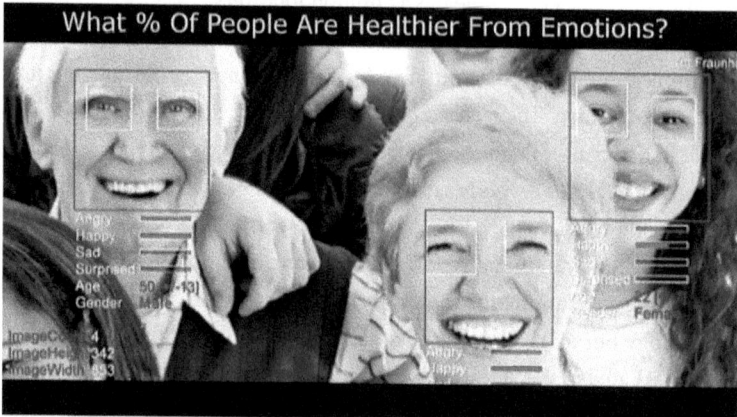

What % Of People Are Healthier From Emotions?

It's about understanding that content. Look at this photo:

©Google

It's a horse, yes. It's a person on a horse, yes. The person on the horse is smiling with a helmet. Many of you are smarter than I am and are going to help us figure out the next 10 layers of this person: that this person woke up late with a hangover this morning after an argument with their parents the night before and that they just had their first kid. That's not in this image but if you look at the trends of the images, you would be able to know that.

Images with blue as a dominant color receive 24% more likes. Images with low saturation generate 18% more likes. We're understanding data in a whole new way. Chobani was able to determine that consumers were putting their yogurt containers in the cup holders of their cars and that was their breakfast. Just by knowing that, they could change their whole marketing campaign.

Recent Computer Vision & AI Acquisitions

2015 Linx > Apple	2014 Madbits > Twitter
2014 Aviary > Adobe	2014 VisualGraph > Pinterest
2014 Jetpac > Google	2013 Lookflow > Yahoo
2014 Sigh.io > EyeEm	2013 IQ Engines > Yahoo
2014 Skybox > Google	2013 DNNresearch > Google
2014 Oculus VR > Facebook	2013 SkyPhrase > Yahoo
2014 KBVT > Dropbox	2013 Anchovi Labs > Dropbox
2014 Vision Factory > Google	2012 Photoccino > Shutterfly
2014 Dark Blue Labs > Google	2012 Face.com > Facebook
2014 DeepMind > Google	2012 Like.com > Google

Here's another list of unbelievable trends that reiterate the opportunity big companies—Apple, Adobe, Google, Dropbox, and others—have to acquire brilliant computer vision AI companies. They're startups, they're viable businesses, and they're brilliant people because they see the value. The ones in green are the ones speaking here today. That's our goal: getting the thought leaders who have experienced the entrepreneurial ecosystem to help the rest of us succeed.

Today we're going to explore many ways to understand visual data. It's just the beginning. There are unbelievable opportunities. But they don't have any value unless you enjoy every single day of your life. And I hope that you enjoy today and meet fantastic people and find ways to help each other. Like I said earlier, I don't have all the answers. I know you're smarter than me and I can't wait to hear all of your wisdom and collaborate with you on how to improve the world we live in.

Thank you.

2

A MOMENT IN SPACE: REIMAGINING PHOTOGRAPHY

Radu Rusu, CEO & Co-Founder, Fyusion

FOR AS LONG AS WE CAN REMEMBER, THERE HAVE BEEN TWO WAYS TO capture memories: photos and videos. One (photography) captures a slice of time and space, and the other (video) captures time. What makes these formats so special? Is there room for one more?

We're going to go through an hour of equations and a lot of mathematical formulations of all these things here. I'm kidding. I usually give talks like this at scientific conferences so CVPR and ICCVs and things like that. This is a different venue, so I'll keep it short and I'll see what we can do. I'm with this company called Fyusion, where we've started looking at the mobile imaging space some years ago. I'm not really sure which ones of you have heard of a company called Willow Garage. Okay, there's a few. Excellent.

A lot of the scientists that are involved at Fyusion actually came from Willow Garage, where we were responsible for developing a lot of tools in the 2D/3D image processing space. For those of you who haven't heard of Willow Garage, this was a very interesting company in Menlo Park in California that develops advanced robotics. Long story short: Scott Hassan, one of the early Google people, founded the company with the hope to build open source software and open architectures for robotics. A few years in we realized that actually building Commander Data from Star Trek might take a little bit longer than initially anticipated, so we started spinning off companies. I think in about two and a half years we spun off nine companies.

Anyway, one of the things that we've done there is this initial difficult point Cloud library, which is a 3D image processing software toolkit. In time, this actually grew quite large. It's one of the world's largest open source ecosystems right now in 3D image processing, so I definitely encourage you to use it. The reason why I guess we started Fyusion is because we got a little bit frustrated of just working on building blocks and research algorithms and never really quite taking one thing all the way. Fyusion was created with a purpose to take all that know-how, all that knowledge that we actually had, narrow it down to one particular problem, and just build something vertically and try to go directly to market and see what happens. Really foolish.

Anyway, the team is about 25 people and I guess there are some statistics here which we've discovered which doesn't really matter so much, but this is where we are right now. As I mentioned before, we come from this space of the intersection of robotics, machine learning, computer vision, 3D image processing, and so on. Here's a few of the projects that we were involved with, from robots bringing your beer to your office if you're lazy and don't want to get up from your chair, to folding towels and laundry, to autonomous car driving, and a bunch of other things. These are things that we tackled in the past five plus years.

At Fyusion, we want to take the opportunity to analyze what's working out there in terms of consumers. One thing I want to mention is that Fyusion is a B2C company, at least for the time being. We've been talking about 3D processing for a long time. It was kind of hard for us to understand why consumers don't actually enjoy 3D imaging as much as we, the scientists, do. We started by analyzing what do they really enjoy and obviously you know this answer already. Something like 99% of the visual data that's shared online falls into two representations: photography and video. One of them is a slice of time and space and the other one captures time. We started asking, "Why are these two representations so powerful? What makes them so ubiquitous on the Internet?"

We couldn't find an answer. The only answer that we found was, well, they were kind of there forever. We've had them since birth. When I was born my mom took a 2D picture of me and I didn't question that, right? "Mom, why, why 2D?" Right? In fact, it's even worse. They pre-date the Internet. They pre-date computers. A hundred years ago someone captured life through a lens and made a film camera. And then when computers came, we digitized the experience. Nothing's really changed since, right, in terms of a capture process? Yes, there's progress in certain small verticals there. You have companies like Lytro trying to reimagine the capture process, but overall the industry hasn't made huge progress.

Same thing with video, right?

We wanted to have motion pictures, Hollywood made them happen, and now we have them as MPEGs on our computers. That's the first fact. The second one is that we already know mobile usage has surpassed PC usage. We have this amazing ability, as Evan said, to just capture data continuously. Obviously he already showed this, this is Mary Meeker from Kleiner Perkins predicting the amount of digital data that we're creating and sharing on the Internet. Basically what this graph shows is that within two years we're doubling the amount of data on the Internet. We're pretty much saying that nothing we've done before matters. It's a new Internet every two years or something.

One of the things that we thought was missing—and again, this is us coming from this 3D imaging perspective—is the notion of space.

You've probably seen this movie. I should actually give credit. This is a snippet from one of the latest *Star Trek* movies and you can see here a young Captain Kirk navigating in space on a tablet. It's all very futuristic, of course, and as computer vision people might think right now that's

quite impossible. How can you do that from a single viewpoint? Well, you probably can't. But the whole idea in itself is quite interesting, right? We started asking ourselves: What is this? Is it a 3D picture? What is it exactly? It's SPACE capture.

We started charting the two dimensions, time and space, and just started looking at the existing file formats. You've got photography in the middle as a slice of time and space. You've got video capturing time. You've got animated GIFs capturing time. Panoramas capture space. We wanted to be over there. We wanted to create a technology that abstracts time completely, meaning that it shouldn't matter how fast you move in the world. I could move around a car in half an hour or in five seconds and the end result should be the same. It should be a graph of visual data with transitions so I can move in that afterwards.

So one of the other things that we had to deal with is: we were always stuck in that space of point Clouds and registration and ICP and 3D and meshes and octrees and so on, and that's not something consumers want, as it turns out.

In fact, what consumers want is this. Quite a shock.

What we set out to do is create a new visual format—for lack of a better name we call it Fyuse (because really it's a sum of advanced technologies and data sources merged together)—and also an application that's exhibiting the visual format. As I said before, the goal is to move around the world and capture data and let technology do its thing. You create this visual graph and fundamentally you're competing with existing formats, like photography and video and panoramas and so on. We think we're the closest to panoramas as anything else from those visual formats, but the goal is to remove all limitations that panoramas have. As you know, panoramas are just 2D pictures warped on a sphere or a cylinder. They can't deal with motion and you can't create the panorama of an object, right?

Maybe the final thing is, we really want to take advantage of these wonderful sensors that we have on our mobile phones, the tap-sensitive screens and gyroscopes, and give people the ability to navigate in that space. Have them be ACTIVE participants for the first time.

For those of you who are interested in the technology behind the hood, there's a lot of interesting things happening in the space of sensor fusion. There's a lot of structure for motion and 3D reconstruction going on for parts of the world that are static. Obviously, as you know, if you're moving a device in space and the world is dynamic, you can't re-reconstruct in 3D. The goal here, however, is not to create 3D models for the sake of 3D printing or anything like that. The goal is to create them for the sake of actually bringing back to the screen virtually-rendered images that look photo-realistic.

For consumers it's basically like this: you move around in space and then we create this infinite band of space that's really nice and smooth and you're just navigating it afterwards. We actually launched this product in December last year right before New Year's Eve. We really didn't want to create another social network, to be completely honest. We really wanted to take the technology and incorporate it in something else. But as you know, if you're creating any file format, it's quite hard to do that because existing software wouldn't be able to recognize it. We set out to create the product, this application called Fyuse, with some interesting UI resembling other existing solutions on the market and we added some of our own tidbits on galleries and channels and messaging and guides. You'll get to see this more if you actually download the application and play with it.

Then once you actually have this powerful new format with 3D information under the hood, you can do interesting things like visual tagging, which comes for free because it's part of the format.

I can go around and create this physical post that I place in the image space and I can annotate parts and I can send people to websites. Where can I buy that wheel? Well, you just click there—and of course they're sticky, they stay and track the objects that you placed them on. They're part of the image. Without realizing, we've sort of created a revolution right now in fashion and retail and eCommerce, because everybody wants to tag things and point people out to the websites where you can actually purchase these things.

One of the things that we also wanted to do is to be able to take the discrete space that our cameras give us and create an infinitely smooth, interpolated space. That's what you see in the upper part of the image.

We make it all very nice and smooth and we deal with motion quite well. The people who are interested—I can talk to you afterwards about it. What you see in the bottom part is this whole notion of loop closure or SLAM, where you can go around and move around the world and then realize that you've been there before and make the whole thing move continuously. It has no beginning and end.

There are really a lot of things to be said here—unfortunately I'm going to be running out of time soon—about this infinitely smooth interpolation, where this resembles a 10,000-frames-per-second video. However, if you've ever recorded slow motion video or with high frame rates, you realize it's actually quite large in size so you have hundreds of megabytes. Fyuses are like one megabytes on average, so they're really, really cool. And of course there are other consumer problems that we as scientists initially didn't really care about, but they became very interesting for us as we started working on them. Like, how do I actually record a selfie of myself and maybe the complete Golden Gate Bridge in the background? Your arms are too short and your head is too big to fit both in one picture. Selfie sticks won't help you, either, as they are just pushing the camera forward, and now you're just a tiny part of the image. So we created this concept of panorama selfies, which are quite interesting. You see one of my colleagues here goofing off in the bottom part.

To our surprise, this actually became very popular as an application early this year and it went viral in the US and Japan and a lot of countries in Europe, surpassing many of the existing solutions on the market—not in this space, just in photography in general. We were trending up to number four in the overall app category in photo/video and in some countries we went all the way up to number one, surpassing Facebook and WhatsApp and Instagram and Twitter and YouTube and so on. My takeaway message from here is that we feel like the world is quite ready to experience something completely different than what we actually had before. This notion of space pops up for the first time in our opinion, where we've been looking at the world only from the perspective of time capture. We as human beings can't really control time—like, how many seconds have passed since we entered this room? We have no idea unless we look at our watches, right? But space is something that we can actually control quite well.

Thank you so much.

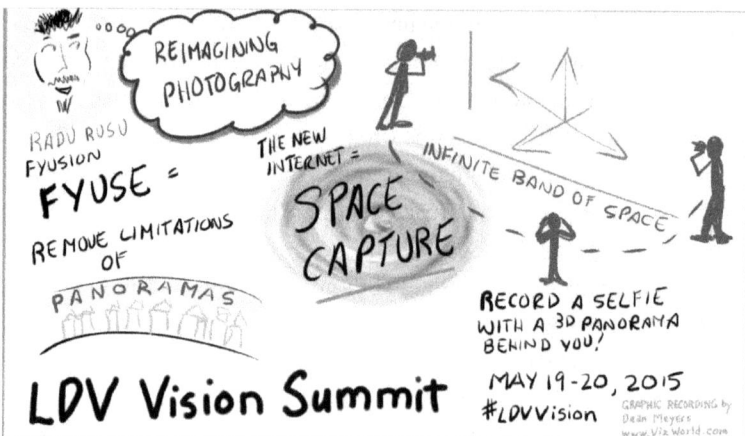

3

PANEL: COMPUTATIONAL PHOTOGRAPHY AND VIDEO

MODERATOR

Evan Nisselson, LDV Capital

PANELISTS

Michael Cohen, Principal Researcher, Microsoft Research
Paul Green, CTO & Co-Founder, Algolux
Ramesh Raskar, Associate Professor, MIT Media Lab

Evan Nisselson: I'd like to invite up our panelists—Michael Cohen from Microsoft Research, Paul Green from Algolux, and Ramesh Raskar from MIT. We're going to have a 35 minute conversation and the last five minutes will be questions from all of you and there will be mics in the audience.

Thank you guys for coming. I'd like to kick it off with a one-or-two minute, at the most: Who are you and what should the audience know in a couple sentences? Then we're going to dig into real details about that. Why don't you kick it off?

Michael Cohen: Sure. I'm Michael Cohen. I'm at Microsoft Research, where I've been for 20 years, but I'm really an academic at heart. MSR is certainly academic. I've been working primarily in computer graphics and now computational photography.

Evan Nisselson: Fantastic. I like concise.

Paul Green: Yeah, good morning. I'm Paul Green. I'm a co-founder and CTO of a Montreal-based computational photography startup called Algolux. At Algolux, our focus is on providing amazing image quality to today's smartphones and next generation smartphones. We enable things like redesigning the optics to make the camera thinner, adding new features like optical zoom, or, more practically, making the camera more manufacturable. Before that I did my PhD at MIT, where I knew Ramesh well.

Evan Nisselson: Cool. Ramesh.

Ramesh Raskar: Hello, I'm Ramesh Raskar. I'm a faculty member at MIT Media Lab. Our group is called Camera Culture. In the group we try to make the invisible visible, and also create new ways to capture and share the visual information. Within that we look at some extreme technologies like cameras that can see around corners or cameras that can create videos of light in motion at a trillion frames per second. We also look at medical devices, medical imaging, and computer vision. Millions of pictures are online and this will change our visual experience.

Evan Nisselson: What's one of the fascinating things here is mixing. We did it last year and we did it this year, and I think it's very critical. Mixing entrepreneurs, executives, researchers, and professors from universities and corporations gives us a lot of interesting dynamics. We've got two research labs, one at a big company, one at a university, and a startup. Polar opposites, potentially, or many similarities.

There's been a revolution at Microsoft in your group. Tell us a little bit about what's changed.

Michael Cohen: Sure. A lot has changed over the last 20 years, as I'm sure you know. Probably one of the biggest things is that for the first 10 years at least, the way we would have impact is by trying to transfer technology out of Microsoft Research into the larger business organizations like Windows or Office, etc. Most recently, we've become our own little startup, I would say, within the company. We're able to build some technology, we're able to build it into an app, we have a designer in our group now who does beautiful designs, and get things shipped out. I don't know if any of you saw last week we shipped out the hyperlapse app on Android phones, on Windows phones, on desktop, and actually a Cloud service, offering all around one vertical with our team, which is about 8 to 10 people within a company of 100,000. That's a very unique undertaking.

Evan Nisselson: It sounds like you highlighted the "get things shipped out." Is that probably the major change over the last 20 years?

Michael Cohen: Absolutely.

Evan Nisselson: You actually shipped a product.

Michael Cohen: We as a small group can actually ship things out to the public.

Evan Nisselson: Fantastic. Ramesh, why don't you take it from your perspective? How is it similar or different from the lab that Michael works in and the projects that you're doing? You've got tons of fascinating projects, but how does the structure work? How do you choose which projects to work on?

Ramesh Raskar: As Michael said, MSR is the economic arm of Microsoft. I consider the Media Lab as the industry arm of MIT, because we work with 80 Fortune 500 companies who are our main sponsors. We are very close to the real-world action. As much as we do very fundamental and applied research at the Media Lab, we are also very, very close to the passion of getting ideas out there in the real world. Right from e-ink that you see in Amazon Kindle to technologies that are far out there, including many visual technologies, [they are all] part of the Media Lab.

Evan Nisselson: Paul, how are you different from that? It's a little obvious but give me your perspective. You used to be at MIT...

Paul Green: Obviously our mandate is a lot different. We're a startup, we need to sell these things as opposed to doing...

Evan Nisselson: Make money.

Paul Green: Make money as opposed to doing interesting, head-in-the-clouds kind of research. Big ideas. We still try and do some of that. I have that background, of course.

Evan Nisselson: That's interesting, because I would say this panel—maybe not in this audience as much—but if we had a survey of the world and asked them what computational photography was, they would say you're doing something far out there. It sounds unrealistic, but for us it's exciting and it's real. Give us an example of more details of what you're actually going to ship, to the extent that you can.

Paul Green: I said we're focused on providing amazing image quality and that's really what the end goal is. We realized pretty early that it's really hard to sell image quality alone. There needs to be some business case around it. Something that the OEMs and the camera module makers really latched onto was the idea of yield. What we're going to ship is basically software that can make the modules more manufacturable.

Evan Nisselson: So you're not making hardware, just software.

Paul Green: Just software. It's really tightly integrated, of course.

Evan Nisselson: For a little foundation, in one sentence each of you—one sentence—define computational photography and video.

Michael Cohen: I'm going to give you two.

Evan Nisselson: Now come on.

Michael Cohen: One of them that I've repeated for the last 10 years is: "Capture, edit, share, and view as a single experience." It's very hard to wrap our heads around that. I think people begin to accept that, but you really need those four things in a single user/consumer experience.

Paul Green: I'd say you know it when you see it. I view it as the intersection of a lot of other areas like computer vision, optics, signal processing, image processing, machine learning.

Ramesh Raskar: In 2005 we taught that computational photography is about creating completely new form factors like thin cameras, a compact flash that can be used as a studio light, or new experiences like interactive photo frames. Or creating impossible photos, pictures that can see around corners, or pictures that feel frozen in time—although the time was different scales. Somehow we lost that dream of computational photography as we defined it 10 years ago. Most of us have been busy for the last 10 years just making the damn camera phone do slightly better. Hopefully for the next 10 years we'll go back to the dream and create amazing new ways to capture and share information.

Evan Nisselson: That's a great lead. It's interesting, you have three different perspectives but they are all really the same. There's different ways of looking at things, which is typical for humans obviously. When you saw my presentation earlier about my first camera phone in 2003, which was a Sony Ericsson P800, and how I took a picture and started using MMS to send it to people... At that second is when I wrote that article. I couldn't believe what was going to happen in our industry. Since my camera bag before was a Nikon F, a Roloflex, a Nikon FM—I've been photographing since 13 years old. I quickly said, "Okay, shitty pictures but the ability to communicate is unbelievable. How long until this camera phone will have 80% of the features of the DSLR so that there is absolutely no need for a DSLR?" We're trending in that direction. People are using camera phones more, but that's my question. Not only for humanity, but from a business perspective. Ramesh, what do you think about how long... Let's say 80%. The reality is that the majority of people that photograph only use 1% of the features on their DSLR so it's irrelevant. The computation, the quality, the output, the quality output—how long is it going to take us to have a camera phone that's the same size, not one of these jumbo crap... Same size as a phone?

Ramesh Raskar: I'm going to look at the question a slightly different way. You're asking me how can I build a horse that can run faster and faster? I'm going to say I'll just give you a car. It's calling a car a horseless carriage. I think we ought to change the conversation now and forget about SLR, like we forgot about LP records, and get into a completely new domain if you are to bring in a revolution in computational photography. You talked about connectivity as being a very important aspect, which SLR frankly didn't care about. If you look to this generation, our set parameters that we thought are very important, like depth of field or bouquet or megapixels, are completely irrelevant. Even post-capture control, which is manipulation of raw and so on, are somewhat irrelevant. I think we are going to change the question. Instead of saying "80% of SLR," can we say "10 times better visual experience with new technologies"?

Evan Nisselson: Than what?

Ramesh Raskar: Than what we don't do now. If we're living age of the horses...

Evan Nisselson: With a camera phone or with a camera?

Ramesh Raskar: Camera phone. Your Narrative Clip.

Evan Nisselson: So anything, it should be 10 times. The challenge there is that 10 times the DSLR we don't need. 99.9% of the world doesn't need that. I'd love to change the discussion, but I think whatever the goal is it's really to communicate. Right? 10 times better communication could be faster. And I question... There's roadblocks. You're saying if you can't build that, I'll give you a car if you want it. Right?

Ramesh Raskar: Yes.

Evan Nisselson: Then translate it before we go to the rest of the panel. Michael, you want to add? What are your thoughts? Add what you're about to say.

Michael Cohen: I'm going to pile on as well and say that you're asking: *When will camera phones have 80% of the DSLRs?* I would say they already have about 800% of DSLRs. It really comes back to that notion of two things. Capture, edit, share, and view—phones can do all of those things.

Evan Nisselson: Let me ask the question better.

Michael Cohen: Let me add one more point. The other point is if we look at the value of those pixels that are captured, and we plotted the value over time, I think most people in here would draw a graph that starts out very high, right at that moment, and drops precipitously and then just wiggles along for a moment. If those pixels have 99% of their value in the first five minutes these days, again, the DSLR is not the place to capture those pixels. Quality is an expectation from a consumer. That quality expectation also is exactly the opposite. We're willing to watch live things in horrible quality because of the excitement of watching it happen. We're willing to look at things and consume things that are five minutes old because it's very fresh. Again, as time goes on our expectations go up. I really think the question is...

Evan Nisselson: How would you ask the question better?

Michael Cohen: When will DSLRs have the capabilities of camera phones?

Evan Nisselson: That's a good way to phrase it. I, however, can't—I only want one device. I am a minimalist and many people like carrying less things around. Some people like carrying huge bags around. I want to be able to have, maybe this is just me and not the majority of people, I want to be able to have all the zoom lenses and all the tools that I used to use as a professional photographer on the same device that I use as my phone and my computer. My portable computer is also my high-end camera. When will we be using a camera phone to replace that? Ever, Paul?

Paul Green: I don't think so. I think it's going to be a while. There are inherent advantages to that form factor that are very hard to overcome in the cell phone. There are some startups even today that are making some buzz. You highlighted one, LinX, that was bought by Apple. There's a couple of others, Pelican and Corephotonics and Light is making some noise, that are going in that direction. As they both said, I think the question is a bit the problem, I guess.

Evan Nisselson: Okay.

Ramesh Raskar: We are making a business case as opposed to research case now.

Evan Nisselson: Actually, I disagree with that. You might be asked in that case, but I guarantee within 10, 15, or 20 years, whatever it is, my view is that 80% of the potential quality output and features—to do zoom and all the other features—will be on my camera phone and I will not need another camera. I might want another camera or another car or another horse. I might want to hold a different form factor, and all those things are right. But I disagree that it's not going to happen.

Michael Cohen: I'll try to answer your question a little more directly. It fundamentally is counting photons. You're asking: *When will this little device be able to capture as many photons as this big device ostensibly in the same amount of time?* The answer to that is maybe never, but computation can do a lot for you. As we saw in the last keynote and as I'm sure we'll see again, by combining not just this instant but maybe a little piece of time that I can capture over, or combining the gazillion photos out on the Internet, I'll be able to do that. I'll be able to go take a picture of my family in front of the Eiffel Tower on a horrible, miserable, misty day and it will come out like a beautiful clear day that I wish I was there on. Is that better than a DSLR? I don't know.

Evan Nisselson: That's another question. I don't know if it is, but I guess for the use case that I mentioned of having one device that's as good as that, there's nobody saying "Well, you don't need anything else." I think that's potential.

Michael Cohen: I was just going to add the phone is a very constrained device in terms of computation and power. The problem is that a lot of people just don't care that much about image quality. They can't detect it. You look at some of the Nokia phones that came out—they are actually pretty good cameras, but they weren't commercially successful.

Evan Nisselson: I think that was less the camera and more the rest of the operating system. It brings up a point and you sent me a great synopsis of a speech you did recently, Ramesh. It relates to the photons. It's superhuman vision hacking physics. I don't know how it's going to be done. I don't really understand photons. I know what the word means, but that's why I'm happy to have you guys here for this discussion. Hopefully it's interesting for the rest. I'll have questions in about 10 minutes. Or if anybody raises their hand for questions... We want this to be as interactive as possible. So hacking physics sounds very similar to directionally what Michael was saying. Well, maybe we could one day do something like that. Is that what you're talking about or something different?

Ramesh Raskar: It goes back to the question you asked about SLR-like. What we have been doing right now is called pixel hacking. We have been trying to take the camera phone and make it look like it's an SLR. But I think our ultimate goal has to be a superhuman experience. Michael talks about being able to see through fog. I want to talk about being able to see around corners.

Roger, by the way... I don't know if you realize but he had big gear taking our pictures and he took out his iPhone and created a panorama. That's why I gave him a thumbs up. I think he was giving his own vote on which one he likes. To get there, I think pixel hacking is not going to take us to this super-human photography, superhuman experiences. We have to start doing some photon hacking to get there. "Photons," of course, means particles of light. When we start thinking about the four standard elements that we, of course, have in a camera, which is optics, sensors, processing, and elimination... All of them involve photons in a certain way, of course... The processing is more on electrons and bits. We have to think about them at the time of capture. We can't just say, "Let's capture good raw and do everything post-capture." Do certain things at the time of capture, and so on. As you know, some of these companies are going in the direction of choosing a camera array, or multi-spectral, but you've got to go well beyond that. In our group, we can read a book without opening it. We can read through the pages using other wavelengths. We are able to create pictures of people behind a wall. For that we are using other spectrums like microwave and terahertz.

Evan Nisselson: Tell us a little bit about that, and tell us about photographing around corners.

Ramesh Raskar: If you are to photograph around corners... If I could look outside this door, it seems like violating the definition of a camera because a camera is supposed to see what's in front of it. But I can just flash light on the floor. It will bleed into what I cannot see. It will bleed back and a very tiny fraction of the light, it will come back to the camera. By analyzing these multiple bounces of light, basically the chatter of photons going back and forth very much like you have chatter for audio and reverberation, the reverberation of photons allows you to actually see around corners. We have demonstrated that and it's difficult to do for photography because these things still cost $1 million. But we're using it for endoscopy and so on. The same thing with seeing through walls. We're able to use radio frequency signals to create full shapes of people that are completely behind walls. Now how will that play as photography? It's not an SLR-like experience. It's going to be a very unique experience where you will be able to see things that are superhuman, literally, and I think for when you watch the movies with superheroes, they always show imagery that looks like something an average person can understand. I don't think superheroes need to see things the same way we appreciate the colors and motion and so on. We need to put ourselves in that mindset of *Hey listen, visual photography was great, but the technological revolution is much faster than biological revolution. We have to adapt to how we experience these things.* I think when phones came around, we understood that we can talk to somebody who is across the ocean. I think when it comes to photography, we are still stuck in the world that I'd rather see a picture that's in front of me.

Evan Nisselson: I totally agree, and I love separating those two things. One was just a use case about that. I didn't mean it to be the whole panel. Exactly like you're saying and I want to talk more about it. That's just one use case I think is fascinating because it has historically, since 2003 when people started doing it, or 2002... Obviously there was the big satellite phones. In 15 years when Algolux is continuing to grow and do fantastic, where do you see the company in several years? Vision-wise.

Paul Green: Ramesh spoke about you have to really... When you design these systems, to think about all the different parts to them. Not just pixel hacking. Our longer-term vision is really around taking a more holistic approach to image reconstruction; having a whole model of image generation and then modeling the physics of your optics, etc. and incorporating those things into our algorithms. It's basically recasting imaging as an optimization problem that models the statistics of natural images and the physics process. That's the longer-term vision and where we're trying to head to.

Evan Nisselson: Michael, you've just released this new iteration of Microsoft Research. You're now releasing products directly. What surprised you the most, that you didn't expect, prior to releasing that you learned after releasing?

Michael Cohen: So it's not the first product we've released. We have a set of them, and I encourage you to come and play with them and give us some feedback. The real surprises are in just what people do with it. The most fun thing is to put something out there and just watch what comes back. There's all kinds of comments that come back and things like that, but the really exciting thing is to see the media that people capture and come back with. You really never know what they're going to do with it.

Evan Nisselson: Are there some examples that you can share with us? What surprised you the most? Like, everybody all of a sudden saw it in the office and everybody started laughing?

Michael Cohen: I think more of the ones that are "ooh" and "wow."

Evan Nisselson: Give us an example of "ooh."

Michael Cohen: The hyperlapse one we put out there was one that came back from some beautiful beach where somebody was walking for miles and miles and miles down this beach in hyperlapse. It's just one of these super smooth, silky experiences that a) really makes me want to go to that beach instead of sitting in our lab, but b) I can just watch this thing forever. It's one of those just wonderful experiences. Then the other thing, basically the kinds of things we see people do. The reception is very interesting. Probably our biggest complaint is that we're not on every Android platform. I can't even count how many there are. The desire for more of those platforms has been really encouraging as well.

Evan Nisselson: What keeps you up at night? Paul? In regards to work.

Paul Green: Obviously I'm in the day-to-day of just making this stuff work, and shipping it.

Evan Nisselson: Is there a certain thing, obviously say what you can, but are there certain things that are the most challenging? Is it hiring? Obviously, each of the three panelists has different challenges as far as hiring and resources. It's obviously harder for a startup. What's the biggest challenge?

Paul Green: Finding amazing people is I think a challenge for all startups. Definitely we're no different. That's one of the reasons we're here.

Evan Nisselson: Hopefully you're here for that.

Paul Green: For sure, yeah, for that.

Ramesh Raskar: Me too.

Evan Nisselson: Are you looking to hire?

Ramesh Raskar: I'm always hiring.

Evan Nisselson: Okay.

Paul Green: That's number one for anything you do at a startup, research labs, wherever—it's the people that are there. More existentially, I think just in the field of photography, it's that nobody looks at their photographs anymore. They are so throw-away, I guess.

Evan Nisselson: People don't look at photographs? After they're created, you mean?

Paul Green: Yeah, they take them, they share them, and they disappear. They stay there I guess but no one looks at them. I think that's why the hyperlapse stuff is actually really cool.

Evan Nisselson: I agree. One of the things recently I think is going to be a huge opportunity is as more and more smart solutions analyze content... There was a story recently of uploading 20,000 images to an online Cloud, at which point I think it was a Google product that brought out images together that told a story that somebody forgot. That was another "ooh aah" moment. I think that's going to drastically change that experience.

Paul Green: There was an interesting thing about a grocery store, I think it was in Sweden or Denmark actually, doing face recognition and seeing all these connections of people who didn't know each other, but by tracking the same time and space you could plot who overlapped.

Evan Nisselson: That's what happened when I was looking for pictures in my horribly organized archive for this presentation. There's a question out here. Yes, right over there, please stand up.

Audience member: Myron Kassaraba of MJK Partners. There's a lot of cool stuff that can be enabled by computational photography—better pet photos, panoramic selfies—but what about significant benefits that computational photography could bring to either business or healthcare? Being able to tell that a motor or a compressor is about to fail, or not only telling if a person is happy or sad, but telling whether they're sick. I guess part two is: Do you see computational photography and drones intersecting in interesting applications there?

Evan Nisselson: Great questions, Myron. Ramesh, do you want to take that?

Ramesh Raskar: I think Myron brings up a very interesting conversation. We think about fusions, which are like photo and video are fusing, or camera and phone are converging. I think what you bring up is a very important debate, which is: Is computational photography and so-called computational imaging also converging? Because so far, cameras have been mostly used for visual memory. Taking photos and seeing them as they are. Computational imaging is more like A to I, analog to information. The goal eventually is not about pixels of photos, but actually some kind of information. You're talking about the motor or a health condition and so on. I think we're going to have a very interesting tension between these two of how can the same device serve both purposes? So far it fits more about A to I (analog to information), then you will make it more rich in capture, thermal, and other moralities, as opposed to what we are talking about. I think that's really exciting. I hope more of that happens, because the computational imaging world has actually moved on very fast when it comes to medical or scientific imaging or military or satellites and so on. Many of those technologies will start coming into a photographic world and we won't even realize it, that these are becoming part of our experiences and also influence the businesses in so many ways. Barcode reading is one of the few examples of how we are treating computational imaging. Recognition is improving, but we still think of recognition for the sake of photos. But recognition for the sake of tasks is also going to creep in.

Evan Nisselson: I think that's a great example. Michael is going to take it, too.

Michael Cohen: I'll try to answer your question a little more directly, but I echo that. What's going to happen is that when you take a picture, there will be understanding of that picture in the device, which offers all sorts of capabilities for business. On your point about health, we saw the work that came out of MIT where you can take a video and actually see somebody's pulse rate by exaggerating the change in color. Many of you probably saw that. That's an amazing thing that we can actually watch and see somebody's heart rate just by watching a video. The next step is by watching that pulse happening in your hand versus your face, literally the delay in time. The phase difference between the pulse in your hand and the pulse in your face will tell you the blood pressure. It just continues from there. I think there is an amazing ability for computational photography plus understanding, etc. to help out in both business and... I think health is a really, really exciting area right now.

Evan Nisselson: I totally agree. We're going to have a couple of companies speak at our Summit tomorrow, Zebra Imaging and a bunch of others, talk about the medical side, the business applications, also satellite imaging. You bring up an interesting point in talking about what could improve my semantic use of images. That's funny. Words! When I think of photography, I think of anything visual. You're bringing up a point of what photography might be, correct me if I'm wrong, photography might be when you make a picture and imaging it might be anything that's not personally captured. Is that the way you define it?

Ramesh Raskar: The way it's typically distinguished is computational photography is about creating a photo out of video, to be consumed by humans. And computational imaging is understanding information from these images.

Evan Nisselson: But when you say imaging and photography, just those two words, are they the same or they're different in your mind?

Ramesh Raskar: Imaging and photography can be confused but computational imaging and computational photography...

Evan Nisselson: Are different.

Ramesh Raskar: It's just semantics.

Evan Nisselson: Yeah, I know, but I'm just trying to understand because I want to make sure it's clear for the audience. If I'm having this trouble, I'm sure others might be having questions. There is a question in the back of the audience.

Audience member: Hi there. I am Julian Green from Google. What do you think will be the next sensors added to camera phones?

Paul Green: I think there's a lot of really cool stuff happening with time of flight sensors. For example, the work that you talked about, seeing around corners, there's work that uses similar sensors, or the same sensors, which are much lower cost. This is the sensor you're seeing now in the Kinect, I think. You can do a lot of 3D and depth and whatever else. I think there's even the Project Tango from Google that's using that.

Evan Nisselson: Any other questions from the audience? We've got about two minutes left. Yes, right here. Stand up, we don't have to wait for them. Stand up. Speak up, Eric.

Audience member: Hi, I'm Erik, and I have a technical question.

Evan Nisselson: He's got a phone right here.

Audience member (cont'd): Ah, thank you. Hey, I'm Erik Erwitt. I was just wondering if any of the panelists have an opinion on if graphene-based photo sensors may one day replace CMOS-based photon sensors in consumer electronics and why?

Ramesh Raskar: I think graphene is a very, very interesting development. It goes back to the conversation we had. We have done pixel hacking, we've done photon hacking now with cameras and so on, then they are moving into the physics hacking, and that's where the sensors come in. Time of Flight is an example of that transformation and changing the sensors themselves—whether it's black silicon or whether it's graphene based—I think as long as we're improving the quantum efficiency, which is you want to convert most of the photons into electrons, it's very exciting. Another interesting thing about graphene is that you can create non-planer sensors. As you know, the human eye, for example, has really simple optics. If you can just use curved sensors as opposed to flat sensors, all the complexity, all the wind that Roger is carrying and the lenses would disappear, and you can create one, two, three gigapixel cameras that are literally two by two centimeters. Two centimeters by two centimeters by two centimeters cubic form factor, if you just use monocentric lenses, which is just lenses like the human eye to concentric spheres, and a curved sensor. I think all the sensor technologies are going to change that. Right now you should invest in Paul's company because he is solving some of those problems. Very soon, I would say 15 years timeline, we'll see new types of silicon, new types of graphene that create curved sensors.

Evan Nisselson: This is great. We're unfortunately out of time. We could talk for hours here, but we've got another 70 speakers and hopefully in the end we'll have more interaction to learn more and spend more time during the breaks. We talked about different use cases, personal, business and other different kind of opportunities. What's the one use case, or scenario, or "ohh ahh"—however we want to define it—that you can't wait to happen? You don't obviously have to say when, but you can't wait for this to happen in regards to computational imaging. Ready? Who's ready? Come on. Who's ready? It's a tough question. I didn't send that in the email. Go ahead, Paul.

Paul Green: Go ahead.

Evan Nisselson: I like asking tough questions. There have to be surprises. It's not a political campaign.

Paul Green: This was the age of information technology. The next revolution I guess is biology, biotech, so maybe bionics. When you can integrate sensors and become superhuman as Ramesh wants.

Evan Nisselson: Is that going to happen in our lifetime?

Paul Green: I think so.

Evan Nisselson: Great. Michael?

Michael Cohen: Very short term I would say it is being able to really tell my story very efficiently and with just the wonderful design that goes into carefully crafted stories.

Evan Nisselson: Great. Ramesh?

Ramesh Raskar: Superhuman experience. We really want to see the world like we haven't ever imagined. The movies, as I said, are showing superheroes how they would see it. Whether they start with Harry- Potter-like photo frames that I can really appreciate, changing viewpoint and lighting and seeing the future... On the other extreme, it creates deep emotional connections with people across the world—especially with cultures we don't understand—to reduce some of the anxiety and tension to empathy, to allow me to learn completely new things.

Evan Nisselson: Fantastic. I can't wait for my retina camera. I just made a picture of all you guys. Thank you very much for that conversation, you guys.

4

BUILDING A CLOUD
ENABLED HOME FOR ALL
YOUR MEMORIES

Peter Welinder, Engineering Manager, Dropbox

MY NAME IS PETER. I'M FROM DROPBOX, AN ENGINEER ON THE DROPBOX photos team. Some of you might be thinking, *What does Dropbox care about photos?* To explain that, let me start by telling you a small story about my own relationship with photos and how it's changed over the last few years.

This is a photo of me. I had just started at University and what you see there on my desk is my first laptop. I just bought it with some money that I saved up and I was so excited about this thing, because I could suddenly store all my photos and all my projects in my backpack pretty much. This was also around the same time as I got my first digital camera, and naturally all the photos that I would take ended up on this laptop.

Eventually, though, the laptop got old and I replaced the old laptop with a new one and every time I do that I sort of start my life anew. I'd move all my new photos onto the new laptop and I'd keep all the old photos on the old laptop, and then I get photos and videos and files through other sources like email and social networks and so on. Eventually I got an external hard drive to try to keep it all organized and backed up, but that made the problem worse, because it meant that I had ever more copies of my life scattered over ever more devices.

Eventually this just turned into a mess. There were so many scattered files, so many different devices, and this general feeling of being overwhelmed. Naturally, this is the kind of problem that Dropbox was built to solve. It's a really easy solution for solving for this, because you can install Dropbox and move all your files into the Dropbox folder and then they're backed up and you can access them from everywhere.

With time, a lot of people have realized this, so we've ended up with a lot of people's files on Dropbox. We now have over 300 million users spread over 200 different countries and together syncing over a billion files per day. If you look at your own most important files, most of them turn out to be photos and videos.

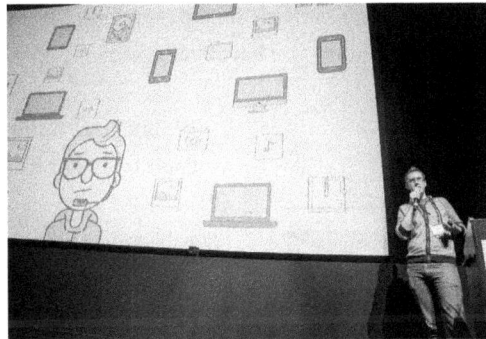

Peter Welinder sold his company, Anchovi Labs, to Dropbox and now leads Dropbox's computer vision team. He shared stories from building Carousel, Dropbox's dedicated photos experience, and lessons learned from being the home for photo collections of millions of people.

In some way, we've ended up being this repository for millions of people's memories. Being this kind of shoe box of people's memories and keeping them backed up and secure—that's one piece of solving this problem. But it's not all of it. There's more to it. What about the second piece? How can we make these photos come alive?

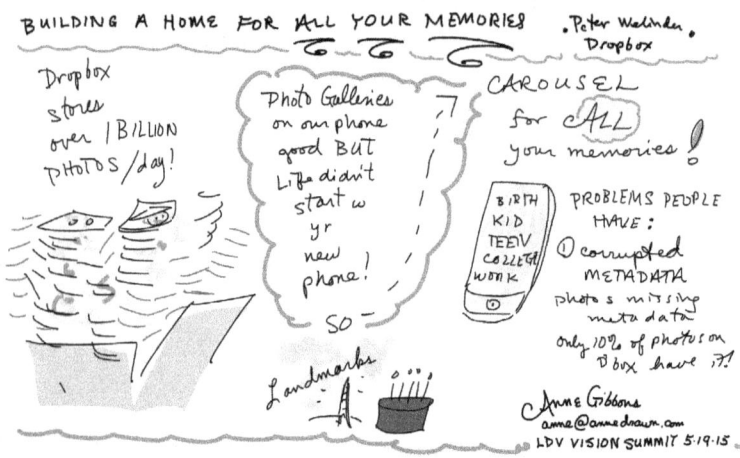

BUILDING A HOME FOR ALL YOUR MEMORIES • Peter Welinder • Dropbox

Dropbox stores over 1 BILLION PHOTOS/day!

Photo Galleries on our phone good BUT Life didn't start w yr new phone!

SO —

Landmarks

CAROUSEL for ALL your memories!

BIRTH KID TEEN COLLEGE WORK

PROBLEMS PEOPLE HAVE:
① corrupted METADATA photos missing metadata only 10% of photos on D box have it!

Anne Gibbons
anne@annedrawn.com
LDV VISION SUMMIT 5·19·15

@AnneGibbons4

For us, it started with a pretty simple idea. It was an idea that these photo galleries that we have on our phones are pretty cool. They're really useful little tools. They're a bit like the photos that we used to keep in our wallets. It's the photos that we have on top of our minds, it's recently taken photos, and so on. It's really fast, it's really simple. We can show photos to a friend very easily and so on.

But there's one big problem with these photo galleries, and it's the fact that it's sort of like life started the last time you got a new phone. None of your old memories are there. You can't get to them. So we thought, *What if we can create a similarly simple experience?* Just as fast, just as simple, but it has all the photos you have ever taken.

We built this thing called Carousel. It's an app available on Iris or Android, and it lets you access all those photos that you have on your Dropbox. It really becomes this timeline of all your memories and you can access it from everywhere. You can keep it in a pocket or you can view those memories from the palm of your hand.

Just to give you one example here, this is one of my own Carousel. All these photos are in Dropbox, they're not on the device, and yet I can do things with these photos—I can zoom in on them, I can view them in high resolution detail, I can browse through them. The really cool thing is since I have all my photos in my

Dropbox, I can use this scroller here and with just a few swipes I can go back many, many years in my photo library and relive those old memories.

That's a pretty powerful thing, because I no longer have to get home to plug in my external hard drive into my desktop to view those old memories. Instead they're all right there in my pocket. In some way that's really cool.

Now we have all those photos accessible, but there's still a long road ahead. I thought since there's a lot of students, researchers, innovators, and entrepreneurs here in the audience, I want to talk a little bit about the problems people have around their photos today (once they have them all together) and tell you about those problems.

They are based on things like: studies we made or analytics that we see amongst our users or things we hear from our users. The first one is a pretty big one, but it's one that's often neglected. It's the problem around missing or corrupted metadata.

If you take a photo with a modern camera today, you have a piece of metadata called XF on it. A lot of apps, including Carousel, use this metadata to organize your photos by, say, time and location. On Dropbox people upload pretty much every photo they've ever taken from every device they've ever owned, so we end up with a lot of different photos. We see a lot of edge cases, we see the trends of photos that people are uploading, and one big issue is that a lot of photos are still missing metadata; they don't have it or it's corrupted. For example, people forget to set the time on their cameras or they forget to switch the time when they're traveling time zones.

GPS metadata is even worse. Just to give you one statistic here, out of the millions of photos that are uploaded to Dropbox every day, only 10% have GPS metadata. 90% of the photos have no way of organizing them by space.

The interesting thing is that there are a lot of cues in people's photos about where they were taken. For example, there's seasonal cues: What's the weather outside? Is it summer or is it winter? There are landmarks you can use to figure out where a photo was taken. Then there are cues in the subjects of the photos. For example, how the age or appearance of a family member changes throughout photos. There's also other things like birthday cakes or Christmas trees or pumpkins to signify events or important holidays.

It's a really exciting opportunity for machine learning and computer vision to use this information to infer or correct this missing or corrupt metadata. We need to do this in order for people to really deal with the photos they have today. We can talk a lot about what will happen in five years, but people are still stuck with all these old memories. It's an interesting challenge to use the techniques we have today and the data sets we have today to solve these kinds of problems.

The second thing I'd like to talk about is the kind of photos that people are taking today. In the age of point-and-shoot, everything was pretty simple. You'd take your camera out of the closet every time you'd go on a vacation, you'd document your vacation, and then when you go back home, you plug that camera into your desktop and you'd organize those files into your folders. Life was pretty organized.

All of that changed with camera phones. Now, we take photos of everything. We still take photos of vacations and trips, but we also take photos of other things, like these fun and spontaneous moments that we want to share with friends. Or photos of products we might want to buy at some point. Or, for whatever reason, the food that we're eating. Or we might be using our phone as a scanner, to scan those documents that we need to archive. There are also these ephemeral photos, we take them and we only want to remember them for a few hours, like: Where did I park my car when I went to do my errands in the city?

The problem is all of these photos today, they end up on our photo rolls and they're not organized at all. How can we organize them? How can we make them indexable and searchable? What are the right interfaces for exploring all these photos? This is a problem that gets worse with every day because we are accumulating ever more photos through ever more devices and it's still largely an unsolved problem.

That's the kind of challenges that we're working on at Dropbox when we deal with all of your photos that you've ever taken. And if you like working with those kinds of problems, I'm going to do a small plug, then you should come see me afterwards. This is a problem that anybody with a camera would have.

Thank you.

5

PANEL: COMPUTER VISION ADVANCEMENT AND CHALLENGES

MODERATOR

Sean Ammirati, Partner, Birchmere Ventures

PANELISTS

Serge Belongie, Professor, Cornell Tech
Andrea Frome, Software Engineer, Nuna Health
Mor Naaman, Chief Scientist & Co-Founder, Seen.co
Simon Osindero, A.I. Architect, Flickr

Evan Nisselson: Sean Ammirati—I twisted his arm—he's a good friend, a great entrepreneur, a troublemaker, and a good investor. Sean, come on up. Thank you, not only for being here for a panel later, but for this—I recently said, "Hey, would you share your wisdom and moderate a panel that you're learning quickly about?"

Panelists: Andrea Frome, software engineer, Nuna Health; Mor Naaman, associate professor, Cornell Tech; Simon Osindero, A.I. architect, Flickr; Serge Belongie, professor, computer vision, Cornell Tech; Moderator: Sean Ammirati, partner, Birchmere Ventures [L-R]

Sean Ammirati: It is a lot of learning, to learn about this for the last couple of days.

Evan Nisselson: Well you know, I had ulterior motives.

Sean Ammirati: Yeah, I assume you want me to invest with you.

Evan Nisselson: I love it, he's a quick guy. I'm going to let you take it away.

Sean Ammirati: Okay. Great.

Evan Nisselson: Thank you, Sean.

Sean Ammirati: So the topic that Evan asked me to moderate a panel on is computer vision advancement and challenges, and specifically this question of: When is computer vision going to become a commodity? Which I thought was an interesting concept, because a lot of the advancements that the startups that we invest in today take advantage of is different types of computing that has become commoditized over time. We're going to dive into that over the next 30 minutes or so. But quickly, just to set the stage, I asked each of the panelists to just do a very quick introduction. So maybe we can start with Serge and just go right down.

Serge Belongie: Sure, is this on? I'm Serge Belongie, professor of computer science at Cornell Tech.

Simon Osindero: Simon Osindero, and I'm an A.I. architect with Flickr.

Mor Naaman: Mor Naaman. I'm also faculty at Cornell Tech, and chief scientist and co-founder at Seen.co.

Andrea Frome: Andrea Frome, software engineer and researcher at Google until last Friday.

Sean Ammirati: Wow, okay. So what we want to do here to start is get synced on vocabulary. I think this has come up a little bit already this morning, that people use the same words and mean slightly different things, so let's agree—at least for the next 30 minutes—that when we use the word "computer vision," we mean: fill in the blank. So. Let's get started again, going down the list. Serge, why don't you start? When we're talking about computer vision, we're going to talk about it on this panel, what do we mean exactly by that term?

Serge Belongie: I think of it as three main things: recognition, reconstruction, registration. This is taking images to recognize things, reconstruct things, or track things.

Simon Osindero: I was just thinking about this on the way in. I'm not quite sure how to answer the question. I know it when I see it, but it's hard to define.

Sean Ammirati: How would you see it… Let's go ahead, right….

Simon Osindero: So on a basic level… computation based on information received from optical photon sensors. But then, that feels overly broad. Typical things that come to mind are, given an image we want to recognize objects within it, and performing scene understanding. But then there are also more exotic applications like inferring audio signals by visually analyzing tiny movements of objects within a video stream. Is that computer vision? I guess I'd say "maybe," but it's probably not what people typically think of. So, I'm actually not sure where to draw the boundaries.

Sean Ammirati: Okay. Mor, got anything for us?

Mor Naaman: Well, to keep on the same level as Simon, I think to me computer vision is extracting information from the content of images from the pixels, as opposed to looking at the context surrounding it—words, whatever else is there.

Andrea Frome,
Software Engineer
Nuna Health

Sean Ammirati: Okay. Anything to add?

Andrea Frome: I think that covers it pretty well, and I was going to say extracting meaning from pixels, from images, from video. We're really trying to get the semantics of what's happening and get to something deeper.

Sean Ammirati: This question, the first question is: Will it become a commodity, and if so, when? We did not get a chance to do our prep call for this, so this panel could go in a number of different directions here. But let's start with a yes or no, and if the answer is yes, when, and if the answer is no, why not? And why don't we, just so we don't always continue to put you on the spot first, Serge, why don't we start on the other end and work our way down here. Andrea?

Andrea Frome: I do think at some point, yes, and I think the deep learning models are coming closer to being a commodity. I think the problem we're going to run into is the data to train these models. And ImageNet really pushed things forward, but where is this data even going to come from in the future, who's going to label it, who's going to pay for it? It might become a propriety, very carefully guarded thing. So that's going to be the big question: the data that's going to be used to train these models.

Sean Ammirati: Great. And just so we have a sense of the audience, is everyone familiar with ImageNet at this point? Yes? Show of hands if you are. About half. We want to just give a quick, what that is....

Andrea Frome: Sure.

Sean Ammirati: For the other half.

Andrea Frome: Gosh, how many millions of images? 10 million? I should know the number off the top of my head, but millions of images that were labeled across—I think, the largest set is 10,000 categories or something like that—but anyway, it was a data set that was put together by Fei-Fei Li and people at Stanford, and I don't know if it was in collaboration with other... I'm not giving good details here at all, but it's a very large object recognition data set, and also now there's a detection data set covering 200 categories as well. And that was used to train Alex Krizhevsky and Ilya Sutskever and Geoff Hinton's model out of Toronto that really pushed things forward recently in the deep learning computer vision models. And that model ended up being the basis for a lot of really great object recognition work that's followed. That's the starting point for a lot of commodity, computer vision models that are being used right now.

Sean Ammirati: So the catalytic moment for this trend, that's great. Mor?

Mor Naaman: No. Well maybe, partly.

Sean Ammirati: So let's go with why, then.

Mor Naaman: The two places I think computer vision will become a commodity soon are places and faces. I think these are two things that we have enough background information to be able to provide recognition capabilities in the Cloud, off the shelf, whatever is needed for a company, because these are two very well-defined questions that can use algorithms that exist and the data sets that exist to provide answers. So, where was this photo taken, who is in this photo? Companies—I'm thinking Canary, who are sharing space with us in a Two Sigma incubator—they're doing computer vision to look at home monitoring. That has to have computer vision stuff and as long as companies that do their own apps will need computer vision stuff—and I don't see a near future when that is not the case—computer vision will not be entirely a commodity. So things is yes, faces, yes, the rest—I don't know.

Sean Ammirati: Not so much.

Mor Naaman: As long as Flickr still has this guy on board....

Sean Ammirati: All right, so what do you think?

Simon Osindero: I would say yes, actually. I guess I'd argue in some ways some bits of computer vision are already commoditized. So if you wanted to, not necessarily state-of-the art, but just pretty good object detection or recognition then there's open source stuff that you could download and use right now. And I think going forward, considering some of the more sophisticated machine learning approaches to computer vision, I guess that pushes the question into to what extent machine learning is commoditized. And there I think as Andrea suggested, that also becomes a data availability question. But I think the data is being put out there. For instance, Flickr have released a dataset of 100 million images. They have their user tags, although they haven't been rigorously curated. But that's still an enormous amount of useful data.

I also think a good analogy, or at least an analogy that comes to my mind, is databases or data storage. You know—to what extent is that commoditized? On the one hand, you could say that that field is totally commoditized, but on the other hand you do still need to hire database engineers and experts. I think it depends a little bit on how close to the cutting edge you want to be, and how good is "good enough" for your use cases.

Sean Ammirati: Sure. Serge?

Serge Belongie: I think I agree with Mor. I lean toward no. It's not because I'm a pessimist, it's just that our field, computer vision, tends to disown success. Something happens where, something starts to work, and then we call it something else. It could be fingerprint recognition—well, that's biometrics—or optical character recognition—well, that doesn't belong in a computer vision conference. That's okay, it doesn't have to be, it doesn't really matter what these things are called. What really matters is success. I think that computer vision doesn't really have the visibility as the name of a field like computer graphics does, for example. Part of that is just that most of what we do involves error trade-offs. Like what Andrea was talking about—the ImageNet experiments, deep learning, and so on—you get these precision-recall curves; you have false accepts and false rejects. Anyone doing research in machine learning or computer vision understands this is a reality. Nothing works perfectly. If the precision-recall trade-off is good enough, and you get flat bed scanned OCR, you get fingerprint recognition on phones. It becomes a commodity. But when it's not? It's not actually the computer vision that's the most important thing. It's actually the design. It's the design of the product, because we have to figure out how to accommodate those kinds of errors and the trade-offs between the Type I and Type II errors. So I think computer vision is vital as a research area, and I'd love to see more and more companies get many computer vision PhDs, but it has something very research-y in its DNA, and when it really starts working, it just gets called something else.

Sean Ammirati: Okay. Excellent. Maybe we can come back to what we need to make the answer "yes" more definitively for two of you. For the moment, let's assume yes, in the following couple of ways: Mor, for you, we'll just stay with faces and places, not things and other stuff, and Serge we'll go with you as "yes" when it gets spawned off as OCR or whatever else you want to call it. So let's kind of agree that there's some of these things that are leverageable. Simon, I want to pick up on something you said and ask you to start, and then maybe others to weigh in. Your analogy of databases was interesting to me because I think about before I became lazy and joined Evan as an investor, I was an entrepreneur at one point, and the second company that I started, I spent a million dollars on servers. Actual, physical servers from Dell. We were a named account as a seven-person startup, and the Dell rep would go visit PNC bank in suits and come see us in flip flops, if they were lucky, and t-shirts if they were really lucky. They'd get similar orders from both of us, which was kind of scary, but when we sold that business to LinkedIn, I had a $500-a-month credit card bill and I owned exactly zero servers. I had sort of pushed all of that content into Cloud storage. Your analogy of databases is kind of similarly provocative. How much of this can you just leverage off the shelf versus write yourself? I guess, if you're a startup out there right now thinking about taking advantage of some of these things today, and then maybe if you fast forward 12 to 18 months, there's more of these catalytic events like the library's become available, whether it's Flickr or the one you mentioned, right? What are the things that startups are going to take advantage of, and just do out of the box that they maybe today would look at, or a year ago would have looked at as a science project?

Simon Osindero: I was just saying, definitely for things like "what are the main objects in this image?," "what type of scene is this?," and "faces," there's already a bunch of companies that offer APIs for that. So, literally, if that's something that is a feature that would make your product better and you don't want to develop that in-house, you know you can, say, import a python library, get an API key, make calls and you have pretty decent computer vision capabilities for those tasks out of the box. I think it's always going to be a question of: How good do you need to be to solve your problem? And as Serge was saying: How able are you to design around whatever limitations exist? I definitely think we're going to see more and more companies offering API vision services. And so as a developer, you can just tap into that and focus on the rest of your product.

Sean Ammirati: Can you give a couple of examples of startups that you might imagine starting? Let's say you had to leave Flickr tomorrow and start a company. Everybody should leave big companies, so we'll pretend like you're going to do that for the sake of this panel. Sorry, Yahoo, if you're actually a sponsor of something. So what are you going to start? How do you take advantage of this to start a company?

Simon Osindero: One cool thing that I saw recently, and one of the companies in the audience—Emotient—they've got some really cool tech for doing emotion detection and recognition. I can imagine all sorts of awesome applications you can have there: you can read, with super-human accuracy, the micro-expressions that cross someone's face. And with that there's an awful lot you can infer about their state and make predictions or adaptations. And so if you had that in a store, say, or if you baked that into your user experience research, it'd be great. So that's one example. Clarifai and Metamind are other examples (of companies) that offer APIs for object detection and classification. Was the question what I would do, or just some examples of what's out there that could be used?

Sean Ammirati: Yeah, I was trying to get an example of some of the startups that you would do. So are there some examples left on the panel? How are startups going to be taking advantage of these in the next couple years?

Serge Belongie: Well, I have a related thing I just want to mention. It's sort of an anecdote, it's sort of evidence of—is it commoditization? Commodification?

Sean Ammirati: I believe that was the term. Commoditization.

Serge Belongie: Commoditization of computer vision. This is kind of a plug for Yahoo, so maybe you can correct the story here.

Sean Ammirati: See what happened there? He saved the sponsorship. You're good again.

Serge Belongie: There's a very well-known web comic called XKCD and one of them was called "Park or Bird." So the web comic shows two characters and one of them… What the comic is about is the subtle differences between impossible and very easy problems in computer science. So someone's taken a photo and is talking to a hacker sitting at a computer and says, "I want to determine whether this photo was taken in a national park." And the hacker says, "Oh, no problem. Just check the geotag and you know that it's in a national park." Then, the other person says, "I also want to know if it's a picture of a bird." And then the hacker sort of stops and says, "Oh, well that'll take five years and a team of PhD researchers," or something like that. Maybe five years ago, that actually would have been true. There's this stark difference between simply looking up a geotag and determining whether a photo contains a bird. I don't know what the turnaround time was, but it seemed like days later, a group of researchers at Yahoo took the challenge and said, "We can do that," and then they released parkorbird.yahoo.com. You upload a photo and it checks the geotag and if it's in a national park it will say so, and then it uses this commodity, deep ConvNet to determine whether there's a bird in the photo. Now, it's not perfectly accurate, but they just did it in what seemed to be a couple days. I think that it really sent ripples, at least through the nerd corners of the Internet, that this has arrived, that this is something that is almost as easy as just looking up a geotag.

Sean Ammirati: That's awesome. Mor, were you about to weigh in?

Mor Naaman: I love this, because when it came up, and the comic came out, I think I posted it on your Facebook, and said that's the difference between my research and Serge's. But I think it also illustrates something about… We had a couple of comments about personal photo collections from the Dropbox speaker. I don't want to sound as old as Evan—I'm not, I don't think so, at least—but when I started my PhD it was early 2000 and we said: What happens when in the future when all the photos are not just digital, but are also accompanied by time and place data, geotagging? And what can we learn from photos taken near each other, what can we learn about the context of the photo—can we build better personal and global photo collections based on that? We have come a pretty long way just by knowing the context of the capture. Looking at the applications we have today, like bird recognition, I still think that the context alone is pushing us 90% of the way there to organizing context-sensitive photo albums and even recognitions. We said back then we had a pretty good estimate of what people are in the photo without even using any pixel. Just by looking at the context of where you are. Because if I took a photo here, it's not very likely that my mother would be in it, for example. We went 90% there, and computer vision takes us maybe a few more percentages towards the optimal photo collection organization, but I just don't see how it's pushing us a whole mile further. Not with bird recognition, at least. Maybe Flickr has different opinions.

Sean Ammirati: You look like you want to respond to that, so go ahead.

Simon Osindero: It seems like you're saying computer vision couldn't, or wouldn't, or wasn't adding much to the context, but in that case I think I disagree. We're not quite there yet, but in terms… One of the goals is full semantic understanding of images, and maybe I'm being overly optimistic, but that seems very achievable within a decade to a pretty decent level. Not necessarily to a full human level, but we're really good right now at figuring out what the objects in the image are. I think there's a kind of crop of new models, some of which] I know Andrea's worked on using active visual attention, and I think in a couple years or maybe less, we're going to see that give us an extra leap. So instead of trying to take things in a single snapshot as we do now, kind of more deliberative and focused processing. I'd say for still images, a lot of the analysis is the kind of stuff that we can take in a kind of snapshot blink right now, and getting that next step. But I agree that part of that is probably going to involve drawing in contextual information. If I know where it was, and I know the people that you know, and as I do that more detailed processing, I can integrate that information in my visual understanding.

Andrea Frome: Yeah, and I think this is really where things are headed, to more integration of information. Video right now—we do really stupid things with video when it comes to detection and tracking and recognition most of the time. I think we do a lot of processing and then we throw a lot of it away and then we post hoc combine things. But to combine more information as you process ongoing and integrate that information as you process the new pixels, I think there's going to be big things coming.

Mor Naaman: These two comments are related to something we do when, in my company, at Seen, where we try to… Seen is an engine that tries to make sense of the real time web. As things go by, things happening, events like today or breaking news events, we collect the data and we try to analyze and present it in different ways, and we started doing that a little bit before streaming content as well, like Meerkat and Periscope. Our algorithms can detect the panel is happening and visuals and then find a representative photo, and when we want it to do that, we only use the context. We only use the text. My suspicion is we already do a lot better than just vision alone.

We can probably do a little bit better when we have vision added. We talked to some very talented people, the Clarify folks and other computer vision companies out there. It wasn't that the price point of the futures and the readability that we were offered weren't quite there yet, to integrate that into a product as we were running. Maybe later on, but not quite there. So that's why... Not commodity, yes, I'm sure quality and integration will be key to doing this in the future, but for now, we're like, "heck," you know, it's just easier when doing contextual features and....

Sean Ammirati: There's been a number of comments in a row here where we started talking about where computer vision is going, and there's a little bit of a discussion in the email thread I had about not only are there some interesting business implications of this, but there are some significant societal impacts. Or sort of impact to society around these trends as well. I think most of us get into computer science because we want to use software to make the world more the way it ought to be. I guess we shouldn't just ignore the societal impacts of these things. As you guys who were kind of pushing this area of study forward and then eventually disowning it once it gets to a level of maturity, as you think about where this is heading, how do you think about how this impacts society and how this fits into making the world the way it ought to be? And then, what are the things that also scare you about where this is heading? I know a lot of you have opinions on this, so does anyone want to start? Serge, do you want to start?

Serge Belongie: There's two main areas with positive societal impact that my group works on and that I find very interesting. One of them is assistive technology for the visually impaired, either low vision or blind individuals. And this is an incredible opportunity. One very visible company that's doing this is OrCam, which is out of Israel. You wear these glasses that have a little camera and you can point at a sign and it sees your fingertip and it can read the sign or identify an object. What's characteristic about a lot of work involving assistive technology is that it's not simply running in batch on some server. You have an engaged, motivated user who may have partial vision or other abilities to assist the system. It's not just the computer helping the user exclusively, it's also the other way around, like pointing at a sign. You can see there's some text but you can't read it. All of us in some sense need assistive technology. We don't speak every language. If you're looking at a Japanese menu, that's, in a way, assistive technology, to use Google's translate to swipe over a word and translate that. I think assistive technology is not a sexy market, it's not something that a lot of people are working on, but it is actually very closely connected to all the problems that we're talking about. It's something that's helpful even when those precision recall curves are not that great. The other one is citizen science but I can come back to that one later.

Simon Osindero: Yeah, I think those are both great points in terms of the benefits. Another one that is maybe a double-edged sword is significant augmentation of human memory. I saw Evan already has a "Narrative Clip." And it's certainly feasible that in the not-too-distant future we're going to be able to record audio and video, pretty much everything that goes on around us, and if we put search on top of that, and semantic search, then you have a situation that, maybe there's a conversation where you have an argument, and you can go back and get a replay, which is not necessarily good.

Sean Ammirati: So I can go to tell my wife, "You know, let's go watch the video." I'm not sure that's a good thing for either of us, to be honest with you.

Simon Osindero: That's the downside, but on the other hand, you can figure out where you lost your keys or... That one's interesting to me because I think it's going to happen. It's not clear to me whether it's something I want or not, but I think it's going to happen anyway, and it's just one of those things that changes the way things are. I remember when cell phones came out, and a lot of people were like, "I don't want a phone with me all the time. I don't want to decide whether I can be reached or not." But now I bet they never leave it. The best example on the side that I think is more clearly negative, particularly in the way a lot of global politics seems to be going... is related to all the recent stuff with the NSA: the ability to do at scale, massive, detailed analyses of people's lives—I think that's cause for caution.

Sean Ammirati: Yeah. I agree.

Mor Naaman: I think these are very exciting implications, and I think the tension is between the personal use and the surveillance application of the same technology. If Flickr does a great job in recognizing faces, like knows where my mother is in the picture, then that's great, it works for me. But if the government knows that, then maybe that's a little more worrisome. Similarly, for Apple to follow to Flickr and recognize the Empire State Building and know where I am so I can retrieve it later, again, that's great for me—the government doing the same thing, I'm now a little bit more worried. Same technology, we don't need anything else.

Sean Ammirati: I'd be a lot more worried, right? We can definitively say, it's not just a little more worried.

Mor Naaman: This would be the same technology, and people may not even realize that. A lot of people don't know that uploading iPhone images with GPS in the Exif metadata would expose where they were. I think there were some high tech criminals that were caught that way in Central America or something. To be able to understand that a bridge in the background of your image can not only locate the general area where you are, but exactly the angle and maybe the place where the photo was taken.

Sean Ammirati: Right.

Mor Naaman: I think that would be an issue. These technologies, in a way, are a double-edged sword.

Sean Ammirati: So we're going to go to questions in a minute and I want you to answer as well. If people have questions just get your hands up. I want to give you a chance to weigh in on this, Andrea.

Andrea Frome: I think they took all my best answers.

Sean Ammirati: I started with you only the time you didn't want to start it, is that right? Excellent job at moderating.

Andrea Frome: I'm hanging in there. I guess one more, just kind of picking up the dregs here. I'm really interested in photo sharing among strangers who share experiences, for one thing. I think that's one where it's a very exciting thing to realize—take the Arab Spring for example, Tahrir Square—to have a way for the people who were there to bring their information together and to share that information for people to get a more holistic view of everything that happened. I think those kinds of experiences would be very cool, and again, there's another edge to it, which is: How do you do that and preserve privacy and not leak private information? Sort of moving to that next level of sharing. It's not just your close friends or your family, but photos and videos in a more public space. And it's not just going by in a stream, it's brought together in a more meaningful way. There's some tricky issues around that, too.

Sean Ammirati: Yeah, absolutely.

Speaker 7: We've got a question out here.

Sean Ammirati: All right.

Audience member: Hi, my name is David Heger. I'm a professor at NYU. I've been doing computer vision and human vision research for 25, 30 years. I would argue that computer vision is a commodity and I think the panel is being overly restrictive in defining what computer vision is, with a focus on recognition. Most of what we do with human vision is to control action. Recognition, really, is overrated. If you think of using image processing and computer vision technology for controlling action, there are a number of really good examples of products that are out there. For example, Mobile Eye makes the technology that many of you use every day when you get in your car, that tells you whether you're about to get in a collision or a lane departure warning or an automatic high beam control. That's all computer vision technology. The recognition part of it is really pretty simple. As a whole package, it's having a big impact, not only commercially but a societal impact. Other than the assistive technology example, which I think is great, I'm wondering if any of you are thinking about or are aware of any other things like that where the focus is on computer vision for action, as opposed to computer vision for computer recognition.

Simon Osindero: There's some really nice work coming out of UC Berkeley at the moment for visually-guided robotic control, and I think that's probably somewhere where I can see this having a big impact in the next five or so years.

Evan Nisselson: Any questions over there?

Andrea Frome: I just wanted to point out that the Mobile Eye example is an example of what Serge was talking about, which is—once we have it working, it's no longer computer vision.

Audience member: Can you speak to what you're learning from other areas, whether or not it's speech or text technology and the Siri-type stuff that potentially can be applied with the vision side that's made...The early work may have been done on other forms?

Serge Belongie: I think I can return to the theme I briefly touched on, which is the importance of design in these products. I think that computer vision, if you think of it as simply a signal, then it is a commodity. It's a signal, but it's not the absolute best thing in any given product. Usually computer vision simply doesn't work that well or it isn't the entire solution. It's a signal along with other things. Something like a QR code, it just works. It's a commodity, but hey, we don't call that computer vision. I think what happens with things like Siri, with any sort of recognition where you have context like Mor was describing, where you know the social network that's in the background. The right way to approach this is really to think of computer vision as one of many signals, with a sort of humility, in a sense, to say it's alongside, the accelerometer, the geotag, the time of day, and all those other things. Then when you approach it that way you can actually do some pretty cool stuff.

Andrea Frome: I think a little bit on that theme, the example earlier of word lens, the translate app that you could hold over a Japanese menu and it'll show it in your language. Something that's really interesting about that is the word error rate on that is quite high, but because it's an augmented reality experience, because it's in real time and you're interacting with it and you're integrating that information as you're using it, it's a really compelling experience. Even though the computer vision behind it is not very accurate, really.

Sean Ammirati: I think we've got time for one more question. Evan's given us a little bit of bonus time.

Audience member: Hi, Paul Melcher from Capture. Could you talk a little bit about how computer vision is going to be affected, or is affected, by what we heard previously on computational imaging? All this pixel hacking, is that a problem for you guys, or is it making your life easier?

Sean Ammirati: Give me one sec.

Andrea Frome: Just give us data.

Simon Osindero: I agree. All of the new senses, when you capture them at the...

It's definitely a great thing, but there's some modifications that we'd want to make, so if you're capturing a light field instead of just a single image, then how you go about analyzing that efficiently has some challenges. But yeah, overall I'd say it only makes the field stronger, so it's great.

Sean Ammirati: All right, so that's it for our panel, we're over at this point. Please join me in thanking them for coming out.

Evan Nisselson: Thank you very much. All done. Thank you, Sean. Great job.

©VizWorld

6

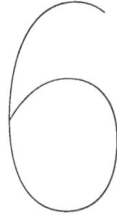

WE'RE TAKING BILLIONS OF PHOTOS A DAY LET'S USE THEM TO IMPROVE OUR WORLD!

Pete Warden, Engineer, Google

MY DAY JOB IS WORKING AS A RESEARCH ENGINEER FOR THE GOOGLE Brain team on some of this deep learning vision stuff. But, what I'm going to talk about today is actually trying to find interesting, offbeat, weird, non-commercial applications of this vision technology, and why I think it's really important as a community that we branch out to some weird and wonderful products and nonprofit-type stuff.

Why do I think we need to do this? Computer vision has some really deep fundamental problems, I think, the way that it's set up at the moment.

Pete Warden, Engineer, Google

The number one problem is that it doesn't actually work. I don't want to pick on Microsoft because their How-Old demo was amazing. As a researcher and as somebody who's worked in vision for years, it's amazing we can do the things we do. But if you look at the coverage from the general public, they're just confused and bewildered about the mistakes that it makes. I could have picked any recognition or any vision technology. If you look at the general public's reaction to what we're doing, they're just left scratching their heads. That just shows what a massive gap in expectations there is between what we're doing as researchers and engineers and what the general public actually expects.

What we know is that computer vision, the way we measure it, is actually starting to kind of, sort of, mostly work now—at least for a lot of the problems that we actually care about.

Kinda-does now…

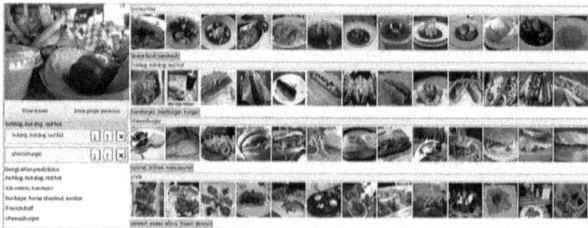

"It is clear that humans will soon only be able to outperform state of the art image classification models by use of significant effort, expertise, and time."

http://karpathy.github.io/2014/09/02/what-i-learned-from-competing-against-a-convnet-on-imagenet/

This is one of my favorite examples from the last few months, where Andrej Karpathy from Stanford actually tried to do the ImageNet object recognition challenge as a human, just to see how well humans could actually do at the task that we'd set the algorithms. He actually spent weeks training for this, doing manual training by looking through and trying to learn all the categories, and spent a long time on each image. Even at the end of that, he was only able to beat the best of the 2014 algorithms by a percentage point or two. His belief was that that lead was going to vanish shortly as the trajectory of the algorithm improvements just kept increasing.

It's pretty clear that, by our own measurements, we're doing really well. But nobody's impressed that a computer can tell them that a picture of a hot dog is a picture of a hot dog. That doesn't really get people excited. We really have not only a perception problem, when we're going out and talking to partners and talking to the general public and talking to people. The applications that do work tend to be around security and government, and they aren't particularly popular either. The reason this matters is not only do we have a perception problem, but we aren't actually getting the feedback that we need to get from working with real problems when we're doing this research.

What's the solution? This is a bit asinine. Of course we want to find practical applications that help people. What I'm going to be talking about for the rest of this is just trying to go through some of my experiences, trying to do something a little bit offbeat, a little bit different, and a little bit unusual with nonprofit-type stuff—just so we've actually got some practical, concrete, useful examples of what I'm talking about.

The first one I'm going to talk about is one that I did that didn't work at all. I'm going to use this as a cautionary tale of how not to approach a new problem that's trying to do something to help the world. I came into this with the idea that...I was working at my startup Jetpac. We had hundreds of millions of geotagged Instagram photos that were public that we were analyzing to build guides for hotels, restaurants, bars all over the world. We were able to do things like look at how many photos showed mustaches at a particular bar to give you an idea of how hipster that particular bar was. It actually worked quite well. It was a lot of fun, but I knew that there was really, really interesting and useful information to solve a bunch of other problems that actually mattered.

One of the things that I thought I knew was that pollution gives you really, really vivid sunsets. This was just something that I had embedded in my mind, and it seemed like it would be something that I should be able to pull out from the millions of sunset photos we had all over the world. I went through, I spent a bunch of time analyzing these, looking at public pollution data from cities all over the US, with the hope that I could actually build this sensor, just using this free, open, public data to estimate pollution and track pollution all over the world almost instantly. Unfortunately it didn't work at all. Not only didn't it work, I actually had worse sunsets when I was seeing more pollution.

At that point, I did what I should have done at the start, and went back and actually looked at what the atmospheric scientists were saying about pollution and sunsets. It turns out it's only at really high atmospheres, with very uniform particulate pollution—which is what you typically get from volcanoes—is what actually gives you vivid sunsets. Other kinds of pollution, as you might imagine if you've ever lived in LA, just gives you washed out, blurry, grungy sunsets. The lesson for me from this was I really should have been listening and driven by the people who actually understood the problem and knew the problem, rather than jumping in with my shiny understanding of technology, but not really understanding the domain at all.

Next I want to talk about something that really did work, but I didn't do it. This is actually one of my favorite projects of the last couple of years. The team at onformative, they took a whole bunch of satellite photos and they ran face detectors across them. Hopefully you can see, there appears to be some kind of Jesus in a cornfield on the left hand side, and a very grumpy river delta on the right. I thought this was brilliant. This is really imaginative. This is really different. This is really joining together a data set with a completely different set of vision technologies and shows how far we've come with face recognition.

Pete Warden, Engineer, Google

But shortly after I saw this example, I actually ran across this news story about a landslide in Afghanistan that had killed over a thousand people. What was really heartbreaking about this was that the geologists looking at just the super low-res, not-very-recent satellite photos and the elevation data on Google Earth said that it was painfully, painfully clear that this landslide was actually going to be happening.

What I'm going to just finish up with here is that there's a whole bunch of other stuff that we really could be solving with this:

- 70 other daily living activities
- Pollution
- Springtime
- Rare wildlife

What I'm trying to do is actually just start a discussion mailing list here at Vision for Good, where we can bring together some people who are working on this vision stuff and the nonprofits who actually want to get some help. I'm really hoping you can join me there. No obligation, but I want to see what happens in this.

7

DEEP REASONING
FOR VIDEO CONTENT
UNDERSTANDING AT YAHOO

Alex Jaimes, Director of Research & Video Product, Yahoo

THE FIRST THING TO SAY IS COMPUTER VISION IS HARD. WE'VE SEEN A lot of stuff going on, and just to show you an example, how many people see things moving? Raise your hand. Oh, great. I can hypnotize the entire audience now and say whatever I want. Obviously there's no motion. It's all in your head. It's just an example. What I want to say with that image is essentially that a lot of the processing, a lot of the computer vision processing that we make as humans is actually in our head. It's not just in the signal. It's in the processing that we actually do. Right? Because of that the approach that I like to take in any computer vision problem is what I call a human-centered approach.

Alex Jaimes, Director of Research & Video Product, Yahoo

Basically what that means is: a very strong focus on more than one area. So it's not just pixels, it's not just computer vision. It's understanding user experience, understanding data analysis, what I call human aspects, and really working at the intersection of those three. That's where the real machine learning work actually happens. Sorry—the timer's not working, by the way. That means I have an extra two minutes. Great, lucky me. How do we do this? And what are the main tasks? One is understanding the users, second is understanding the content, understanding how they go together, and then creating new algorithms and new applications—this through large-scale data analysis and machine learning.

The reasons that these dimensions are important is because, as has been said before today, most computer vision techniques don't work perfectly all of the time. We need to really understand how they can be applied in a specific context and that depends on the content and the users—how the users will use it to design the application and so on and so forth. Recognizing things in baseball videos is not the same as in music videos or as in any other kinds of sports videos or surveillance, right? The technologies might be similar but the actual application really varies significantly. From that point of view, everything you do is really a cycle where the user has to be at the center of it all and we start with either the data analysis, design, or a hypothesis. That's how we kind of develop it and the basic idea is that we cycle between these three areas within a specific context. The more that we do that, the closer we get to the user and that's how we innovate and get to a point where we make a difference for users.

Alex Jaimes, Director of Research & Video Product, Yahoo

What I'll do is... I'm going to do something really risky. I'm going to show some demos. Hopefully some of them will work. I'm not going to show all of them, but these are some of the areas in which we're doing some work. I'm not going to show anything adult because I think everybody knows what that is. In essence, I think there are kind of two main areas. One deals with algorithms and the other one deals with insights. It's really important to do both because, as I said before, you need to understand how users are using the content, how they're viewing it, and how those insights and content go together. What is the main computer vision task? It's basically this: given an image or video, recognize objects, actions, or events.

In essence, that is the core technology that most of us are working on. I would say that that's not enough. In most cases, that's definitely not enough. And from this what we usually get is labels, which historically a lot of people have referred to as "image understanding" and "video understanding." I don't like that term ("understanding") because we're not really understanding anything. We're just labeling. It's not the same. Labeling and understanding are not the same and I'll say a little bit more about that.

Just some examples of stuff. We're doing things like celebrity tagging so the basic idea is that we want to find images of celebrities and find not only images of celebrities but find the best quality images and often the most representative of images. For a single celebrity we want to find one image that is the best one for that particular celebrity. This is an example of something that's done automatically from a very large collection and if you look carefully you'll notice that there are a lot of possible issues in doing this automatically, which are very subtle. They're really difficult to deal with. We want to have the highest quality images. The one at the bottom where you have beer mugs in the guy's face, that's not really acceptable and it's really hard to detect automatically. Or if you look at Sarkozy at the top—ironically, I think he's next to Bush—you can see the way he's using his hands. For him that's actually a pretty good portrait because he uses his hands often, right? But in other cases, you don't want to have a hand in front of the face. You want to make sure the eyes are open, the lighting is good.

Here's another example of an image where the quality is maybe not that high, but again we're looking at a particular set of images. We're picking the best one. The applications of this are many, of course. If you search for a celebrity, for example, you get a little box with a photograph of that celebrity. One of the things I discovered working on this project is how many celebrities there are. There are thousands and thousands and when you work at a global scale, of course, it's not just the ones that I know. There are famous people in India, in Africa, in Indonesia, and all these places make it a pretty hard task. Obviously it can't be done fully manually. The other issue is obviously that when you have so many, you have different celebrities with the same name. A lot of interesting challenges. In relation to that, one of the things that we need to look at, not only in that context but in the general context of image and video, is aesthetics. All of the quality and creativity metrics help you find the most creative and most interesting content. This is quite a big challenge.

Now I'm going to jump into the demos very quickly. Okay.

This is an example of deep learning framework for annotating a video. What we're doing here is... Essentially when you take a video and you tag it, automatically all you're left with is a bunch of tags. Now the problem is the tags. Alone they're usually not interesting enough and they're not sufficient to determine what's in the video, so what we're working on is going multiple levels deep into the tag hierarchy and reasoning on top of that classification. I'll show you some examples of what we're doing with this. In this particular example we have a number of tags and then how frequent they are at a particular point in time. Again, some of the ideas and the challenges that we have is at the video level. What I'm saying with this is that individual tags are really not going to be enough for what you need. We need a reasoning layer that takes the tags, expands them, solves contradictions, etc.

Here what we do is match videos with ads based on the tags. It's basically a reasoning engine on top of the deep learning framework, so that this is the best ad for that particular video. This is also useful for search, so if you search for furniture, you get a video where furniture is more prominent and so on and so forth. Another important area is the automatic generation of video summaries.

This is one example of that. The 15-second summary generated fully automatically from the original video and then a summary animated GIF. I'm not going to have time to show the live demo of that but it takes a few seconds to do a full video.

Let me just finish by saying, in terms of video understanding, there are many levels of classification, labeling—and humans do this. We have what we call "basic level categories." As an example, there's some research in psychology where they've done studies where they find that human accuracy in recognizing objects varies depending on whether the background and the foreground match.

There's a paper cited there. One of the biggest challenges, I think, is what to index. We have these deep learning frameworks that can easily label stuff, but that's usually not enough. We need to go a level higher. How to match that to user interests is the second one and the third one is aesthetics and creativity.

Thank you very much. Oh and that's an example of some of the videos that we are able to automatically classify as creative versus not creative. Thank you.

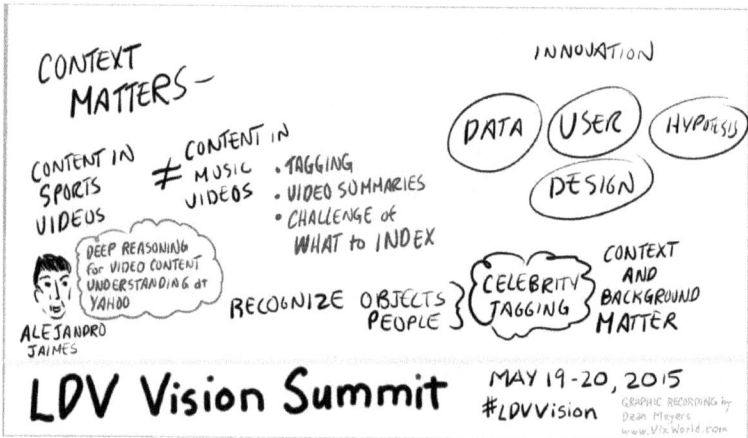

© VizWorld

8

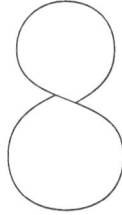

PANEL: LEVERAGING COMPUTER VISION TO SEE IMPORTANT SECONDS OF OUR LIVES FROM HOURS OF SECURITY VIDEO FOOTAGE

MODERATOR

Andy Parsons, CTO, Kontor

PANELISTS

Samir Kumar, Sr. Dir. Business Development & Product, Qualcomm Research
Carter Maslan, CEO & Founder, Camio

Serge Belongie: Lots and lots of people are setting up video cameras in their home. It could be webcams, it could be iPods that have been retrofitted to be security cameras, it could be baby monitors. All of these different technologies have emerged to monitor home spaces, retail spaces, and so on. That creates a lot of data and what are we going to do with it? Andy will be moderating this panel and he will introduce the panelists. Thank you.

Andy Parsons: Hello, thanks Serge. I'm going to invite the panelists up here with me to introduce themselves. This should be pretty interesting. I'm going to change the topic, just ever so slightly, which we'll get into when we talk about this. I think a lot of video intelligence is going to end up being about decision making, of which security is a piece, a component. But there are many more aspects that we're going to get into. Carter, we'll start with an introduction.

Carter Maslan: Sure, my name's Carter. I'm co-founder and CEO of Camio, and we do smart video monitoring that's accessible and easy for everybody. That's a quick one line summary on what Camio does.

Samir Kumar: Hi, I'm Samir Kumar, I look after product and business development in Qualcomm Research and specifically our efforts in computer vision and machine learning.

Andy Parsons: Good, I'm going to ask you guys to be much more long-winded than that because there are only two of you and this is not a subject of deep knowledge for me. I'd like to organize things in three parts, roughly. I'll keep an eye on the clock. I hope this inspires a few questions from the audience because it's a fascinating topic. First, I'd like to talk about state-of-the-art, kind of where we are in your respective companies and what you're working on now. Then, of course dive a little bit into security with the obvious privacy topic being one when we talk about security and video and ubiquity of cameras. Then, I'd like to spend the majority of the time on what's coming in the future and how optimistic or pessimistic we are about what's happening. Samir, can you say a few words about—more than a few words, actually— about what you're working on and what Qualcomm is up to?

Samir Kumar: Sure. You may or may not have heard of something called Qualcomm Zeroth, which is our effort to create a platform on which you can run deep neural networks on embedded systems, specifically Snapdragon-powered systems, and being able to do applications like on-device classification. You've seen examples throughout the day of image tagging or being able to classify videos. Now, imagine being able to do that fully on-device, on an embedded system, without requiring a Cloud connection.

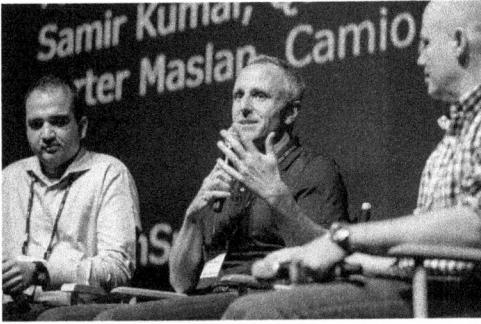

Panelists: Samir Kumar, senior director, business development & product, Qualcomm Research; Carter Maslan, CEO & founder, Camio; Moderator: Andy Parsons, CTO, Kontor [L-R]

We've demonstrated some of this already. For example, in camera preview, being able to do concept detection. You're looking at buildings or cars or planes or bicycles or things like that and really taking state-of-the-art convolutional neural nets and showing that you can, in fact, run these on existing mobile and embedded devices with the power constraints or the thermal constraints that they have and get really good performance. That's really the focus of our research effort, is how do you do this on an embedded platform? How do you make it go fast without destroying the user experience or killing the battery of the device? Then, creating a platform on which you can deliver lots of use cases using a common on-device platform on which people can build different types of machine learning classifiers and specifically deep neural nets. And then you can do it for audio, you can do it for video, you can do it for imaging, you can do sensor fusion, and things like that.

Andy Parsons: Great, Carter?

Carter Maslan: Just as background, when we started Camio it had harkened back to memories. Luc Vincent is here, director of Street View engineering. When we launched that at Google, one of the things that people did right away, when they were first exposed to Street View, every usability lab, was they would go to their house. That's not surprising, but the shocking part was that at least one person in every usability study would say, right after that, "Wait a minute, why isn't my car parked there? I left it there this morning." All the engineers laugh at this, but it got us thinking, *Well why isn't the real world indexable and searchable in real time?* What would it take to make that happen? A lot of the innovation that we're focused on is doing it at a scale and price that's accessible to everybody. Traditional computer vision has a lot of brute force computation and stuff that is expensive. How do you do that across millions of cameras where you're processing in real time and labeling in real time so that I get an alert on my phone that just says UPS has a large package on my doorstep at 2:32?

Carter Maslan (cont'd): The innovation that we were focused on was blending the image processing part with the machine learning part so that every single camera has its own set of neural nets, actually, that are learning what's interesting in that scene. Approaching the problem differently: in the sense that a little gray, furry animal on my walkway that's a squirrel is not particularly interesting, but to my co-founder, who has a place in Tahoe, that's a coyote in his backyard—with the same dimensions, same kind of object—that he would want to see. Learning, even without knowing whether it's a coyote or a squirrel, that the squirrel is uninteresting to me, not because we know it's a squirrel, but just because we know that pattern for that camera and that user is uninteresting. Whereas the coyote, also not known to be a coyote, but is very interesting because he saves it or shares it or we have the little thumbs up, thumbs down kind of like a Pandora music rating. It learns over time quickly so that we can train a camera within the first 30 ratings and really get a clean signal so you can see a digest of a day on one screen. A lot of the work that we're doing in terms of current computer science is just doing things cheaply. We'd love for more capable devices to come out where our app-based stuff tries to push as much to the edge as possible, but we also want to have a uniformity and accessibility so that it's open to any camera source. We're in this zone where we'd love to have more capable clients but they're not evenly accessible and so that whole shift in what's in the Cloud and what's on the edge in the client, I hope just becomes a really nice collaboration to take things to the next level in learning.

Andy Parsons: That was the almost perfect segue into my next question, actually.

Carter Maslan: Okay.

Andy Parsons: It's almost like we rehearsed this. We did have a short phone call. Around the topic of what logic resides in the Cloud versus what can be on-device, and think a little more into the future, where some more of Qualcomm innovation will appear on phones. If I can take the liberty of situating you guys on either side of the debate, if it even is a debate. Carter, what you guys are doing at Camio is much more, for the time being, Cloud-based, Cloud intelligence, lots of cheap, virtual machines doing a lot of interesting logic at a low cost. Samir, you guys spend a lot of money on research to get more of that onto our phones with the obvious subtext of what does that mean for security implications—the more intelligence I can get for monitoring my home at the cost of having more of it in the Cloud and accessible to prying eyes. I'm talking in extremes just to set up a little debate here. The question is: Over time, what are the right cognitive capabilities to reside in the Cloud and what are the right ones to be in my pocket?

Samir Kumar: Yeah, it's a good question. I think I would break it up into two pieces. I think there's clearly evidence that you can take trained classifiers— let's say that you trained them in the Cloud with lots and lots of data—and they're able to detect things like cats or that's a mountain goat or whatever kind of thing that you're looking for and bringing that trained classifier and then running that at the edge or on-device and therefore being able to extract that intelligence or salient information immediately and then make smart decisions about what aspect of that do you want to send to the Cloud for more processing versus blindly sending everything to the Cloud without first doing some filtering or edge processing. I think hopefully everyone agrees that over time, as we're able to do more and more classification on-device, it's clear that it's a hybrid model. Now, the other piece of this is: In the long term, the whole bit about having to train models and having to have all of the data in the Cloud—does that also have to be true? One area that we get excited about that we hope to see realized in the future is this notion of distributed learning. There are so many of Snapdragon-powered devices out there, and no single one of them will have the capability GPU cluster in the Cloud to train models, but now I've got billions of these and people are contributing both training data and a little bit of compute to also collectively build models and then ultimately decide which part of that model makes sense to have on their device versus your device or somebody else's device. I think that's where we would love to see the world go to, but there's a lot more R&D that needs to be done to make distributive learning, including on mobile hardware, commercially viable. The vision is there and there's definitely evidence that some elements of that are, in fact, possible and are worth investing in.

Carter Maslan: Already, today, just as an example, one of the biggest problems with video monitoring is all the false alerts. If you point a camera, like your old Android phone, out the window, just shimmering leaves or changing shadows or the lights coming on will trigger a lot of false alerts. As an example, Camio in the Cloud will develop an adaptive motion filter that says

"Oh, it must be breezy for the next two minutes" because it's started seeing a lot of activity in a particular region, so "ignore that as significant motion for the next n number of minutes." That's something that we can push to the clients. The Cloud has learned that this is not interesting, so apply this mask locally and don't send everything to the Cloud. That's like a simple version. But then some of the learnings—for example, rain night streaks on infrared cameras—the Cloud is uniquely good at being able to see that it's raining across a region, so it gives you a little more confidence to say that's unlikely to be something interesting. It's probably just among the pattern of rain. I think this hybrid thing is key, even as we get great edge capabilities, you'll still want to have the supplement and the ease of interaction across different guests that are accessing the video. Our cameras are trained by everybody that's a guest on the account so that you can increase the density of feedback. The things that people play or share can be known to be good and the things that people down-vote or delete are known to be bad. That kind of ease in having coordination across people is always Cloud. The part that I dispute about your "prying eyes" comment on the Cloud is that delineation in security—there's the machine learning and capability part— but the security part, I feel much better having it in the Cloud. You look at what happened with the FTC, who I think fined a whole bunch of camera makers for having 50,000 cameras accessible. Because even major vendors of security cameras—in Comcast and ADT and all these—they'll say step one is enable plug-and-play port forwarding so you can see live video remotely. Then, no one puts passwords on their devices, the firmware's hackable, and you end up with more risk than I think having just a secure connection with encrypted storage, distributed data center. I think the natural emotional reaction around "Wait, my stuff is in the Cloud" is maybe going to change if people really dissect the elements of the security. Because I feel safer with two-factor authentication encrypted on the disc. Anyway, that was a side point on the Cloud versus...

Andy Parsons: No, it's good. You can't dispute it because it's not actually my opinion. I'm just trying to stir up some controversy where there is none.

Carter Maslan: Oh, okay.

Andy Parsons: You guys are both agreeable folks working on amazing stuff. There may not be any controversy. Well, we'll talk about security in a minute. Again, mostly from a lay person's perspective, to do the kinds of things that we're talking about cognitive analysis. You talked a little bit, Carter, about how to filter out noise with the terrain or leaves. Given the amount of video that people are generating and the amount, exponentially, that we expect it to increase, as tools like what you're developing become more commoditized, more available, and more generally used, what's coming next? What do you need to think about in terms of storage? Just things that explode very quickly like storage, like noise, uninteresting frames and video. How do you think about that?

Carter Maslan: The first thing that we're asking camera makers—we can control the part where we are running the iOS or Android app or web browser app—is, *Can you please send down-sampled stuff to the Cloud for analysis while keeping the high-res local, as an example, because then you could get enough known immediately?* If, within 10 seconds, you could know whether an event is interesting, then you can cue the request for the high-res stuff and leave it on-device otherwise. I think some very simple first steps will easily handle our bandwidth and amount of video problems just with that level of tiering—apart from any improvements we have in connectivity and bandwidth speeds. Just that basic stuff. Most of the image analysis you can do on 120-pixel images. In fact, sometimes it's better because the noise is less when you have a lower res image. You could potentially have just a trickle of data and still get a lot of the information content out only to have the high-res stuff subsequently fetched when you really want to see the full HD.

Samir Kumar: Yeah, I fully agree with what was just said. I think the other thing I would add is, in order for us be able to deal with the explosion in the amount of video that's being collected and it's not just going to be true security cameras. There's cameras in cars, there's cameras in drones, there's wearable cameras. How do you decide what's interesting and what's not? Generally, for us to get better at doing anomaly detection, because those are the things that I didn't know about or I did not predict and I do want to capture that, versus right now it's all about we want to detect things like people shaking hands or hugging or my dog jumping through the door or something like that. I pre-define what these are, I can train on that and create a classifier and then look for those events. When I think about security, what's the most interesting is when there's an anomaly. How do you define an anomaly? How do you train to detect an anomaly? I think that's a very interesting area and there's a lot more research that needs to happen there. I think as we talked about on the call, this notion of, and you're seeing this in deep learning going to unsupervised learning, you don't have labels for the data, you just have a lot of data and you're able to look for patterns that are relative differences and then decide these are the ones that are interesting and I need to look at these further. I think getting to a place where unsupervised learning advances far enough that it becomes commercially viable, that will be another big breakthrough in doing really robust anomaly detection.

Andy Parsons: Great. Another awesome segue.

Carter Maslan: That's one of the most popular features. We have Camio Daily that just highlights the top four events of the day. It tries to find the unusual stuff, but it's both successful and failure because the highest ranking events sometimes have the people that were already there in the event. That's another example of Cloud influencing the ranking, where if you're at the same network location as the camera, it's unlikely that you think that was important... There's a whole bunch of signals you can use based on where the user is, where the family is, or the co-workers are, and ranking anomalies longitudinally across an extended period of time. What is the weekly pattern? What's the daily pattern? There's a lot there, but there's high demand for it, which is what I was leading with. People love seeing the highlight of anomalies every day. That's like a top feature, it's like 60% open rate and 52% subsequent search rate, just on this one daily email.

Andy Parsons: Let's jump into sort of security in society a little bit. That was a good segue because anomaly detection, obviously, is one essential aspect of security in general. Maintaining safety and some of the things that society is supposed to guarantee us that can be enabled, but also there's the darker side of security and surveillance. I want to raise the Boston Marathon bombing because it's so close in time proximity, the events of last week I think have it burned in everybody's mind right now. Taking a positive spin on that, what could be different were that, God forbid, to happen in the future or not happen? Talk about how it might be prevented, but more clearly, more succinctly, how your technologies and the combination of Cloud analysis and on-board cognitive intelligence on phones could have been applied.

Carter Maslan: That was a galvanizing thing for us, when that happened. Even if only one out of 10 residences in Watertown had had something like the video monitoring, then the kind of collaboration that could have avoided people getting killed with the pursuit of those guys I think is so powerful because you could say something. One of the things that I love is the idea of private convenience and public benefit. Where the private convenience is you get to see your package delivered, your kids are home, is my mom okay? Those kinds of things that are just private to you, no one else can see it, but then if something happens where there's a public safety issue, private citizens can marshal a coordinated response instantly, which is just so powerful. In fact, there was a tech reporter for the local *San Francisco Chron* news that did a segment on Camio and afterwards he said, "You know, I think it's people's civic duty to have a camera facing the front." I was like, "That's a pretty strong statement because it's controversial and stuff." But I get where he's coming from because you could have police broadcasts saying we're looking for a person in a black hat and a white shirt last seen at 2 p.m. All of the limitations that were mentioned around computer vision are so painful, but humans can glance at a thing so quickly to identify. One of the things that Camio's doing is enabling a broadcast like that to just say, "Hey, you have three candidate matches for a recent request. Would you be willing to participate?" And then you just browse and say yes, no, maybe. Being able to instantly marshal that kind of coordinated response I think is awesome. There's no kind of Big Brother surveillance eye that sees across all cameras. It's predicated on the opt-in support from people. I love that as a model and it was specifically in response to everybody seeing how effective video was in the Boston bombing.

Samir Kumar: Yeah, the Boston bombing one is really interesting. Maybe some people have seen it, but PBS *NOVA* actually did a special that talked about all of the technology that was used to find the Boston bombers. They highlighted a couple things that are very relevant to this discussion. One was, what is the current state-of-the-art in facial recognition technology that the FBI and different law enforcement agencies in the US are using? Then the second piece that they highlighted is the fact that they have so many different cameras in so many different locations and getting a centralized feed where they have analysts that are looking for anomalies. Today, it's a manual process. One of the key ones was just over and over looking at one particular corner in Boston and seeing one of the bombers drop a bag next to some mailbox or something in the street.

Samir Kumar: Now, imagine if we could automate that process. If I were to build an engine that said, "I'm looking for this type of anomaly. If someone drops a physical object or a bag in this location, I want the camera to actually trigger an alert or automatically do that versus a human being." That's one example. The other piece that they highlighted is when you look at face recognition, even with state-of-the-art, there are still many challenges in getting it to work. Dealing with occlusion, for example. I got a really grainy, partial shot of a face and the FBI is doing some interesting things like basically generating or creating a face model just with partial information and then trying to do matching on that. If you think again about machine learning, there's clearly more opportunities to advance facial recognition, doing things like generative models where a neural network could hallucinate what a face looks like and not necessarily have a bunch of labeled examples of *this is the face that you're looking for.* That's another example.

The last piece was the crowd sourcing piece, which is I think what you were also talking about. Everyone, whether it's on Facebook or different social networks, contributing their photos and then aiding the law enforcement agencies to find these anomalies. Again, being able to do some of that automatically where you can use my camera or you can use my device and I can do some interesting things on-device and then send that up centrally to law enforcement or to a Cloud. I've already done some first-level filtering, first-level processing, and then the experts can look for the more detailed information to make a decision if it's interesting or not. An analogy I like to think about is SETI@home. SETI's out collecting tons and tons of data from different radio telescopes. You can contribute your computer to process some of that and then send that back to the folks at SETI and then they'll decide if the data's interesting or if maybe there's a signal here. Now, imagine doing that for computer vision, with all of the wearable cameras or phone cameras or car cameras that we'll have and applying that same model.

Andy Parsons: Cool. Something in the future, two big questions. Same order of magnitude. I'm going to ask the first one and then open it up to the audience for a few questions and then follow up with the final one. It seems like a fair assumption, based on the discussion we just had, is that connecting everybody's camera, security aside and Cloud being as secure as it can possibly be, is something that will enable some of the innovations, home security, larger scale national security, state security, cities—but it's a huge task. And I don't know what protocols exist or what standards exist that would even enable connecting everything. How do you guys think about approaching that problem? Imagine that you two are at the forefront of making that happen in the next decade.

Carter Maslan: There's the simple version, which actually has happened already in my neighborhood. We had a robbery, that was rare. The guy had $35,000 cash in the back of his car. He owned a bar in San Francisco and would do check-cashing on a certain time of the month. There were three people that were casing him and they carjacked and robbed him. It was as if a director had said, "Cut to cam one, cut to cam two, cut to cam three." They robbed a street where the Camio co-founder lived, so all of the neighbors had Camio. It was a social experiment because within five minutes of the robbery, every neighbor had donated their five minute window, I remember 4:41-4:45, and with a single click, they just donate and now the police have seven neighbors with links to that specific time facing the street, showing scene by scene the three people that did the robbery. It's not future, actually. You can do this right now.

Andy Parsons: So your answer is that everyone should subscribe to Camio?

Carter Maslan: No, the protocol is, you should be able to have a link and the link should link to something that people can use. Yes, Camio produces URLs and they can go to vetted video clips that are distilled just to the parts that are relevant. I think the standards are there in the sense that it's going to be human collaboration. There's no machine good enough to prosecute a crime or investigate a crime, so it really comes down to using the standards we have for already great information exchange and collaboration and just not having it all locked up in DVRs and things that you can't access. I would just say web links is the standard.

Andy Parsons: Samir, anything that we should be thinking about on phones or beyond?

Samir Kumar: I think, again, this comes back to a little bit about as much as there's the social good aspect of sharing your camera data, there's also the privacy paranoia. The question is how do you balance those things? I don't know if there's a perfect answer to that, but one place to look at is when you think about things like peer-to-peer networks and headless type of deployment. Things we've seen with BitTorrent, and BitTorrent, as many of you know, recently has launched a messaging service where I don't need to have a Cloud back-in to do IP messaging. I can really go just peer-to-peer. What if I could do that with cameras? What if I could just say, "I want to share my camera feed with my neighborhood, as a starting point?" Maybe it's five, 10, 15 people. Maybe the second element to that is that when we collectively decide elements of that we want to share with the Cloud or with the authorities, we would also enable that. A starting point would be maybe not immediately jumping into my camera and my camera information is immediately going to law enforcement or to the authorities or to the Cloud, but I feel comfortable sharing it with my neighbors and participating in this local web for finding anomalies or finding threats in my area.

Andy Parsons: Great. Any questions from the audience, before we wrap up? There's one over there. Run, Evan, run!

Audience member: I'll talk louder, if you want. I was just curious, Carter, if the police apprehended those criminals in short order?

Carter Maslan: They identified them but did not apprehend them. They were in Hunter's Point and the Burlingame police said that they knew who they were. They came and deposed me for admissibility of evidence and things and frankly I don't think they even apprehended the guys, even after knowing who they were. I don't understand and I've pinged the police because I wanted to actually close the loop on that story and I still have to do the follow-up to see what ever happened. They said that they had ID'd both the car... But there were so many embarrassing things. The car that they hijacked was left parked in a neighbor's driveway and the neighbor was calling the police saying "There's a car parked here that shouldn't" and they were saying "That's a private matter, sir. You need to hire your own towing company." But that meant they had the license plate on file, so there were just a series of embarrassing prosecution things, but that I just need to close the loop to see what happened.

Andy Parsons: I believe the mic's not on or not working. You can just shout. I bet we can hear you.

Audience member: The question is concerning CCTV. We all know England has the highest saturation of CCTV. I know Time's Square, as a location, has the highest saturation of cameras. But recently *The Guardian* did an article on CCTV concerning that it does not prevent... Street crime has not decreased in England as a result of CCTV. Also, prosecution of criminals hasn't increased as a direct result of CCTV. The only thing that CCTV helps with is planned crimes, like jewelry store, et cetera. To your point that saying that's civic duty of everybody to have a camera, isn't that kind of absurd? How's it my civic duty? The government already has cameras everywhere and it doesn't help, anyways.

Carter Maslan: Andy, can you repeat the question?

Andy Parsons: I can try. To summarize, the question is given the density of CCTV cameras, stationary cameras in cities, there's a sense that there may be plenty of cameras. Why is it our civic duty to provide more?

Carter Maslan: I know the study you're referencing and I think the counter-argument is that it's not the severity of punishment or anything that matters, it's the certainty of being caught. We live in this fear state like we've got to lock doors, have guns. To me, cameras are an alternative where there's more good people than bad. If bad actors can't outrun the perimeter of people that are somehow going to contribute to the certainty of their being caught, then that's the civic duty part. You don't have to share it at all. It's all private to you, but if something bad happens and there's a collective deterrent just by knowing that there's a certainty of being caught as a bad actor, even though no one can see that... Part of the problem with central government monitoring is nobody's looking at it. It's like black box recordings that are ex post facto forensic evidence, as opposed to, say, a door-to-door solicitor that is casing a neighborhood and the neighbors say, "This guy tried my door without ringing the doorbell." Immediately the police are there at the time that the person is planning the burglary. I don't think it's apples to apples comparison with a state-sponsored, central CCTV system versus a distributed network of private citizens that are just trying to say, "Let's relax knowing there's more good people than bad and we can coordinate if anything does happen that's bad."

Evan Nisselson: Some more questions right over here.

Andy Parsons: Yeah, let's take one more and then I want to make sure, yeah.

Evan Nisselson: Okay, one more question.

Audience member: I don't know if you guys heard recently that BAE Systems said they're going to build an automated drone that could go bomb on its own. In that case, who bares the responsibility? Is it us, because we're developing technology that can recognize things while it's flying? It's like, "Oh, this is a terrorist" and just goes and bombs something. Who's really responsible there and should we even be building these things?

Samir Kumar: Let's make it even more personal versus a drone that goes and bombs. Let's think about self-driving cars and let's say they're using computer vision to run auto pilot and also then make decisions about "Do I stop, do I turn left, do I turn right?" I see a pedestrian, I don't want to hit them, I want to hit the wall. Even more controversial, you may have heard of this scenario. I've got a self-driving car. An accident is going to happen no matter what you do. There's no way to stop it. You can either hit those four children crossing the street or drive it into the wall and kill all the passengers in the car. What's the moral dilemma? Whatever the car does, it's going to cause the loss of human life. What should it do in that particular example? I think these are all very good questions. Whether it's for the drone use and it's for the military application. The one that we're all going to be facing very, very soon is these robot... These autonomous vehicles or semi-autonomous vehicles that are going to be making decisions on our behalf and in some cases, those could be decisions that have an impact on human life. I don't know if there's a good answer, a good, clear answer on that. There is a lot of legal debate happening there. Let's take cars, again: Where's the liability? Is it with the driver? Is it with the manufacturer? What is the insurance policy going to look like in that type of situation? Very, very active, open discussions, but I don't think there's a clear, defining answer there.

Andy Parsons: Okay, we're just about out of time. I just want to ask you guys one quick one. Given we have an audience of entrepreneurs and scientists and engineers, what's the number one innovation you think we need to see in the next few years to move the ball forward?

Samir Kumar: I think there's more than one.

Andy Parsons: Your top one or two.

Samir Kumar: I'm going to give you two answers, anyways. The first one is, if you want to do this out in the real world and in lots and lots of devices—whether it's your wearable camera or drone or car or your mobile phones—how do you do always-on at very low power and run sophisticated machine learning algorithms? We're seeing lots of noise about how we can do these kinds of hardware accelerators to do this really fast or at very low power, but the big white elephant in the room is right now, the way we've architected things, the biggest bottleneck is memory. Moving things in and out of memory, the cost of doing that from a power perspective and from a compute perspective. We need novel architectures to deal with very low power-embedded systems that can do things like machine learning algorithms, always-on. The second piece, when you think about video and understanding what's happening in a video—I think Simon from Flickr made a point about this. This notion of not just labeling and tagging things, but we're seeing evidence of things like top-down attention control, where there's one field in machine learning, which is language understanding and language modeling, and there's another field in terms of image classification or object recognition. Combining those two together to be able to actually direct and say this is the salient information, this is what's happening over time versus just what's happening on a frame-by-frame basis.

Andy Parsons: Great. I feel Evan's presence behind me, all around me. I want to thank Carter and Samir for an excellent panel and thank you guys for listening and asking great questions. Thanks.

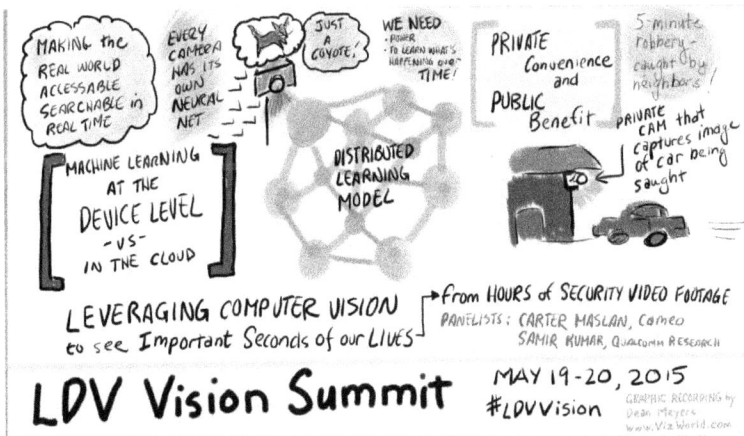

© VizWorld

9

FIRESIDE CHAT WITH SERIAL ENTREPRENEUR LANE BECKER

Lane Becker, Serial Entrepreneur & Author
Evan Nisselson, LDV Capital

Evan Nisselson: This is Lane Becker. He is fantastic.

Lane Becker: That's a great opening. Hey, everybody.

Evan Nisselson: You are a phenomenal serial entrepreneur... Loves parties but only when there's certain goals, which we'll talk about. We talked the other day. Actually, I think I was flying 30,000 feet in the air and you were somewhere else. I saw a Twitter stream and I responded and then we started talking via email. We haven't talked in awhile. This discussion is about the life of an entrepreneur, the roller coaster life of an entrepreneur. The good, the bad, and the ugly.

Lane Becker: Mostly about the ugly.

Evan Nisselson: Is it?

Lane Becker: I think we're mostly focusing on the ugly today.

Evan Nisselson: We can focus on the ugly, but also in that ugly, there's a discussion about what's ugly and how we can do better. It's a two-way street here. For the audience to know, we've both been entrepreneurs for about the same amount of time—18 years.

Lane Becker: Yeah, around that. Since...

Evan Nisselson: 1996, like I said?

Lane Becker: Yeah, 1996, 1997.

Evan Nisselson: I went to the other side of the table after my last company did well and then had an unfortunate disaster, for which I blame myself. The company got to about $3 million in revenue. It was a SaaS platform and we were trying to raise money or sell in June of 2008 with an investment banker. I was chairman at the time and we had about a dozen companies interested in June. In September of 2008 the economy crashed, the potential leads all closed their doors and said that they were no longer interested when the market crashed. Diablo Management was hired to liquidate the company by the majority shareholders. That was actually their name, "Diablo." Several companies were in discussion with them to acquire the company, including myself. I tried to negotiate with the devil, honestly, to buy back the company to keep it alive but it was not possible during that economic crisis.

Lane Becker: That's truth in advertising right there.

Evan Nisselson: It was really amazing.

Lane Becker: They know what they're doing.

Evan Nisselson: The bizarre part, and then enough about me—we're going to get to you and then back and forth... Actually, Andy, who was our CTO, who's in the audience, was at the last board meeting. Our typical board meeting was about seven people—normally investors, myself, and a couple executives. And this last board meeting was about 25 people, including lawyers and others on the phone. It was announced about 12 hours prior to the board meeting that Diablo was invited to come to the meeting by the majority investors. And all of a sudden it was a discussion of liquidating the company—of what to do, what not to do, and how to manage liability. Let's transition to you. So you've done a bunch of companies. How many?

Lane Becker: Three. I mean, if you want to go back to the stuff I did in college, four or five, but I'd say three in that kind of classic, Internet startup space. One was a design consultancy. One was an analytics tool, an early analytics tool in the mid-2000s that we sold to Google, and then my most recent company, Get Satisfaction, which I think will be the subject of this discussion.

Evan Nisselson: It will be the majority of the subject, absolutely. Talk about Get Satisfaction. When you were starting it... Or before you start, what was your goal? Why did you want to start that one?

Lane Becker: Oh, it's a good question. Well, it was 2007, which was a lifetime ago at this point in Internet years. It's funny. While getting ready for this conversation, I was thinking back to all the stuff that didn't exist in 2007 when we were starting Get Satisfaction—like Facebook as a platform that anyone used that wasn't a college kid or Twitter or SAS. Even the concept of SAS. I remember when we were originally pitching Get Satisfaction to First Round Capital, actually Rob Hayes from First Round Capital, who was our first investor early on in his investing career. I remember that I had a slide where I was showing the buying page from Basecamp and I was like, "I think we want to sell like these Basecamp guys do, where there's three pricing tiers and there's this one in the middle and we just think that's a really great way. We've never seen anybody sell like that before. I think it would be a really great way to buy online." I remember Rob, who is a wildly successful investor, going like, "Yeah. That will never work."

Evan Nisselson: And your response was?

Lane Becker: He is a much better investor now. We actually listened. I would say my entire experience at Get Satisfaction... I also want to take responsibility for the mistakes that we made. I could talk about the mistakes that we made all day.

Evan Nisselson: We'll talk about some of those, too.

Lane Becker: When it comes to our investors, I would say that the mistake that we made was that we listened a little too often.

Evan Nisselson: How do you choose when to listen? I had similar challenges as well. We all do. It's life.

Lane Becker: I don't know. Thor, Amy, and I came up with something... Thor and Amy are a married couple, the Mullers, and we came up with what we called "The Muller-Becker Rule of Investor Advice," which is that you should always assume that the approximate percentage of advice that your investors give you that is correct is equivalent to the percentage of your company that they own. Good luck figuring out which percent that is. I don't think there's a great rule.

Evan Nisselson: There's not, that's why I asked—because you're smarter than I am.

Lane Becker: I talk to a lot of people about how to manage boards since the experience of Get Satisfaction—and particularly the experience of not managing our board particularly well at Get Satisfaction—and I think it really comes down to just really knowing what it is that you believe in or what you care about and making sure that is represented and then weighing all the advice that you get relative to that. These are people who are spending part of their time looking at what you're doing. You're spending all of your time looking at what you're doing. You're the one with deep knowledge and deep experience in that area. It's valuable, but you always have to gauge it against *What is the core of what it is that I'm doing?* and *How do I apply it against that and decide?* I realize that's really abstract. It's also almost impossible to do, especially relative to your investors who definitely have a sort of power or authority position over you.

Evan Nisselson: I think that makes a lot sense but I just realized, we should probably take a second and back up. What was Get Satisfaction in the beginning? The goal of it, the funding that came in, and the recent outcome, which is what sparked our discussion when I was flying in the air and saw a Twitter stream discussion that I thought was very valuable.

Lane Becker,
Serial Entrepreneur
& Author

Lane Becker: Get Satisfaction was a customer service community platform. We came up with the idea in 2006, 2007, looking at online forums and seeing all the conversation that was happening about people who had products and sharing and having ideas and sharing problems and getting solutions from other customers. So the original idea for Get Satisfaction was a consumer site that was basically customer service without companies. How do we create a space where customers can share ideas, answer questions? That sort of thing. It did pretty well initially and early on, but very quickly, one of the things that we had done is we'd created a mechanism for employees to... This is our little accidental growth hacking thing, what would now I guess be called growth hacking. We had created a way for employees to self identify so they could come and they could say "Oh, I'm an employee of Yahoo, I'm an employee of Apple" or whatever. There were no Apple employees that showed up, at least officially. There were plenty that showed up unofficially.

Evan Nisselson: Maybe they're here unofficially, too.

Lane Becker: We would give them a badge and they would start answering questions alongside the customers. It was actually very effective. Again, this is prior to Twitter and Facebook becoming customer service platforms. That was still just a twinkle in someone else's eye. I think we were very direction-ally accurate in that sense about what the product needed to be, but kind of basing it on old technology forums instead of thinking about where it needed to go. But the thing that did happen that was so interesting is that once the employees started finding out about us... We'd done a really good job on SEO on the pages. Mid-2000s, SEO is still kind of a thing, and it turned out that all of these marketing and customer service types in all of these different companies had actually set up Google alerts for the name of their company and we had made sure that the name of their company was very prominent in the page. Basically, what we had developed is that as soon as someone came in and asked the question about, say, a Samsung product, it was like having a direct line into all of marketing and customer service people in Samsung because they'd all set up these Google alerts and it was SEO'd so it would show up relatively high. We developed this really fantastic way to get access to all these people and so suddenly we had all of these employees from all of these companies, some fairly high up, who are self identifying and answering questions around these products. That was the point at which we started having a conversation with Rob and some of the other folks on our board about, *Maybe we should start to turn it in this direction. Maybe we should go back to that original idea we had about selling this as a product, as opposed to the place that we had initially pushed it based on our investor feedback, which was towards more of a consumer environment.*

Evan Nisselson: At that stage, you were making how much revenue?

Lane Becker: Oh yeah. We were making no revenue. This is Digg-era days of trying to get big fast with a consumer platform. So much of this ends up being subject to the sort of vagaries of the moment. Do you know what I mean? What's the hot thing? What's the direction that everybody else seems to be going? Where can we pattern match today? We were kind of flip-flopping around a little bit based on that. Personally, I think if we just stuck with our initial idea of following the Basecamp model and getting into the SAS approach earlier, we actually probably would've been in much better shape in 2008 when everything tanked. It's funny, listening to your story and my story...

Evan Nisselson: Well, not really funny.

Lane Becker: Well, I'm going to go with funny because I have no choice but to laugh at this point. My childhood dream was to be a cautionary tale for others. Here we are up on stage. In 2008, the economy tanks and there are these macro conditions that have fuck to do with anything that any of you in the room are doing. We can blame 2008 entirely on a bunch of guys wearing ill-fitting suits sitting about a mile and a half that way. That way? That way? I'm lost, orientation-wise. Again, what they were doing was terrible. It had nothing at all to do with what I was doing or with what you were doing, but those macro conditions totally influenced our ability to succeed. In your case, it was your ability to sell. In our case, it was our ability to raise a series A, which had been going quite well up until that point. Suddenly, the money's not there. So I'm scrambling. I end up asking some of our better-off friends... I mean, I did all sorts of things to keep the company going through this period, like asking people for money.

Evan Nisselson: Tell us a couple of those things that you did.

Lane Becker: Actually, one of the things that I did which was fantastic advice that I got from one of the guys who would later become our investor, Josh Felser from Freestyle... Josh is amazing. I would totally recommend taking money from him and that's a very short list of people that I have for that.

Evan Nisselson: You're an advisor to them.

Lane Becker: Yeah. I was for their first fund.

Evan Nisselson: They're also an investor in one of our companies, Camio, and the CEO, Carter, is in the audience. There he is, right in the back.

Lane Becker: That's right. Everybody likes Josh. Josh has this thing that he says, which I think every investor should say but very few do. He says, "We will always have an agenda, but we promise to share it with you." So when he feels like his interests are going to diverge from your interests, he will point out how that is happening and why he is lobbying for something separate from you, which is terrific because there are always times that your investors are going to diverge in their interest from you. Always, in every company.

Evan Nisselson: Is that solely a personality trait or is that because he's a serial entrepreneur?

Lane Becker: I think it's a personality trait but I also think it is both.

Evan Nisselson: Both.

Lane Becker: Yeah, I've watched them win all sorts of investment opportunities based entirely on the fact that he's able to empathize with the people that he is talking to far more than somebody who's never actually run a business in their life. He suggested that we have a dinner party and that I invite all my high net worth friends to come to the dinner party and basically just tell them, "Hey, I'm pitching you an investment and dinner will be terrific, but if you're not interested in investing, we'll have other dinner parties. Don't bother to show up." He was like, "80% of the people won't show up and then 20% of them will and the 20% that show up are all going to invest." Which was completely, 100% accurate. That is exactly what happened.

Evan Nisselson: Wow. That's great.

Lane Becker: It was. This is like December of 2008.

Evan Nisselson: Did you believe him or was that like, *I'll give it a try. I don't think it's going to happen?* At that stage, it was kind of like *I'll do anything to try and raise capital to keep the company alive?*

Lane Becker: I was already in that *I'll do everything to raise capital to keep the company alive* place. I was very much there. In fact, this story has a very sad ending in which we sold Get Satisfaction to a company called Sprinklr, probably like six to eight weeks ago. And every early investor, every employee, all of the founders got completely washed out, so none of us see a dime from the sale. The only people who make anything off of the sale are, interestingly, all the people who are still sitting on the board: later investors, the current CEO, and the former CEO. It's funny how that works. Actually, the mechanism by which that works is one of the things I want to talk about. We should get to that. One of the things I have observed through the experience of being more public about this than most people are when their company sells (but not really) is how much opacity there is. We have so much transparency and we talk about our world as so transparent and open, and it certainly is relative to the way investing worked in the '90s, for example, but I want to be transparent and open about things. Like in the beginning, there's still this surprising shroud of secrecy and uncertainty around how things end when they don't end well, which is most of the time, or at least they don't end fantastically. It turns out there are things you need to know about that part, too—including, for example, the way we got screwed, which is one of the many ways you can get screwed in a sale where not everybody's going to see something from it.

Evan Nisselson: We'll talk about that in a second. Let's lead up to it. You evolved your role with the company from what to what and when?

Lane Becker: Well, I was always co-founder and chief product officer. I didn't actually take the CEO title but Thor and I used to joke... he was the CEO of the company. We used to joke that he was the peace-time CEO and I was the war-time CEO. He was really good when things were going well. I was really good when you needed somebody to get kind of pissed off at board meetings. I sort of ran that piece of it and the fundraising piece and everything that happened post 2008. Thor was still very much the CEO of the company.

Evan Nisselson: A couple of years ago, you switched again. Correct?

Lane Becker: Yeah. So what happened to us post 2008 is that we were basically told by our board, "Okay, it's great that you guys are having fun with this little consumer toy but it's time to bring the adults in." This is kind of classic Silicon Valley behavior prior to the Zuckerberg-Sandberg-Andreessen-Horowitz era, where their one-two punch of Sheryl Sandberg coming in underneath Mark Zuckerberg instead of on top of him, which is what they would've done in years previous and what they would've done, frankly, if Zuckerberg hadn't had Peter Thiel advising him on how to structure his board... It's true, he was 20. He would've gotten screwed if he hadn't locked into some good advisors; Andreessen Horowitz opening up and saying, "We believe in founders. We think founders need to stay in charge of their businesses. We think a good VC teaches a founder how to become an investor." That is all absolutely, 100% true and came a couple of years too late for us. So in late 2008, they're basically like, "You need to raise money and prove to us that you can raise money in this totally fucked up environment or we're going to lose all faith in you. And oh, by the way, also we've lost all faith in you and we think you need to get an adult in here." Those are sort of the messages...

Evan Nisselson: It doesn't sound like it was an option. It was a way of phrasing.

Lane Becker: Right. The thing is, if I could go back and do it again...

Evan Nisselson: What would you do differently?

Lane Becker: I would just tell them to fuck the hell off. Seriously. And you know what? I think that's kind of what they wanted me to tell them. I actually ended up having a conversation with Rob Hayes years later about Travis Kalanick, who, let's be honest, is like clearly the most successful asshole billionaire in the industry, right? Really plays it to the hilt. This is long before Uber is in any other city besides San Francisco. Rob Hayes, to his credit, was one of their seed investors at First Round, so nice work Rob. He told me the story about Travis Kalanick that really stuck with me, and this is after I had left Get Satisfaction, so it was probably in 2010.

Evan Nisselson: You left and you were still on the board or you weren't?

Lane Becker: I left. I got pushed off the board first with the series A round and then Thor got pushed off the board in the series B round. Rob tells me the story about how there's a board meeting he needed to reschedule because he had a conflict so he has his assistant call Travis. Travis probably didn't even have an assistant at that time. Call Travis and say that Rob needs to reschedule the meeting and Travis says, "Fuck you. We're not rescheduling that meeting. I'm putting my time and energy into this. He needs to put his time and energy into this, too." Rob was like, "I have so much admiration for Travis for doing that." And I realized, "Oh, that's how we fucked up. Rob's a bottom." Clearly, that's what he wanted. He wanted me to dominate him because that's what venture investors want from you. Now my attitude towards this sort of thing is to go to a BDSM metaphor. Thor, who was always the sort of more politic of the three of us, Thor's take on it is that your investors are always testing you. And that was the test. In that sense, we failed that test because in that moment, we weren't forceful or aggressive enough. We weren't doing the things that they needed us to do to see that we were passionate or committed. And I actually think, as fucked up as that is, it's also totally, totally true.

Evan Nisselson: I actually...

Lane Becker: I usually don't swear this much.

Evan Nisselson: The subject is relevant for swearing. I'm sure I could loft some out there as well, depending on the question you're asking me.

Lane Becker: Apologies if you have delicate ears.

Evan Nisselson: No, the kid left earlier so he's no longer here. That was Serge's kid that I invited and gave a little name tag, if anybody didn't know. What is he, three weeks? A month old, a month and a half? So that brings up the point. Let's finish the story of the situation and go back to talking about investors and the crux of how you got screwed. All of a sudden it sells—the company. And you'd been out.

Lane Becker: I'd been out for awhile.

Evan Nisselson: You probably had limited knowledge of what was going on.

Lane Becker: Limited.

Evan Nisselson: You're still the shareholder.

Lane Becker: Limited and as we'd argued, deliberately incorrect knowledge of what was going on.

Evan Nisselson: And you were out for how long?

Lane Becker: I left in 2010. Thor left in 2011 and I believe Amy left in 2012 or 2013.

Evan Nisselson: So over the last two to five years...

Lane Becker: Left.

Evan Nisselson: Right, and still had equity. Not a lot.

Lane Becker: Still had equity. Sort of ever decreasing.

Evan Nisselson: Decreasing, recapping, and other things.

Lane Becker: By the end, probably the three of us collectively owned between 7% and 10% of the company, depending on the day.

Evan Nisselson: Then all of a sudden news hits. Get Satisfaction is sold.

Lane Becker: Right.

Evan Nisselson: Tell me the story just before that, because I think there was some behind-the-scenes to that. How did you find out and then how did it evolve to all of a sudden having an extensive discussion with entrepreneurs around the world on Twitter?

Lane Becker: This is where the story gets kind of gross and ugly.

Evan Nisselson: And that's why I asked.

Lane Becker: Yes.

Evan Nisselson: I'm sorry. That's why we're here.

Lane Becker: I don't mind. What the hell. We're here. We found out because we had maintained, even after leaving the organization, we had actually maintained pretty close ties with a number of employees, which is what I would recommend to everybody even if and especially if you end up getting shoved out of your own organization. Employees on the ground usually know what the hell is going on. In this case, we actually found out from an ex-employee who had also done the same thing, who had maintained tight relationships. And apparently, the employees at Get Satisfaction had been explicitly informed not to tell any of the founders that the sale was happening because the current management, I won't be more specific than that, had decided that we were a liability in this situation and they weren't going to tell us that the sale was closing until—it's unclear to us—either exactly the day it was closing or perhaps the day after it had closed. So they had just cut us out of the loop entirely. But this ex-employee had caught wind of it and he had no reason not to tell us so he just called us up and he was like, "Hey FYI, your company's selling." That is just the shittiest way to find out that your company is selling. At the time, though, we assumed, *Okay well, we're probably getting washed out, right?* Because why wouldn't they be telling us if they were going to get us even a little bit of capital? Now, up until this point, our understanding has been that revenues had sort of leveled off. We knew that they were struggling. We understood that they'd done a convertible note with really onerous preferences towards the end in an attempt to keep things going. But again, they were like *happy, happy, rosy, rosy* in their conversations with us.

What I understand now is that it was 3x... A convertible note with a 3x liquidation preference was there primarily to ensure that only the people that participated in that note were going to see anything from it, because they had done a successful job of hiding how badly, frankly, they had managed the business, how off-a-cliff the revenues appeared to have gone. So they were all kind of scrambling to make sure that they were going to get their money or they were going to be able to get something out of it. That's what that last year actually was—not an attempt to get the company back on its feet, but an attempt to basically steer it in a direction that was going to guarantee the maximum outcome for the people that still had some insight into what was happening: the people sitting on the board.

So we find this out. Thor emails very politely their CEO and CFO and says, "Hey, you mentioned awhile back that you were thinking about maybe acquisition or fundraising. How's that going? Can we maybe talk to you about it?

Come in and talk to you about it?" One of them, the CEO or the CFO, writes back and says, "Yeah. We're really busy this week. Why don't you come in next Wednesday?" Thor was like, "So, funny story. We actually know what's happening and we think you should bring us in much sooner." And all of a sudden they're like, "Oh yeah, you should come in on Friday." So we go in on Friday and we know that we're going to get washed out. There's no way they would've been screwing with us as much as they were if we weren't going to get washed out. But we go in basically with an argument like, "We think, ideally, you should at least recognize that the common stock is getting completely washed out in this instance. We still think you should give something to the common. Even if you're not going to give something to the common, even if it's just like pennies on the dollar of the sale in recognition of all the people who have put so much time and effort into this…." This, by the way, is not uncommon behavior even in the situation where common gets washed out. Frequently, in order to maintain relationships or in recognition of the work that's been done or to just not look like an asshole, the preferred stock will actually throw something to common anyway. That's actually something that does happen. Not in this instance. We were like, "Okay, even if you're not going to give it to common, at least recognize that we are the founders of this fucking business and we put a ton of time and energy and capital into it and we would like to see some small token thing just in recognition of that. And hell, even if you're not going to do it because you're nice people, you should do it because why the hell else would we support this sale? What is the point of us supporting this sale?" And the CEO's like, "Oh, but it's your baby. Don't you want to see your baby make it out into the universe?"

Evan Nisselson: It's his argument that if you don't support it, it goes out of business?

Lane Becker: No. His argument was "suck it up," basically. He basically says, "Okay, I'll go back to the board." Oh, so the other thing we learned in this meeting on Friday, besides the fact that we're getting washed out, is that the sale's closing on Tuesday. And we were like, "Wait a minute, you told us you didn't want to talk to us until Wednesday?" Like, what? Jerk!

Evan Nisselson: You probably used a stronger word than that. You don't have to use it here.

Lane Becker: No. There was this really funny part. The worst part about this meeting is that I go in, Thor goes in, Amy goes in, and we know the whole point of this meeting is for him to deliver us this shit news that we've been washed out. But before he gets to that, the first 20 minutes of the meeting is spent with him grilling us on which employee told. "Which employee told you? Which employee told you?"

Evan Nisselson
@nisselson

Genuine & transparent entrepreneur roller coaster stories @monstro @LDVVisionSummit yesterday #narrativeclip thanks!

Evan Nisselson: So that situation was horrible. Let's jump to...

Lane Becker: He says, "We're going to ask the board." The board basically comes back and says, "No. We're not going to give you anything. We're not giving common anything but we do expect you to support the sale." And it was in that moment that I realized this happens. It's so frequent. I end up going out on the morning of the sale... It's Tuesday morning and this congratulations note hits my phone and wakes me up at 6:00 in the morning because this news has gone up on the wire that Get Satisfaction is sold. So all of my friends start doing the thing that you would totally expect them to do because they're your friends, which is they start sending congratulatory notes. I was just not in the mood for it so I wrote this thing on Twitter, which I have to say I did not expect to get the kind of pick-up that it did. I was basically just like, "Hey everybody, I appreciate the sentiment but don't congratulate me on the sale because the founders got totally washed out and we got nothing."

Evan Nisselson: As you are a very genuine and sincere person. That's what the message was, but to everybody else, that's unusual in Silicon Valley.

Lane Becker: I know.

Evan Nisselson: Unfortunately—or in most of the ecosystem.

Lane Becker: No, I know that's accurate actually because one of the very, I don't know if "amusing" is the right word, but one of the really fascinating outcomes of this is that I got a lot of private messages from a lot of successful entrepreneurs who all said something along the lines of, "Wish I could pretend I didn't know what you were talking about." It just made me realize in that moment: this is absurdly common.

Evan Nisselson: I was on the plane and I saw the thread, which had many, many comments—I don't know if you have any numbers, I mean dozens, hundreds... It just started going and going and going and going.

Lane Becker: Yeah, it was great for my follower count.

Evan Nisselson: Anyhow, it started and I felt so bad. I know what it's like. A friend was going through this and I actually belabored for like 20 minutes on the plane: Do I post publicly? What can I say? What would be appropriate? What would be right? What would be helpful? And then I just sent you an email and that's where we started a back and forth thread. A lot of people voiced that they thanked you for sharing that.

Lane Becker: Yeah. It made me feel great, actually.

Evan Nisselson: But now that it's out, and you look back... So talking about the investors. First question actually, most importantly: You mentioned earlier there's a lot of negative results. There's only a small percentage of successes. Our audience is all trying to build businesses or build technology. Are you going to do another one?

Lane Becker: Oh, yeah. I would totally do it again.

Evan Nisselson: Perfect, I was assuming that was your response because there's only one answer for a true entrepreneur. But now, looking back at that, what would you do differently next time?

Lane Becker: We covered the one point, right, which is that I would have had a lot more independence and this is something I think comes with age.

Evan Nisselson: Independence?

Lane Becker: From the board. I would've set things up better so that I would've been able to maintain control because I understand how to do that now. And then I would've actually maintained control because it's not just about the percentage ownership or how much money has been invested. It's also about the more subtle social ways in which investors can create pressure on you. I mean, at the time that we gave up the CEO role of Get Satisfaction, technically we still owned more than 50% of the company. We didn't have to do that. It was far more the intimidation factor of it that made us do it than anything else. I just feel much more prepared for that sort of thing these days.

Evan Nisselson: I had a similar situation where, unfortunately, we almost had a large financing and it blew up in the 12th hour. And all of a sudden there was a transition where the top investors and the new CEO said, "It's best if you don't come into the office after you transition to chairman. It's best for everybody." And actually, it was a very difficult period because I wanted to do what everybody thought was best for the company and at the time I was not sure that was the right decision. Learning hindsight is 20/20, but now I realize the decision of not going into the office after that transition situation was a disaster and wrong on many levels. How else can we learn if it's not from our own challenges? Is it advisors? How do we know which one's the right advisor? Is it just that we have to go through this crap and become better at the other side and hopefully it's not disastrous?

Lane Becker: Well, I definitely think learning from other people's experience is better than learning from your own experience.

Evan Nisselson: That's a fucking rhetorical question. And see, I cursed. I knew it would come up when I started talking about this. How would you answer that differently?

Lane Becker: If I were doing this again today, I would definitely have a much stronger support network. There just weren't as many of you in 2008, 2009 honestly. I would definitely build a much stronger support network. Or I would have come up through a much stronger support network, like say an accelerator-type structure that gave me access to other people who I could talk to and work with. I think that is huge. The other thing I would do differently today is I wouldn't panic as much as I did then.

Evan Nisselson: I panicked all the time.

Lane Becker: Yeah. And now, I'm like, "Well, it's just a company."

Evan Nisselson: One of the things I try to do now after successes and failures... I am a mentor to about five accelerators and try to help others avoid the mistakes that I did. You can never dictate. I think it is best to share stories which can share perspective for others to make their own conclusions. That's the same thing I do as an investor. You've met some investors now who you would definitely work with in the future and you probably have signals that say, "Oh no, not that one." Tell us a little bit of how you would choose those signals. I know we've got a couple of minutes left and need to wrap up, but this is a fantastic discussion to help others avoid the mistakes that we might've made.

Lane Becker: I just think your investors are essentially your boss. For all the *blah* in this industry about how you get to be your own boss, that's really not true. You have people that you report to, right? And your investors are those people, and so like any situation where you're going to have a boss, the question is: Do you like and trust this person? At the end of the day, that's really what it comes down to. Are you able to communicate with them in a way that feels meaningful? Have they done things, have they said things that are indicative to you in some way, shape, or form that you can trust them? Do they talk about things as if they were experienced? Do they have an experience that you can relate to, because that's one of the fastest ways that you can form a trust relationship with somebody? I think that's the reason why Josh and probably you were successful with a lot of the companies that you invest in is because that resonates with them. I just look for that human connection in this situation and some sign that they're not just some autonomic bottom feeder and just trying to sort of take what they can and then run off, which unfortunately, the business world has quite a few of.

Evan Nisselson: I think that's great advice. We could talk for a long time, but I want to end on this: Is there anything else that from your experience you've learned that you would give as concrete advice to anybody in the audience who is either building or wants to build a company?

Lane Becker: I think you're better off when you think about venture capital as a game or taking venture capital as a game. I recognize that it has all sorts of real-world inputs for a lot of people. Actually, starting companies is one of the ways that they aspired to class-jump, which is very hard to do in this country. We have a surprisingly static class system and Silicon Valley is still very aspirationally one of the ways that you can do that. So you're thinking about your future, you're thinking about the people that care about you—there's all sorts of reasons. I recognize the gravity of it. At the same time, you are going to be better off if you can treat it as if it wasn't the most important thing in the world. You know what I mean? Entrepreneurs are always at a disadvantage in negotiating situations with investors because the person who wins a negotiation is always the person who has less of an emotional investment. Always. You can go into any negotiation and if you have more of an emotional investment, you will lose. That's just kind of the deal. You have to figure out how to manage that and to me, managing that means pulling yourself away from it as much as possible, looking at the situation dispassionately and understanding that if it's a game, there are rules. It is a system. It's kind of a weird game because some of the rules aren't necessarily as apparent to you as others. If you can, treat it as a game—understanding the system and the system dynamics, understanding the person and their motivations and intentions, and understanding how they're going to align or not align with yours. On the one hand, I want to say, "Get to know that person as a person and trust them," and on the other hand, I want to say, "Step back from the situation and recognize that it's not just about the two of you as individuals." It's actually about the system that you're participating in and the expectations that you have and that your LPs have and your returns on capital and the margin macro economic situation.

Evan Nisselson: So it sounds like you have to do both and there's a fine line between getting to really know them and taking a detached perspective. I think that's fantastic advice. It's probably everything in life. But in this situation, investors, co-founders—it's very critical to have that view without blinders on.

Lane Becker: Yes. I don't think it's surprising that so many of the startup founding teams that manage to build these sort of wildly successful companies, they're on their second or their third or even their fourth company before it kind of hits. The Twitter team is a great example of this. It is the act of working with these people over time, learning their strengths and their weaknesses, figuring out how you can trust them, figuring out where you can lean on them, that makes you quite successful. I have no doubt that that's true over repeat investments as well.

Evan Nisselson: Right. This had been fantastic. Thank you very much for sharing the ups and the downs with transparency. You're going to be here for the rest of today?

Lane Becker: Yeah, I'll be around.

Evan Nisselson: Others, if you have questions, we don't have any more time right now but Lane is here. He's fantastic. Thank you for sharing and enjoy the rest of the Summit.

Lane Becker: Thanks for having me.

Evan Nisselson: Thank you, Lane.

Lane Becker: Thanks, everybody.

Evan Nisselson: A round of applause, guys.

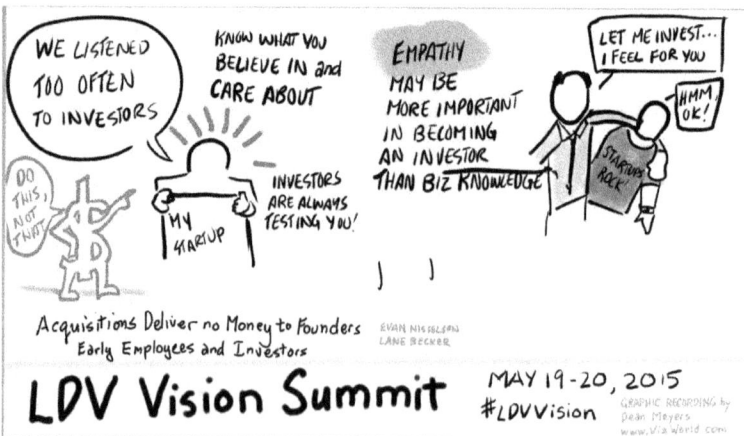

© VizWorld

10

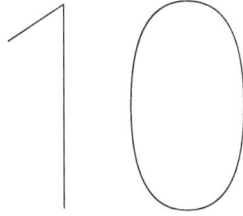

PANEL: FUTURE INVESTMENT OPPORTUNITIES IN BUSINESS

MODERATOR
Erick Schonfeld, Co-Founder, TouchCast

PANELISTS
Sean Ammirati, Partner, Birchmere Ventures
Patrick Eggen, Head of Seed Investment Program, Qualcomm Ventures
Fran Hauser, Partner, Rothenberg Ventures
Geoff Judge, Partner, iNovia Capital

Erick Schonfeld: So this is the Bottom Feeders Panel.

Erick Schonfeld: We're going to talk a lot about where you guys are investing in this space, but I found that a very powerful story, Lane. That does happen a lot. Would anyone care to respond to what happened to Lane? Does that happen in your experience? How can we avoid those type of situations?

Sean Ammirati: Before I became a bottom feeder, I spent a dozen years starting and selling some companies. Part of the reason that I got into this is that I wanted to be a better investor for entrepreneurs than some of the experiences that I had gone through. Some of those have been more successful than others, but I think one of the things I really appreciated about that panel was I feel like—you were once a part of a TechCrunch or Read Write Web that I was at—but I was like, "It's easy to get people to go on the record and talk about the good stuff. It's hard to get the honest conversations about what happens when shit hits the fan." I thought that was just a very helpful conversation for everybody to have. No offense taken to being a bottom feeder. I would point out just for the record, Evan, that you are also a bottom feeder now by definition. We're all in this together, but trying to be good partners for entrepreneurs.

Patrick Eggen: Quickly one lesson learned from that, and that was a wonderful talk, was when you do a reference check on investors, actively seek out companies where that investment didn't work out previously and see how that investor shows their true colors when things go south. Because when money's involved and it's a distressed or very modest outcome, their actions oftentimes deviate from the norm. You really need to see how they act in that. Do they have the moral consciousness or are they just pursuing their own agenda? And if they are, it's okay if they're at least transparent about it.

Erick Schonfeld: One thing that I think that came from that story is that we've been seeing a lot of people talking more publicly about just the whole funding process, the startup process, and just founders understanding more what the different terms of the term sheet mean, and it's not just about a valuation but about these other terms. Lane mentioned the 3x liquidation preference, and I imagine that in these instances when you're on the other side of the negotiating table and you're trying to get the best terms as an investor, it's not incumbent upon you to explain to the founder what might or might not happen three years down the line. But are there rules of thumb? What are the signs in that negotiating phase that founders should take as a red flag?

Geoff Judge: The 3x liquidation, usually you see that when a company is already in trouble. It's called: *If we're going to get any more cash in here, we might have to offer that up because there's probably a lot of the risk involved.* Most of us don't like to see any liquidation preference besides our original investment because then the next guy who comes on, then they want that same liquidation preference. If a lot more money is coming in, then guess what: there's going to be a lot more money going out before you get your share. But usually those type of things do show up in bad times. Somebody is taking a risk and putting more capital in. We certainly have all had companies that go belly up. As an angel investor, I've made over 80 investments, so plenty have gone south. But I think you want to build a good relationship with the founders. Only in one instance have we ever removed a founder. We've had founders who may have decided they wanted to do something else, but as Lane was mentioning, you really have to have a good level of trust with your investors and really feel that they're going to do the right thing. And to your point, talk with some of the founders of the businesses that go south because they certainly have a good picture for what the investor is like.

Erick Schonfeld: The topic that we're supposed to be talking about is leveraging visual technology. That's what this whole Summit is about. Can you just tell us a little bit about the areas that you guys are currently investing in? What are the areas that you're most excited about? The way that I think about these technologies is that they're enabling technologies, they're foundational, but those may or may not turn into good businesses. How do you decide when to become involved?

Sean Ammirati: We tend to be one of the first investors into a company, so pretty early. I've made about a dozen investments personally in the last three years. I would say that three of them have leveraged visual technology, but none of them would I have said, "That was a visual technology investment." But they were part of bigger trends. Visual technology was an enabling technology to let the businesses solve the problems that they were solving. Just to quickly illustrate that: we're investors in the seed round at Peloton that just raised $17 million about 18 months later, but Peloton is basically a mobile eye for trucking companies. If you guys have adaptive cruise control on your car today, it's basically adapted cruise control for trucks. We have a company that leverages OCR to do workflow automation. We have a company in the e-commerce space that we haven't yet announced that uses visual technology to create a better e-commerce experience.

Erick Schonfeld: If we can just go down the line, give us a sense of what your investment is.

Patrick Eggen: In terms of the visual technology area, our portfolio is heavily weighted on that side. At Qualcomm Ventures, we thrive on the complexity of technology, but at the same time, we want to ensure that there are commercially-viable use cases in the longer term. In the augmented reality world, we have Blippar, which is the leading campaign management platform for brands in the world. "AR" was a four-letter word for years. Blippar was probably the only one out there on the low end that's monetizing at scale. On the high end, we're investors in Magic Leap, which is very stealth but next gen. On the computer vision side, we have an investment in a company called Matterport, which does 3D image scanning for indoor areas, the reconstruction of indoor areas. We also have investments in the deep learning side in Clarifai, which I know they're represented here as well. What we look at is we really aim to find very disruptive technologies that take a phased approach, that in the near or mid-term, they're a viable use case. It's very focused. So with the issue of Matterport, it's real estate. We're early investors in Skycatch—they do aerial imagery for drones, laser-like focus on construction sites. You have to have this phased approach, prove it out, validate, raise more growth capital, and then expand in other adjacent markets. That's our philosophy.

Erick Schonfeld: Magic Leap, that was a high-profile company. They raised how much?

Patrick Eggen: It was a small round of about $500 million.

Erick Schonfeld: Right, $500 or $600 million. When a company like that raises that much money, is that the amount of money that is required to bring that next generation technology to reality? Are they spending that on salespeople? Why do they need that much?

Patrick Eggen: That's one company I really can't talk much about, given how stealth they are, but it's such an ambitious idea and disruptive. For those that don't know Magic Leap, it's focused on cinematic reality through emerging physical and virtual worlds. It is a super ambitious idea that redefines augmented reality. Think of something like Microsoft's initiative, HoloLens, but very different than Oculus at the same point. In fact, we think better than both and by an order of magnitude better. But that's all I can really say on that one. They're obviously very stealthy. If you go on YouTube, you'll see.

Erick Schonfeld: But that's not your only augmented reality investment. You're a believer in augmented reality.

Patrick Eggen: One of the few

Erick Schonfeld: Any of the other investors here believers?

Geoff Judge: A believer, yes, but an investor, no. We are much more B2B investors. Personally, I'm primarily a B2B investor. We have a lot of investments in the ad tech space, where mobile is growing at a faster rate but video online as a form for doing advertising is growing enormously fast. It is becoming a very big business at this point in time. We also have an investment—I was talking with you earlier—in a company called Vidyard, which is up in Waterloo, that helps mostly B2B companies market themselves when they are going after institutional buyers. They're using video, tracking it, knowing who the user is, being able to reach them later, and basically allowing you to take a prospect through the funnel using tools like Marketo or Eloqua or Salesforce. As I was mentioning, Salesforce is by far and away our biggest client. They have over 4,000 videos on their sites, all giving you short, three-minute instructions on how to use Salesforce, how to make it effective. Video is a very, very effective sales tool. It's easier to watch and listen than to read, not that we don't all do plenty of reading.

Erick Schonfeld: I totally agree with that. I have my own video startup called Touchcast. We can go back to the visual communication as a theme, but I do want to get back to this AR and VR. Are you an investor in VR?

Sean Ammirati: We do not have any investments.

Erick Schonfeld: You do not. Is that by choice or you just haven't had the opportunity to invest in something you like?

Sean Ammirati: I think we just tend not to think about the world that way. We would think about it more like, *Here is a top-down market problem that we're interested in solving.* For example, we're interested in automating routine cognitive professional tasks. If AR was part of the solution for that in a specific vertical, we would unpack that technology early to serve that market. We don't just tend to say, "Let's go make five AR investments"—or really any kind of tech as a filter for deals.

Erick Schonfeld: Right. There have been lots of AR startups throughout. The technology is cool, but it's never taken off in any meaningful way. We could argue about what needs to happen for it to take off. I think some of it has to do with the devices that are required to view the AR. Part of it's maybe behavior. What is your view on why it hasn't taken off? Why do you think that would change?

Patrick Eggen: In full disclosure, I think AR is an abysmal failure historically. I looked at about 35-plus AR companies around the world. I have colleagues in six different regions around the world. Our only two AR investments are Blippar and Magic Leap. We feel like we have a lot of inside edge and domain expertise in that area that we can really leverage from the mother ship. We make investments independently. We don't need a BU. We don't need spon-sorship on that front. But my clear takeaway in AR land, after seeing 35 companies (about three years ago), was either they had great technical chops and couldn't sell a thing or they were just complete Madison Avenue and zero technical acumen. Blippar was the rare breed that we saw out of the gates using a computer- vision-based approach. The CTO was brilliant. 200 tier-one brands signed up in the first year. We seeded that company. We couldn't get a VC to put money into that for the following 18 months. And they just raised a $45 million round. Ambarish Mitra, the CEO, has a chip on his shoulder because at the time no one believed—and people still don't believe. People still think it's a gimmick.

The beauty of Blippar is they've moved away from AR. In fact, they don't call it AR. They call it a "blippable moment." Any object is a media format. They have tweaked the AR stigma, but their next product launch was at South by Southwest. They did a visual search through the camera phone leveraging off the strategic assets that they had built through their computer vision AR platform. That's the next generation of what Blippar is doing. They keep on innovating. I would say they bought Layer for pennies on the dollar. Intel Capital invested in about eight AR companies. The money in your wallet is probably worth more than the asset value of those AR companies. It is a space filled with dead bodies.

Erick Schonfeld: But you're saying they found a way to make it work for marketers?

Patrick Eggen: Yes. Selling to Nestle, Unilever, General Mills—they all closed global deals south of 12 months. It's unprecedented. They have real technical chops and they can sell. That's the beauty of a co-founder partnership from the CTO and CEO.

Erick Schonfeld: And consumers are pulling out their phones and engaging with the units?

Patrick Eggen: Absolutely. Would we want higher engagement? Sure. But I think that is still a shift in consumer behavior that's required. That's the challenge with AR.

Erick Schonfeld: Let's take a step back. I think that if you have a thesis in this space, I would hope it would consist of the fact that humans are visual creatures. We communicate visually. Therefore, new technologies that allow people, humans, to communicate in a visual fashion is a strong trend to bet on. Certainly, we've been seeing just with the growth of video on the web and on mobile, that that's a preferred media type going forward. Let's just talk about video. You mentioned a little bit, Geoff. What are the big underlying trends or metrics that you look at to see where we are in that curve? Do you have any thesis on it? Is this just something that's been growing steadily over the years or is there some fundamental change in the underlying infrastructure that is now allowing video to really take off?

Geoff Judge: I'm far from an expert on the science itself, but I think for most of these companies that are finding great success, especially in the online advertising world, data is very important. It's tough to pull data out of a video, but there are certainly some companies doing it. We just sold a portfolio company, Chango, to Rubicon Project, which we're very pleased about. Basically, an advertiser is looking for an audience that is looking to buy something, is in the market—whether it's for an auto or a camera, whatever—and that's when the advertiser wants to get in front of that individual. It's been a little bit easier in display based on: *What's the copy on the page? What's the subject matter?* In video, it's much tougher. I couldn't tell you what technology they're using, but I will tell you it's quite effective. Very, very effective. The performance for an advertiser is fantastic. That's why we're all seeing so much more video advertising, because the agencies have been making TV commercials forever. They get it. It's a very effective story.

Erick Schonfeld: Right. Certainly we're seeing a lot of previous text publishers moving more and more into video. We see that just in terms of our business. People who aren't necessarily broadcasters who are really interested in video. The economics behind that is that there's just a huge gap between CPMs, banner ad CPMs and video CPMs, but I don't think you can just take the 30-second ad or even compress that to a 15-second ad or a 5-second ad and put that on the web. You certainly can and people do do that, but I'm a little bit more excited in hybrid forms that are either rich media or interactive where you can actually tell that the consumer is engaging with the ad in one way or another. Are you guys seeing any of that either on the content creation tool side or on the advertising side, where they're taking the best lessons from online advertising and applying them to video?

Sean Ammirati: I think that's really interesting as somebody who sold ads for a number of years. The thing I think when I meet startups and they talk to me about that and which I think they underestimate, is the scale they need to have in terms of reach for that to be interesting. They'll tell me they're going to do this amazing interactive ad unit that intuitively seems like that would be the most powerful thing and would be much more powerful than a 5-, 15-, or 30-second spot, but then they don't realize that media buyers are lazy and that they want to buy from a small number of people and that it's really hard to be one of those people they buy from. If you think about the people who bought ads on... When I was running Read Write, you'd think they might buy the half-dozen big Internet brands and a deal with Federated, maybe, or a deal directly with one or two of us. To convince them that you could be one of them and to do a unique ad unit was not an easy thing. I realize this is different than a content vertical but it's the same conversation. When you say "I'm going to build this video platform. It's going to be this really cool, interactive experience," you better have the ability to get through the J curve of reaching that scale to make it work. I do feel like that's where the world's heading for sure. If you were to say, "Yes or no? In 24 months from now, that's how we'll buy it," I'd definitely bet yes. But I think most entrepreneurs, most of the ones I've talked to—maybe people in the audience are doing this better—are not raising enough money, thinking about the right kind of economics to get from where they are today to when they have that scale to actually have those conversations.

Erick Schonfeld: Right. Another barrier too is that all the media buyers, agencies and brands, they know to buy a banner ad or a pre-roll.

Sean Ammirati: Sure. They have a bunch of 30-second spots sitting on the shelf.

Erick Schonfeld: Right. They're almost working against themselves in that they want better performance but they're not willing to try a novel format that isn't at scale immediately.

Geoff Judge: Certainly native advertising is growing at a very rapid rate at this point in time. We have an investment in a company here in New York called TripleLift that basically will create an ad for you using the agency's graphics that they're planning on using. It will just size right into the publisher's website so it looks just like another editorial brick that the publisher is using. That's the way their site looks. We have an exchange where you can buy these ads. You can say, "I'm Procter & Gamble. I want to be on food sites. I want to be on parenting sites." Whatever. You give us the pieces and each one of these ads will be of a different size on every single publisher, but they'll all look uniquely like they belong, as if they are part of the editorial.

Sean Ammirati: Are they doing that via video or just images and text?

Geoff Judge: They are doing video. I don't think video's a big part of their business today, but everything is very graphical. You're not seeing a lot of words. It's mainly photographs and images. The click rates and engagement is just over the top.

Erick Schonfeld: Sean, you mentioned I think earlier that one of the companies you're working on is using vision systems for autonomous driving trucks or for cruise control?

Sean Ammirati: It's more like adaptive cruise control. It is part of the stand for an autonomous vehicle team. That is their background. But they're working on more... Think of it just like if you got into your car tonight. You turn on the adaptive cruise control driving down the Jersey Turnpike. It's the same kind of thing, but it's a truck. The benefit is fuel savings. They help large trucks, what they would call platoon or draft, communicate with each other while driving across the interstate.

Erick Schonfeld: That sounds frightening.

Sean Ammirati: It turns out those trucks are safer actually because when the front truck brakes, the back truck automatically brakes. You actually end up with safer vehicles—we have now done enough miles on the road to know that—and significant fuel savings for the front and the back trucks.

Erick Schonfeld: They're communicating between two trucks?

Sean Ammirati: That's right.

Erick Schonfeld: Over what?

Sean Ammirati: They're communicating back and forth through the systems in the car. The back truck sees the video monitor of the front truck. Then they use lidar to understand what's going on in the images around them as well.

Erick Schonfeld: A lot of attention has been placed on self-driving vehicles because of Google and other programs around that.

Sean Ammirati: Right. To be clear, there's a driver in both cabs.

Erick Schonfeld: I understand that. There's a driver. But this is on the way towards ... This is a step towards autonomous vehicles. But the impact for trucks, whether semi-autonomous or maybe in 10-15 years self-driving trucks, arguably is as great or bigger than self-driving vehicles, right?

Sean Ammirati: The great thing about it from our perspective is that there's a clear-cut, huge business case for the truck companies to deploy it today. You can imagine 10% of a logistics company's fuel savings being saved as soon as they turn this on. There's a clear business case to use it today. UPS, FedEx, places like that are very excited about it, but I do think long-term there's a lot of interesting upside to the business beyond that for everyone.

Erick Schonfeld: So the truck driver becomes a pilot where really they only touch the wheel at takeoff and landing?

Sean Ammirati: It really is more like adaptive cruise control. The driver is still steering the truck, they're just controlling how close they are to the truck in front of them.

Patrick Eggen: Here's another example because I'm very familiar with the company. I looked at it. It's a fascinating company.

Erick Schonfeld: What is it called again, the company?

Patrick Eggen: It's Peloton.

Sean Ammirati: Peloton.

Patrick Eggen: If two trucks are maybe a kilometer away all day for eight hours, they don't know they're near each other. But with the Peloton solution, they can communicate and know that they are near each other, so one speeds up, one truck slows down. That will save, for eight hours of that day, a ton of fuel efficiency, drafting, etc. It's a fascinating company. The beauty is it's exploiting the existing fleet management system, as opposed to new cars or new trucks. People forget the new car/truck market is relatively small in the US. Used cars are 10X the size of new. They're hampered by a four- to eight-year design cycle. They're a median age of 10 years. The existing car or truck market from an investor's standpoint is actually a lot more interesting so let's not get too crazy with autonomous cars and quasi-autonomous. Let's think of immediate investment opportunities that we can take advantage of right now. We invested with a company called Navdy, which offers a heads-up display in your line of sight using optical imaging. It translates what's on your smartphone into the line of sight. So it's convenient and context-aware which prevents you from driving with your knees, but in a similar manner it exploits the existing market and empowers drivers today as opposed to being beholden to some sales cycle of the OEMs or autonomous vehicles in the future.

Erick Schonfeld: I love these examples because they're examples of where vision technologies can really impact existing industries in novel ways. This gets back to the original question of: Do you invest in new enabling technologies or in new businesses? This I think answers the question. The answer here was both. What are other answers to that where there's a really great vision technology that's been in the labs and in startups for maybe the past 10 or 15 years? And now you're just starting to see a really interesting business application for it?

Geoff Judge: We have a company based in New York called WorkFusion. I don't know what vision technology they are using. They basically have an AI platform that learns how to get things done in small chunks. For example, we just signed a contract with a large global bank that is settling trades on a daily basis in big numbers. What we do is we use platforms like Mechanical Turk and oDesk [now Upwork] to get laborers to come in to do assignments. One of the things that we're then able to do is to learn: How does the human do it? Where do they go to get the information? We pay attention to that. Then we start automating. You take the human out of it more and more and more. We're working with Thomson Reuters. A process that they were doing costs about $3 to do. Now it costs about $.25. With this bank using some visual technology, we're able to look at a PDF and get what the settlement data is. These PDFs come in lots of different forms. The information is in different places. It's really saving the bank a ton of money by being able to use that.

Erick Schonfeld: I think we have about two minutes left. Does anyone have any questions out in the audience? You can just yell it out and I'll repeat it. Yeah, right there.

Audience member: I have a question for Patrick. In the world of AR when it does hit the mass market, will a person have to hold their phone, will they put it on their face, or do you think they'll take another device with them and put it to their eyes?

Patrick Eggen: That's an interesting question. Part of our core thesis with Matterport, which is a 3D scanner that creates rich 3D models of your home or indoor areas, and will move into VR... Ultimately, it will move to the mobile phones. Today it's relatively expensive hardware but still cheap compared to the existing market. But think of the proliferation of 3D sensors and phones, which will become a reality in the next 12-18 months. Through the phone you can scan an immersive 3D model. Essentially, the vision of scanning every home in America could be enabled by a company like Matterport. They would be the underlying 3D software platform. The big issue in AR and VR is there's a bottleneck on the content creation. There's just not enough content creation out there. A company like Matterport... Next time Justin Bieber destroys a hotel room in Vegas, his entourage can scan it, throw it up, and monetize that in some way. That's what's going to enable a really disruptive technology that today has sleepy B2B use cases, maybe B2B2C. Then we can leapfrog to D2C use cases because there are 3D sensors in the phone, whether it's Mantis in Israel or whatever Apple's doing with PrimeSense... Activate those sensors on the phone and then anyone can take a rich 3D model of an indoor area. That's what we get excited about.

Erick Schonfeld: One more question. We have time for one more out there. That sensor is using just the camera or using lidar? What would it be using?

Patrick Eggen: There's a number of options. Apple obviously acquired a company called PrimeSense on the 3D sensor side. We have an investment with a company called Mantis. But there's a number of technologies out there right now. Obviously, they have to be integrated by the OEM, so there needs to be a lot of pull from the OEMs, but that's the ecosystem issue right now. We see these sensors becoming more prevalent in Q1 of next year and hopefully, ubiquitous in the next 18 months.

Erick Schonfeld: Terrific. A big round of applause to our panelists. It looks like we're going to get an answer about Magic Leap because Magic Leap is up next.

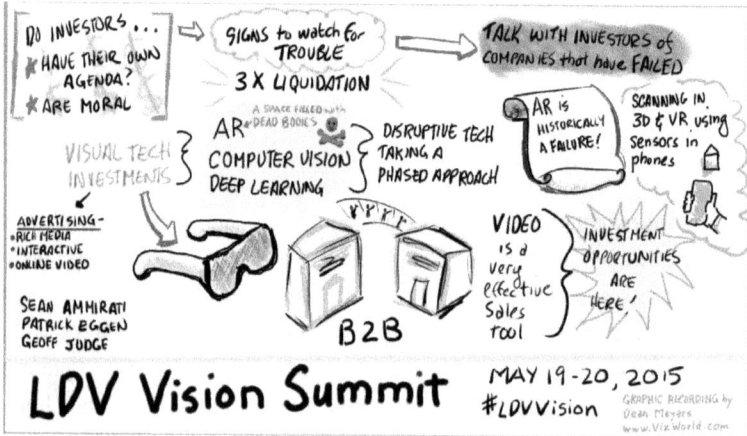

©VizWorld

11

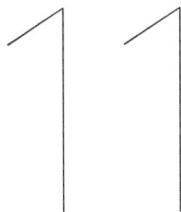

AUTOMATING THE BLACK MAGIC IN DEEP LEARNING

Andrew Rabinovich, Principal Engineer, Magic Leap

I JOINED MAGIC LEAP A LITTLE BIT AGO, BUT I WON'T TELL YOU ANYTHING about Magic Leap. But I will tell you a little bit about some other magic that's involved with this deep learning revolution that started a few years ago.

Let's look at the spectrum of machine intelligence over the last 50 or so years. We started out by carefully teaching machines as to what we wanted them to do by programming transistors to all these gating functions, designing features, classifiers, and so forth. While in the back of our minds, the ultimate goal is to ask the computer to solve some problem and have it figure out what exactly it needs to do. How to design features, how to build intelligent pipelines and essentially solve the problem and teach us as to what the process is like.

Andrew Rabinovich,
Principal Engineer,
Magic Leap

SION SUMMIT

Machine Intelligence Spectrum

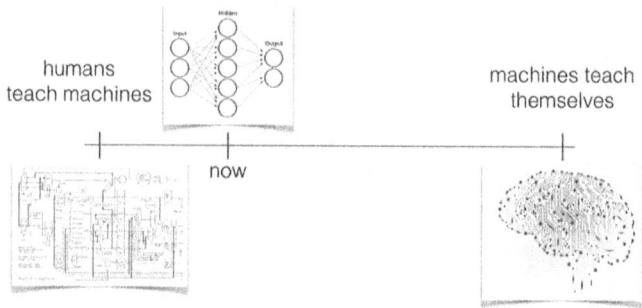

humans
teach machines

machines teach
themselves

now

The reality is, we're about here right now. Maybe this is even a little bit too optimistic. The latest and greatest in machine intelligence are in fact these deep learning models that allow us to achieve such great advances in computer vision, object recognition, speech, and natural language processing. Let's have a closer look at what deep learning has become over the last few years.

This is one of the original LeNet-5s, the convolutional neural networks introduced by Yann LeCun. This is from 1998.

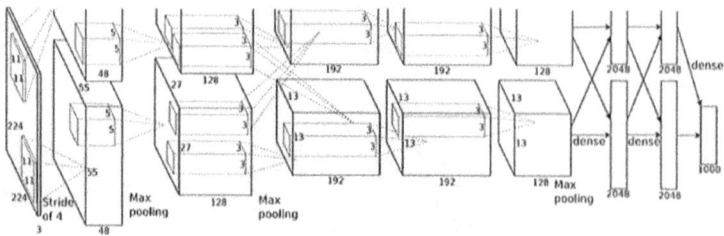

In 2012, a so-called AlexNet, developed by Alex Krizhevsky and Geoff Hinton in Toronto, revolutionized this ImageNet competition and blew traditional computer vision models out of the water.

This is the GoogLeNet, the mother of all networks. It's a highly complicated system that we developed at Google about a year ago.

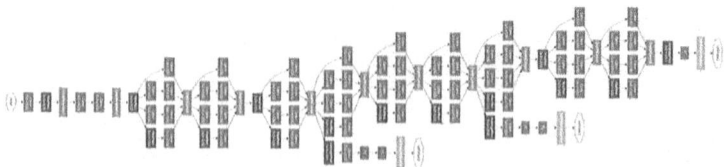

On one hand, these models are quite different, but on the other hand, they're all about the same in the sense that they're all very complex, they're all designed by humans who tried to project the insights that they have about problems into these difficult architectures, and finally, they're all designed without a particular problem in mind—the problem that they want to solve. All these networks essentially solve a general object recognition instance, without having any data in mind.

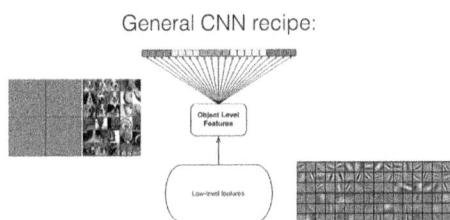

General CNN recipe:

So, essentially, all these networks boil down to the system, where you have a network that learns some low-level features that may be Gabor functions or some representations thereof. Then some higher-level features that a generalist system knows; these are object parts (parts of faces, cars, and so on). Finally, at the top, there is a layer that returns probabilistic labels. What is important in this slide is that there's some structure to the output that's produced by this network. However, this structure's never utilized when these networks are designed and trained.

The proposal is the following: Why don't we learn a system such that we start out with some network—whether it's LeNet, AlexNet, GoogLeNet, or some other Oracle—and we give it the data that we want to use to solve the problem with; we look at the structure of the output produced by this Oracle network and then modify the network yet again to solve this specific problem? Rather than having humans come up with these highly complicated systems, we allow the data to drive the architecture of the network that will be used for this specific problem.

To reiterate, the constraints are: we want the model to have depth and breadth in knowledge, essentially a mixture of experts, we want the model to understand the difference between a cow and an airplane, yet it needs to understand the difference between all possible airplanes and the same way for cows. The design of these systems should be such that it needs to optimize learning in a way that reduces computation because you need these models to run in real time on very low-power compute devices. However, you have to provide these models the capacity that they need to learn, but the only capacity that needs to be provided is in the places where it's required. The general paradigm of just building networks deeper doesn't apply for the reasons of overfitting, lack of data, and many others.

To go back, if we look at the actual structure that you can learn from this Oracle network, you see this graph of labels in the three-dimensional space where all the classes and correlations between the classes are shown in color. You can actually see that the network is learning dependence between the predictions that it's making. However, in traditional deep-learning architecture design, this is not utilized. If one looks even closer at this structure, it is obvious that the system is learning concepts that are actually visually similar to one another.

On the left, you see a bunch of different ships that are in the top left corner. In the top right, you see all sorts of different-looking foods that the system also co-confuses with each other, yet things are similar. Then on the bottom, you see some dog-specific breeds that are also looking the same.

The idea is, you start out with the original Oracle network, whatever it may be, then you learn some structure of the problem that you're trying to solve and automatically you augment the existing network to produce this beast, which in fact is produced completely automatically.

There's a generalist part of the network that knows the difference between cows and airplanes. Then it's augmented with these specialized pathways—I won't say this is close to the way the brain works because I don't know how that works—but these specialists, they come in different sizes and shapes. Some have more features, some have less. Some have more capacity while others don't, because the problem they're trying to solve is much simpler. At the end, the generalist and the specialist compete with each other and the preference is given to the one who has more confidence in the predictions they're making.

This kind of network works essentially with any Oracle and works for any data set that you can think of. We tried this on a number of examples and here's just two of them on the classic

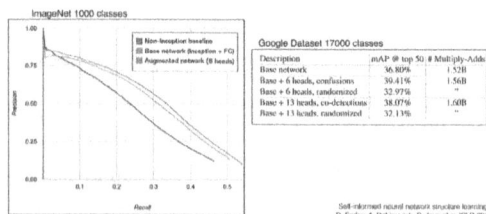

Results using a learned architecture

ImageNet Challenge where you have 1,000 classes. The red line is this old AlexNet, then the orange line is the GoogLeNet. The green line, although the result may not seem significant, is actually rather important because you see an improvement over the entire range of recall. Furthermore, we don't increase any computation as shown in the right plot. We go from over 1.5

billion multiply-adds and we increase the computation by 2% because we can't have models operate at a larger lag. We essentially take the exact same network, we reorganize it using the structure of the data of the problem that we're trying to solve, and we gain automatic improvements.

Similar exercise was done with a Google internal data set that has, in this case, 17,000 classes, so it can tell you everything about everything. Here you want to look at the first two lines. We go from 36.8%, this is the GoogLeNet network that won the ImageNet Challenge this year, to almost 40% without increasing the computation of the network at all, yet by providing capacity only to the pathways that are more challenging for some problems than others.

What's in the future? So far, these are very intuitive and hand-wavy explanations. There's a little bit of theory that supports all of this, but a lot more is needed and we're working on the theoretical bounds of these new architectures. How much can they improve the existing results and how much further they can be pushed? We're expanding this further to analyze video using recurrent neural networks and other architectures, not just for classification but for detection, attention, and so forth. Finally, we want to build a system that actually teaches humans how to solve problems. Ideally, you approach a machine as a black box. You tell it what the problem is, the system learns, and then it tells the humans what the problems are and how to solve it. With this, I'll thank you. By the way, we're hiring at Magic Leap, so talk to me if you want.

Thanks.

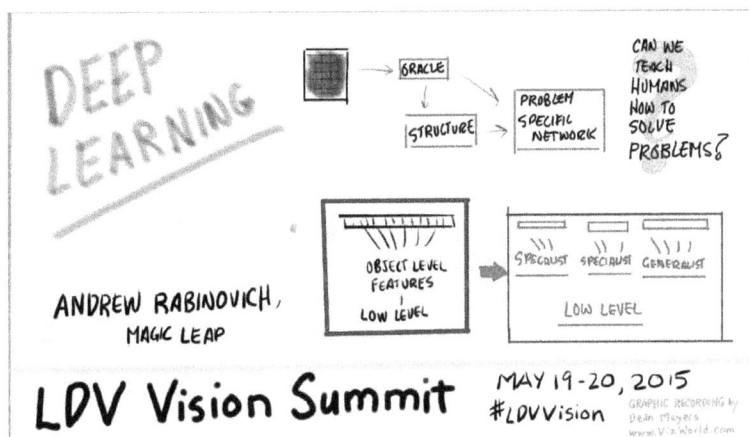

© VizWorld

12

ENTERPRISES ARE LEVERAGING SMART GLASSES TO BE MORE EFFICIENT

Jay Kim, CTO, APX Labs

MICROSOFT AND MAGIC LEAP AND THE BABY... I DON'T KNOW. THESE ARE three really, really tough acts to follow. I'm going to do my best. A lot of what's been presented, I think, really deals with awesome content and awesome stuff. What I'm here to talk about is what my company, called APX Labs, is doing in the Enterprise AR space and specifically drilling down into a form factor of devices called smart glasses. Just to start, I'd like to actually start with showing a really short clip on how one of our customers is using Google Glass in AR in their wire harness assembly operations.

Video Voiceover: Okay, glass, start a wire bundle. Number 2-0-1.

This is a very high-level example of a real-life use case of how these things are being used today. Obviously, with that dark screen at the top corner of the user's eye, this is far from *Minority Report*. This is far from *Terminator* vision. But as Microsoft and Magic Leap are also showing in the forms of videos, it's not too unreasonable to think that there's a lot just around the corner from where we are today. Right now, where businesses are finding the most amount of value in smart glasses and more broadly within the AR context, is delivering the information they already have in the systems that they have spent billions and billions of dollars and decades of time building. In the case of Boeing, imagine how much work-flow actually exists within their databases. Getting that to the people where the work is being done—for people to be able to access that in a heads-up and hands-free fashion-- is a really, really powerful concept.

And from a market opportunity perspective—and these are numbers just within the US across four representative industries, obviously there are a lot more industries where this can scale up to—we are talking about 12 million people who can access technology like this. Even in the crude and rudimentary way that I just showed you in the previous short clip, we're able to deliver, again, information based on the user context, which is where the vision piece comes in. Vision plays a role in being able to drive enhanced knowledge of user context. And to be able to deliver things like next-generation user interfaces and heads-up and hands-free access to information.

You can do that in logistics settings. For example, a picker in a warehouse. Similarly, in a field-service type environment—if I'm out there servicing wind turbines, I don't necessarily want to have to go up to each of the different panels and systems to be able to access data. I can now do that by looking at the different kinds of devices that are out there. And then, of course, in health care, the savings are obvious. The obvious upside is that you could be saving lives. With automotive manufacturing—complex assemblies and things like that—is where we as a company have seen the most amount of traction, because the return on investment associated with this kind of technology can be most easily quantified. So, if you are saving seconds off of a simple task, if you are reducing the error rate of a complex assembly that you are doing so you don't have to go and get re-work done, there is a very obvious dollar amount that's attached to all of this.

We are big fans of HoloLens, as far as the devices that have been announced, and I have been fortunate enough to recently try it on at Build. From a sensor technology, this is one of the most powerful kinds of sensors spanning hardware, software, and the integration of both into a wearable form vector. I can't stress that last part enough. It's really, really impressive what Microsoft has done. Essentially, jamming in a couple of Kinects' worth of sensors along with advanced cameras, IMUs, and other kinds of radios and processors and actually make it wearable. That singularly is the biggest challenge that a lot of the industry players who have tried to have a product offering in the smart glasses space have faced. It is really, really hard to cram in the requisite amount of sensors, to gather the proper user level context, and then to be able to have it be somewhat comfortable. Maybe this isn't a mainstream consumer device just yet, but certainly within the context that we play, which is in the industrial applications, there is a lot of appetite for this specific form factor and the capability that this offers. This is tremendously exciting for us. We basically consider this, as far as state-of-the-art goes today, as the most advanced device that's out there. Look at the number of cameras that are there. It is impressive.

From an optics perspective, optics have been a little bit of the chicken and the egg problem in the sense that some of these things that I'm about to show you have existed—it's just been really hard to drive the price and scale to a point where these can be deployed en masse. So, today what we have is really simple prisms, like you find in Google Glass, where light bounces off a reflector and then a polarizing beam splitter basically just mixes that external light with the light that's been being driven from the projector.

This is the most common and probably the cheapest form that you can get to—a heads-up display or an AR kind of device. But obviously you are getting Coke bottle kinds of glasses. You probably don't want that in front of your eyes. And then you've got a product like Epson, for example, which is a reflective wave guide. Still somewhat thick, but at least you are able to collapse the lens and effectively drive the field of view to be able to get something that is a little bit more over your eyes. If you think about these optics as something that's going to have a 13-to-about-23-degrees field of view, that's still not a compelling user experience any way you look at it.

So we go to HoloLens, then you've got now stereoscopic displays with the right kinds of sensors that are able to do accurate overlays over somewhat of a limited environment. Of course specifications around optics and the parameters are not public yet. The overall user experience that is being driven, coupling with vision and optics technology, is a generational leap over what we have seen so far to date. Then, of course, you've got Magic Leap and the very neat fiber-based modulation they are doing where they are able to portray depth.

Jay Kim, CTO, APX Labs

This is where the technology is going, and really at the end of the day, the goal is to be able to drive a lot of these optics and collapse it into something that is not too unlike the set of glasses that I'm wearing.

Let's talk about what this all means, more broadly. Smart glasses from our perspective is basically just a way to add a human element back into industry buzz words like "Internet of Things" and "big data analytics." IoT generates a glutton of connected sensors spewing out real-time data at orders of magnitude higher than what we are dealing with today. Of course, the analytic systems are going to have to keep up to be able to make sense out of all of this.

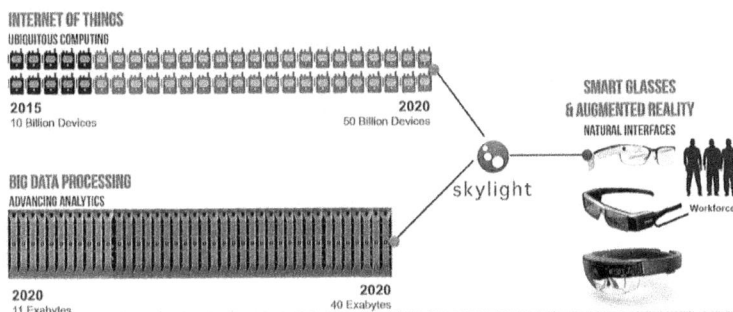

INTERNET OF THINGS
UBIQUITOUS COMPUTING

2015
10 Billion Devices

2020
50 Billion Devices

SMART GLASSES
& AUGMENTED REALITY
NATURAL INTERFACES

Workforce

skylight

BIG DATA PROCESSING
ADVANCING ANALYTICS

2020
11 Exabytes

2020
40 Exabytes

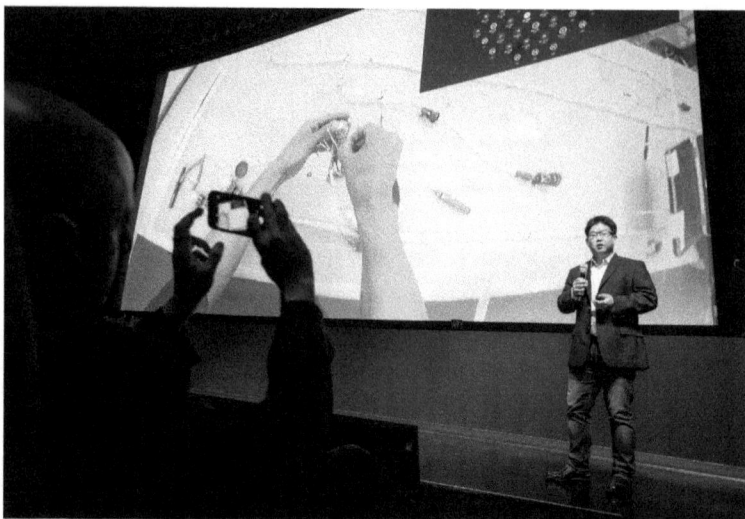

Where we see AR and smart glasses writ large coming into is being able to provide an interface and a mechanism for users to be able to interact with those objects and to be able to do that using the most natural user interface. Fundamentally, there is value in that enterprise today around the form factors that are existing today and around the workflow that exists today. Not too far from here, we are also talking about these form factors from large companies that are getting to the glasses-type of fashion. There is no question in our mind that consumer adoption of this technology is just around the corner. It's a start of a very, very exciting market.

Thank you.

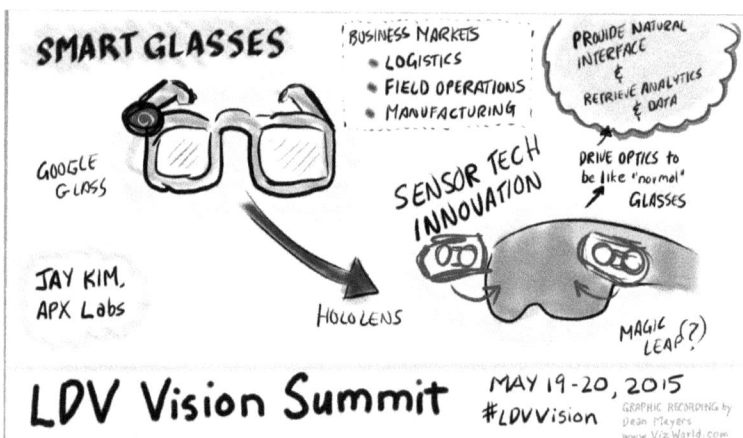

13

THE POWER AND PROMISE OF EMOTION AWARE MACHINES

Marian Stewart Bartlett, Co-Founder & Lead Scientist, Emotient

TECHNOLOGY THAT CAN MEASURE EMOTION FROM THE FACE WILL HAVE broad impact across a range of industries. In my presentation today, I am going to provide a picture of what's possible with this technology today and also provide an indication of what's possible for the future, where the field may be going. But first I will show a brief demo of facial expression recognition. You can see the system detecting my face and then when I smile, the face box changes blue to indicate joy. On the right, we see the outputs over time. Okay, so that's over the past 30 seconds or so. Next, I will show sadness.

That was a pronounced expression of sadness. Here is a subtle sad. Natural facial behavior is sometimes subtle but other times, it's not necessarily subtle. Sometimes, it's fast. These are called micro expressions. These are expressions that flash on and off your face very, very quickly. Sometimes in just one frame of video.

I will show some fast expressions, some fast joy. Then, also surprise. Fast surprise. Anger. Fear. Disgust. Now, disgust is an important emotion because when somebody dislikes something, they'll often contract this muscle here, the levator muscle, without realizing they are doing it. Like this. And then there is contempt. Contempt means unimpressed.

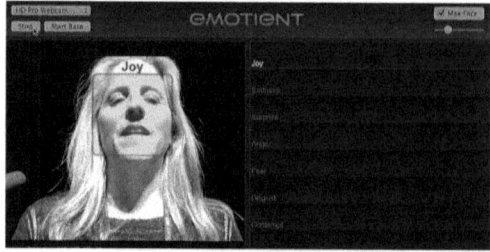

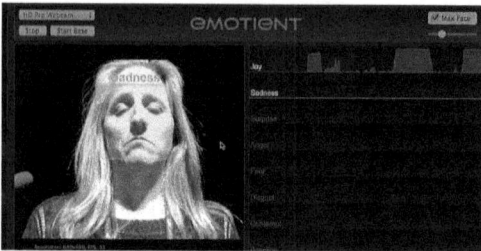

Some things that are possible today are to do pattern recognition on the time courses of this signal. If we take a strong pattern recognition algorithm, capturing some of the temporal dynamics, we are able to detect some things sometimes better than human judges can.

For example, we've demonstrated the ability to detect faked pain and distinguish it from real pain and we can do that better than human judges. Other things that we can detect are depression, student engagement, and my colleagues have also demonstrated that we're able to predict economic decisions. I will tell you a little bit more about that decision study.

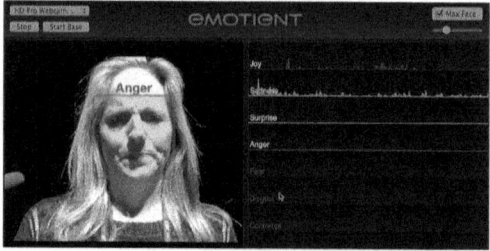

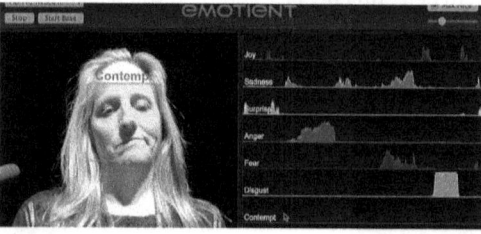

Facial expressions provide a window into our decisions. The reason for that is that the emotional parts of our brain and particularly the amygdala, which is part of the limbic system, plays a huge role in decision making. It's responsible for that fast, gut response, that fast value assessment that drives a lot of the decisions that we end up making. One of the other co-founders of Emotient, Ian Fasel, collaborated with one of the leaders in neuroeconomics, Alan Sanfey, in order to ask whether they could predict decisions in economic gains from facial expression.

Here is the game. It's the ultimatum game. In this game, Player One is given some money and then Player One has to offer some of the money to Player Two. They can offer none, all, or anything in between. If Player Two accepts, then both get the money. But if Player Two rejects, then neither one gets any money.

The optimal solution, according to most economic theories, is to always accept because you will get more money if you always say yes. However, humans don't always behave optimally in this sense. They get mad. This guy is a jerk. I am going to punish him and reject his offer and nobody is getting any money. Rossi, Fasel, and Sanfey asked whether they could predict decisions in this game. They used our system to measure individual facial muscle movements and then they gathered dynamic features of these facial muscle movements. They passed these features to a gentle boost classifier trained to detect whether the player would accept or reject the offer.

They also compared the machine learning system to human judges looking at the same videos. What they found is that the human judges were at chance. They could not detect who was going to reject the offer. However, the machine learning system was able to do this above chance. It was 73% accurate. It was able to predict decisions in this game. They could also find out which signals were contributing to the decision, being able to detect the rejection in this offer. What they found was that facial signals of disgust were associated with bad offers, but they didn't necessarily predict rejection. What predicted rejection was facial expressions of anger.

Player 1 offers some of the money ($0 - all) to Player 2 ...

Player 1   Player 2

Marian Stewart Bartlett, Co-Founder & Lead Scientist, Emotient

They could secondly ask which temple frequencies contain the most information for this discrimination. That is where they found that the discriminative signals were fast facial expressions. These were facial movements that were on the order of about half a second cycle. On the other hand, they found that humans were basing their decisions on much longer time scales. The way they did that, was they trained a second general boost classifier. But this time they trained it to try to predict the observer guesses. Then they went back and looked at which features were being selected. The observer guesses were being driven by facial signals on time scales that were too long.

There are many commercial applications of facial expression technology. Some of you may remember the Sony Smile Shutter. The Smile Shutter detects smiles in the camera image and that was based on our technology back in UCSD prior to forming the company. That was probably one of the first commercial applications of facial expression technology. What I've shown here on the screen is one of the more prominent applications at this time. This is an ad test focus group and here the system is detecting multiple faces at once and is also summarizing the results into some key performance indicators: attention, engagement, and sentiment.

Now where this is moving in the future, is that we're moving towards facial expression in the wild. We're moving towards recognition of sentiment out in natural context where people are naturally interacting with their content. Deep learning has contributed significantly to this because it has helped provide robustness to factors such as head pose and lighting to enable us to operate in the wild.

This shows some of the improvement that we got when we moved to a deep learning architecture. Blue shows our robustness to pose prior to deep learning and then green shows the boost that we got when we changed over to deep learning with an equivalent data set. Here is an example of media testing in a natural context. What we have is people watching the Super Bowl halftime in a bar. Watch the man in green. He shows some nice facial expressions in just a moment.

Next, we have the system aimed at a hundred people at once during a basketball game. Here we are gathering crowd analytics and getting aggregate information almost instantly and it's also anonymous because the original video can be discarded and we only need to keep the facial expression data. Here we have a crowd responding to a sponsored moment at a particular basketball game.

There are also a number of applications of this technology in medicine. The system is able to detect depression and it can be employed as a screening mechanism during tele-medicine interviews, for example. It can track your response over time, your improvement over time, and also quantify your response to treatment.

Another area where it can contribute in medicine is pain. We can measure pain from the face. It's well known that pain is under-treated in hospitals today and we have an ongoing collaboration with Rady Children's Hospital where we have demonstrated that we can measure pain in the face postoperatively right in the hospital room. Now this contributes both to patient comfort but also to costs because under-treated pain leads to longer hospital stays and greater re-admission rates.

Education is another area where this technology will have broad impact. This image shows three facial behaviors related to learning.

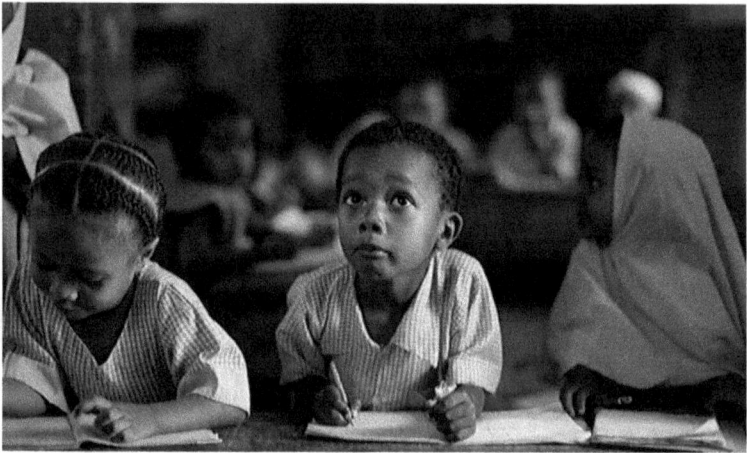

The girl in the middle is distressed. The one on the left is engaged in her task and the one on the right is moving away. These are behaviors that can be detected right now with this technology. We can also take this a step further and we can make online education and adaptive tutoring systems adapt to the emotional state of the student the way good teachers do.

In summary, facial expression technology is enabling us to measure sentiment in locations and scales that were previously not possible. It has the potential to predict consumer decisions and behavior and will have broad impact across a large range of fields. I showed you some in advertising, ad copy testing, medicine, and education. It will be a game changer.

Thank you.

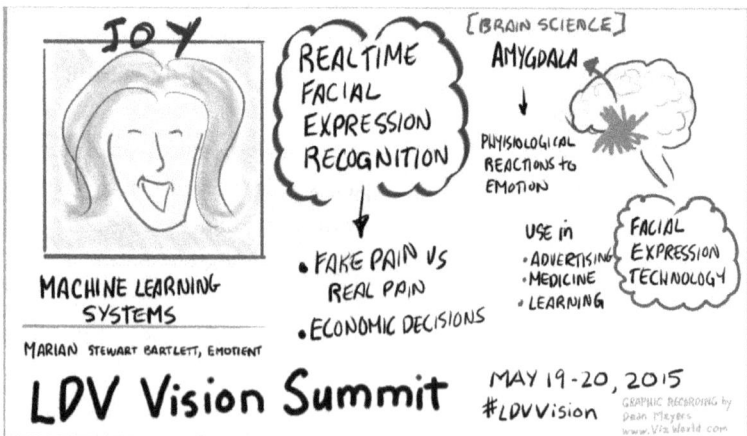

©VizWorld

14

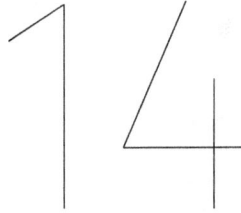

ANALYZING SATELLITE IMAGERY

James Crawford, Founder, Orbital Insight

I AM JAMES CRAWFORD FROM ORBITAL INSIGHT. OUR COMPANY IS BASED on—as many software companies are -- a fundamental change in the hardware world. That fundamental change is expressed very nicely by this image.

This is what satellites are going towards, where Planet Labs launches them 20 at a time. In the past, satellite imagery was interpreted by armies of people going back to the days of the Cuban Missile Crisis. And there are actually still thousands of people in Virginia whose job is to stare at satellite images every day. But as we start to fly not one but tens and eventually hundreds of satellites and UAVs, that doesn't scale anymore, which is why we are here, why we are excited about machine vision, and why we founded the company.

We are interested in processing petabytes of satellite imagery and, more than that, determining what we can learn from them. The most a human can

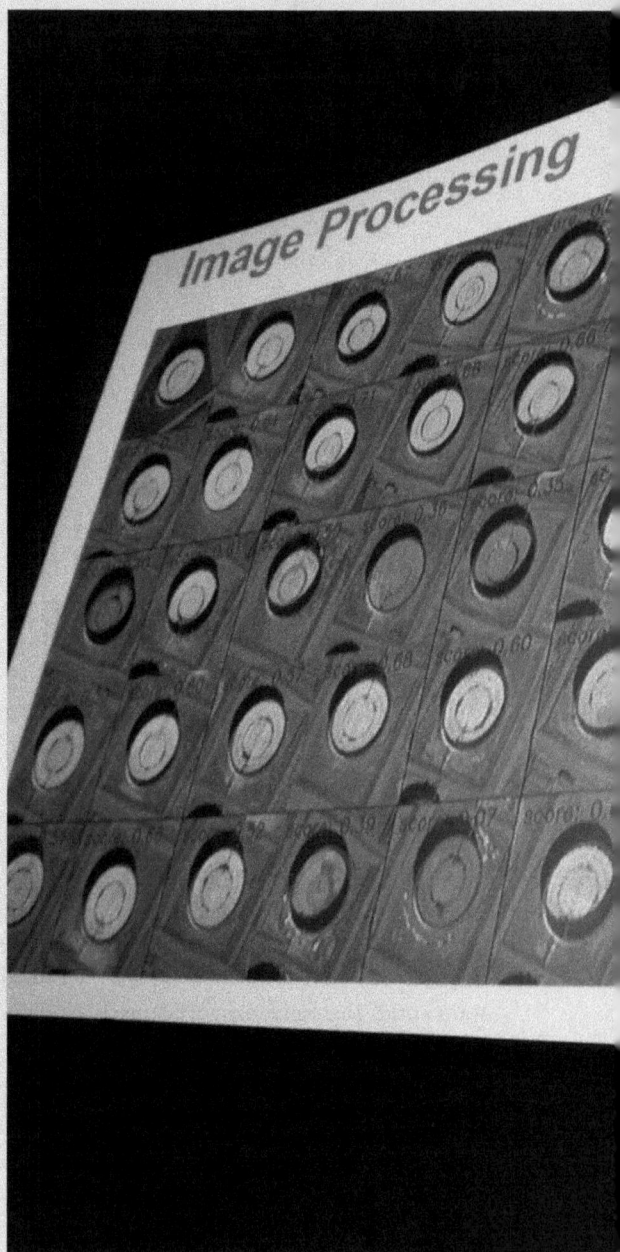

James Crawford,
Founder, Orbital
Insight

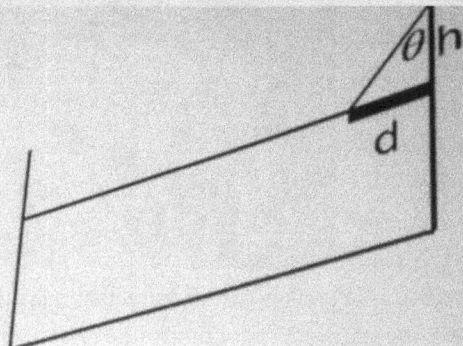

1. Find tank rim and shadow area.

2. Compute height h from sun angle

look at is about a million pixels but we have already run cases where we've analyzed 4 trillion pixels. We can learn things about the world that we never knew before.

If you are a startup and you just got your seed round and you have to show customer traction right away in order to get more money, you start with hedge funds. The reason we did this is that they really want to know about the world and they are like race car drivers. They want to be one second faster than the next hedge fund. If you can tell them something they don't already know about the world, they will pay you really soon. So we've had sale cycles of two days to get contracts signed with some of the hedge funds. In the longer term, there's a ton of other applications that I am going to talk about.

One of the large projects that we've done is looking at US retail. Here we analyze a million parking lot images, which is about 4 trillion pixels, and we counted 700 million cars. This system is still running on a daily basis with imagery being uploaded from Digital Globe and Airbus and we are counting cars on an ongoing basis. We are now tracking 50 US retailers.

We can look at the data—700 million cars—in two different ways. One way is to look at a single parking lot. This is a heat map over six years of where people park. You can see they mostly park by the entrance doors. You can cut the image the other way longitudinally and look over time, and here you can see when people shop as a function of time.

This is Christmas, this is the summertime, and then if we zoom in on just Christmas you can see Black Friday and then the Saturday after Black Friday. All this comes out because we are working at scale. You wouldn't get this if you were looking at a few parking lots, but because we are looking at literally a million parking lots we can pull out these sort of signals.

We found out the worst day to shop at Ross, of course, is Saturday, but also Wednesday is pretty popular. First, we thought this was a bug and then we realized they have a senior discount day on Wednesdays. All seniors are shopping on Wednesdays and that created the secondary peak. Since this is a machine vision conference, I wanted to show you just as much as I can in 10 minutes about how we actually do this.

This is done with deep learning. We had humans go through about 200 images and put little red dots on the cars. This is actually what the tiles look like. You can see at 50-centimeter resolution, the cars are not that obvious. You can make them out but they are not plain as day. We had humans go through and mark them. And then the neural network uses a sliding tile to actually create a heat map of how likely each pixel is to be a car pixel. Then we take that heat map and turn it into a count. In this case, the result says there's 47 cars in this parking lot.

Fortunately, since we are using so many images, we don't have to be exactly precise on every image. We can tolerate single-digit errors in precision on single images and still be able to get these kinds of results—accurate results— at scale. From an investor point of view, a stock market investor, you also want to look at pairs trading. This is the kind of thing that really speaks to the guys a few miles north of here, who are actual traders.

You can see Home Depot versus Lowes, with Home Depot in the yellow and Lowes in the black. Back in 2009 they were almost neck and neck, but if you look in 2014, Home Depot was ahead almost every month of the year. Right after we put this chart together, Home Depot had their best quarter ever and the stock went up 5%. We didn't predict that much of an extreme result, but we were more accurate in our predictions than the consensus of the Wall Street forecasters.

We also look at longer-term trends and we have tons and tons of slides on this, but I will just share one. If you remember the Polar Vortex of 2014, when basically winter failed to end. Usually when we look over multiple years there is a trough in January, but it's pretty shallow in time. But this one was much broader. This is sort of the wisdom on Wall Street, that bad weather in the northeast depresses retail sales. We actually have enough data now to start to see that effect occurring.

It's not all about car-counting, though. We are also looking at a half dozen other applications and I am going to talk briefly about two of them. One of them is construction in China. If you talk to the guys in the investor community, there's a huge division about China. Some people say they are building these things the size of lower Manhattan and nobody is living in them. Other investors believe that they are appropriately building because a large population of a billion people are starting to move in from the countryside. Our role here is not to resolve that. Our role here is to count the buildings and count how big they are.

This is actually really hard from a machine vision point-of-view. You're looking from above, you can't really tell how tall these buildings are from above. We've been focusing on counting shadows. I just pulled out a few—these are actual tiles out of our system—to show you how challenging that is from a machine vision point-of-view. Those are shadows, those are shadows, this is a shadow with a hole in it. This is a shadow in a really dark image. This is a shadow in a really light image, which is actually lighter than the lightest pixel in the really dark image. Nevertheless, the neural networks are getting to be pretty good at this. This is actual output, where we start from the image, we create a heat map of how likely we think each pixel is a shadow pixel and then we create our actual assessment of where the shadows are. We've been running this on a set of commercial developments of known gross square footage and we found a 0.96 correlation between the shadow pixels and the known gross square footage. We're just in the process of rolling this up to run it at a city level and then national level. The data on China is amazingly bad. One of the biggest challenges in this project is there is no good objective data to tell how good we are doing. That also means the data source, once we build it, will be one-of-a-kind in terms of how quickly people are putting up buildings in China and other countries as well.

The last one I am going to talk about is oil. As you may have noticed, the price of oil has been going down quite a lot. Part of the reason for that is the world is literally running out of places to put crude oil. What we've been doing here is we've been looking at these floating lid tanks. Crude oil is almost always kept in tanks with floating lids. This is a cutaway picture.

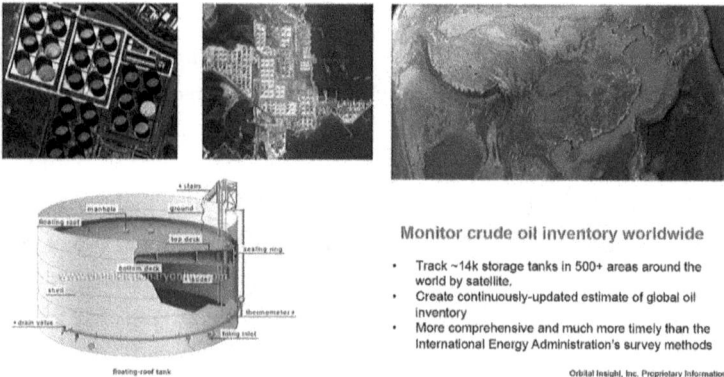

Monitor crude oil inventory worldwide

- Track ~14k storage tanks in 500+ areas around the world by satellite.
- Create continuously-updated estimate of global oil inventory
- More comprehensive and much more timely than the International Energy Administration's survey methods

Floating-roof tank   Orbital Insight, Inc. Proprietary Information

The reason they do this is that if they don't do it, the volatile fumes from the oil will get out into the environment and they may actually explode inside the tank. The lid floats, it sits right on top of the oil, and because the lid sits right on top of the oil, it creates these nice little crescent moon shadows for us. One of our machine vision leads, Boris, has been working on this and he is here so you can find him and get him to try to tell you the secret. I am not going to tell, but he has a new algorithm that gets amazingly good results. All the red lines on that picture are machine-generated using the algorithms. And we can actually find both the outline of the outside of the tank and the line of the shadow demarcation in the tank. By looking at the distance between these two lines and a little bit of trigonometry based on the sun angle, we can compute how far down that floating lid is.

There are about 14,000 of these tanks around the world. Nobody actually knows how much oil is in all of them. When we put this signal together as we start to scale it out, we'll have the first real picture of how much crude oil is in the world. This is one of the biggest drivers, as with any commodity; the amount of the commodity in the world is one of the biggest drivers of the price of that commodity. This will be the first time we have a real-time picture of the amount of crude oil, especially in places like Asia and the Middle East where the data is just not good.

As I said, there are many other things we are working on: predicting crop yield using multi-spectral analysis, projecting deforestation, looking at road-building as a predictor of deforestation to identify areas that are at risk, measuring commodities. It's an extremely long list and so on the strength of that, we raised an A Round, which has been in the press, earlier this year of about almost $9 million lead by Sequoia Capital and we are in the process of building out these and a number of other applications.

Hope that all made sense. Thanks for listening.

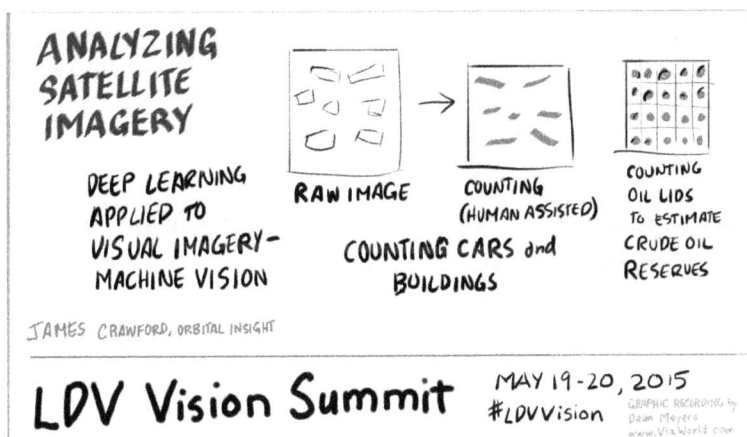

ANALYZING SATELLITE IMAGERY

DEEP LEARNING APPLIED TO VISUAL IMAGERY - MACHINE VISION

RAW IMAGE

COUNTING (HUMAN ASSISTED)

COUNTING CARS and BUILDINGS

COUNTING OIL LIDS TO ESTIMATE CRUDE OIL RESERVES

JAMES CRAWFORD, ORBITAL INSIGHT

LDV Vision Summit   MAY 19-20, 2015
#LDVVision

GRAPHIC RECORDING by Dean Meyers www.VizWorld.com

©VizWorld

15

CAN COMPUTER VISION HELP RETAILERS BETTER UNDERSTAND CONSUMER HABITS?

Alexandre Winter, CEO & Founder, Placemeter

I WILL SHOW YOU HOW WE INDEX HUMAN ACTIVITY IN THE PHYSICAL world using computer vision, very simple and inexpensive sensors, and of course, our computer vision algorithms. I'll show you why that matters and how this is already helping businesses and people in changing some lives.

That's the big picture, right? You probably know that. By 2050, 3.5 billion people are going to be added to the world population, almost all of them in cities. This means that cities' populations will double in the next 35 years.

If you don't take action, this is what's going to happen. We have to change the way we use cities and we have to be data-driven and smart about that.

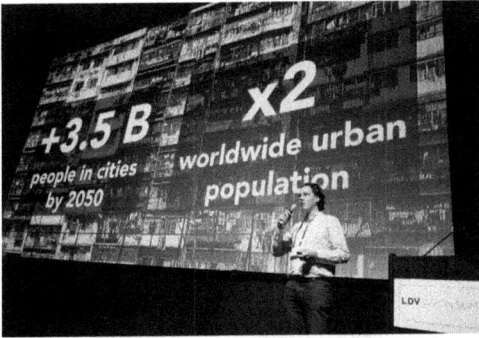

This is what the $1.5 trillion smart city market is about: smart retail, smart transportation, smart buildings, you name it. It's all about optimizing the way people use these amenities, but to optimize you need to measure first, you need to quantify. There's different approaches to that. Most cities are not ready for that.

You can imagine physical sensors. There's lots of physical sensors out there, they're great, and they do the job, but they're single task. So when you want to do something new you have to install a new sensor. Satellites, we just saw, are very powerful. They are amazing. I know these things as large scale but they can't be stationary and high resolution. There's also cell phones. You can locate cell phones and people who hold them. That scales pretty well but it's not data-accurate. You don't know exactly how people interact, where they go, how far they go, et cetera. Plus, it has some privacy issues. Not exactly that, but you get my point.

At Placemeter, we are really convinced that this is the solution: inexpensive video sensors that already blanket our cities and if needed, we can easily install new ones. By running computer vision algorithms on that we can extract everything we need. We can extract everything we need. We can count people, we can see where they go, we can see how they interact with each other. We have a statistical calibration algorithm where we can infer and estimate dimension and speed, so we can count how many people come in and out of these places. We can make predictions based on weather and calendar data. We can do everything we need to quantify activity in the physical world.

At Placemeter, we are all about scaling because we need scale, we need that data everywhere. We don't process these high resolution videos I just showed you for demo. We process tiny video like this one and we extract the same data from them. It's not easy, but that's our challenge. We do that because then we can process a lot of videos, either on our backend or on the edge of the network. We are actually going to release an embedded sensor that is inexpensive, always on, carries a little bit of computational power, and a little camera. We're going to release that in June to ease the deployment of our technology. I'm sparing you our sales pitch, but everything is included in an amazing product that's easy to use and very seamless. Having things that are automatic to set up and don't require any expertise is extremely important, again, for scaling.

So what do people do with our data? We have plenty of applications, just like for satellite imaging, there's plenty of people who need that data, and we talk to hedge funds also. Transportation, urban planning—I don't have time to go over all of them, but I wanted to show you one that's interesting. Retail, physical retail, is a world where co-occurrence and causality are often mixed. There's more people in my store, how much of that is due to marketing efforts, how much is due to my new window, and how much did you adjust to more traffic to the neighborhood? Well, we help them figure that out.

We also recently helped a customer I will not name select a new location for their candy store not far from here in Union Square. They were considering two places and they did an analysis of foot traffic.

One place had more foot traffic (in yellow). It was a lot more expensive, of course. The other place had fewer traffic, so they were about to go with the first one, but then they looked at the details. The extra foot traffic was mostly at commuting time and most of these people were walking pretty fast. You know, people commuting, looking at their cell phone, angry and in a rush—that doesn't really matter. They're not potential buyers. So they took the other location and will get more bang for their bucks. A detail that's interesting that they noticed is that at night there's a little peak in conversion rate, what they call the "boozy late night candy crush," so they're probably going to package something for drunk people at night.

There's plenty of other applications. I won't go over all of them. I just want to highlight the fact that when I started computer vision, images took about 30 minutes to load and hours to process. This with the era of 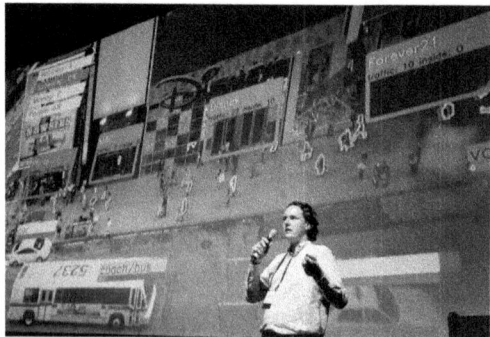 the expert computer vision. Today is the era of the point-and-shoot computer vision analytics. In place of a platform, you just submit a video, you click where you want to measure, and get the data. So it's extremely scalable and this is really what's going to change the way we do business and the way we use our cities.

16

PANEL: WILL COMPUTER VISION EXPONENTIALLY INCREASE CLOTHING COMMERCE?

MODERATOR

Jonathan Shieber, Editor, TechCrunch

PANELISTS

Alper Aydemir, Co-Founder, Volumental
George Borshukov, CTO, Intervisual

Moderator: Jonathan Shieber, editor, TechCrunch; Panelists: George Borshukov, CTO, Intervisual; Alper Aydemir, co-founder, Volumental [L-R]

Alper Aydemir: Cheers, everyone.

Jonathan Shieber: Yeah. Thank you all for sticking around. Hopefully we've saved at least the nominal "not worst" for last. I've got two esteemed gentlemen with me, one of whom worked in a jet propulsion laboratory for NASA, and the other one's won, I think the technical term is a "shit ton" of Academy Awards, right?

George Borshukov: Something like that.

Jonathan Shieber: Yeah, a shit ton. More than three, right?

George Borshukov: Yeah.

Jonathan Shieber: Yeah, but let's be humble. George and Alper, thank you so much for joining us. Alper is from Volumental, which does 3D imaging for clothing and retail, and George is in transition to a bunch of new things, which hopefully he'll be able to tell us about. Let's start it off. When you think about the visual learning space, or the visual space, visualization, it's a technology that's been around for awhile. So as a new startup, how do you differentiate yourselves? How do you stand out from the crowd when there's been so much research and deep technology that's been developed around this stuff?

Alper Aydemir: I think one way to answer that is that, sure there has been a lot of technological advancements and stuff, but that doesn't really translate into successful products. Today's hot new tech is tomorrow's GitHub repo story. That's a given and that's especially true in computer vision and software. Whatever that is today, clothes—we can't talk about it—but we had a bajillion amount of investment and so on. Once it's released, someone will reverse-engineer it, put it on GitHub, and then that's it. The real value, I think, in computer vision is counter intuitive. All good things are counter intuitive, but the good way of building a computer vision company is not to call yourself a computer vision company.

Jonathan Shieber: It's about the application, then. It's not about the technology itself. You can have the whizziest, bangiest, newest technology, but if you're not solving a real use case, then what the hell's the point?

Alper Aydemir: Right.

George Borshukov: You heard nobody believes computer vision works, right? The moment it works they try something else.

Alper Aydemir: That's true. That's true.

Jonathan Shieber: How has that impacted this tension between the technology and the application? How has it impacted the way that you've thought about your own business, George, and the way you're transitioning?

George Borshukov: Absolutely. I have great stories. My background is in visual effects, actually. Steve and I go a long way back. What he showed today that's done at Microsoft was kind of a dream back in 1996 when we worked on the first *Matrix* movie. There were these scenes called "bullet time" that we were challenged to make, and we had no idea how we were going to do them. We looked at different ways of technology and John Gaeta, the visual effects supervisor, challenged us to capture the scenes exactly like Steve showed us today. This was in 1996, and we were supposed to film with multiple film cameras, capture Keanu Reeves dodging bullets, whatnot, and then be able to replay those scenes in the movie, in the slow motion that someone showed earlier. We looked at this for months and the technology just wasn't there.

Then for the backgrounds we used photographs. But long story short, it all goes full circle. I like change. I am in technology. I have applied it in movies and video games, and then six years ago I decided to apply it to fashion eCommerce. I got involved with a startup. I was a CTO for six years. We wanted to do virtual fit technology.

Jonathan Shieber: Let's telescope down into that a little bit. We've gone from the general, what's going on in visualization, broadly how you need to think about it in terms of an application rather than a technology, to specifically the ways in which this can be applied for retail and fashion for consumers. If you project out the next five years, how does your visualization technology get integrated into the retail experience? What does that store look like?

George Borshukov: For me there's two things that are big in eCommerce right now. There's the trend of mass customization, which is still a niche, but it's increasing and some people believe it's huge. The problem there is, before you buy the product, it has actually not been made. So you need good visualization. The second problem in eCommerce for clothing is that most of the things you buy are not going to fit you when you get home; therefore, you'll be disappointed. There'll be high returns. Both of these things require really good visualization, really good computer graphics. They require computer vision so we can capture the clothing because CAD doesn't exist in the apparel industry.

Jonathan Shieber: Right. Alper, how does that relate to what y'all are doing over at Volumental?

Alper Aydemir: Right. I think when we think about the retail experience, it's really related to these sensors that are not yet quite in your pockets. So that's one place to reach these consumers and you would rather reach them, I think, at home if you can. Of course I think the retail stores will never go away. You will always have a need for those. But then again, even at the store environment, you need the right kind of experience. Most people, for example, are not comfortable with stripping down and going through something that looks like an airport scanner, and so on and so forth. I already feel awful about trying on three pairs of pants. I don't want to go through this extra step again. I think a lot of that depends on nailing that user experience where the technology is invisible. It's not even part of the conversation. You're there to buy a product and this happens to be the best way of doing it. You don't care if it's computer vision or x-ray.

George Borshukov: Also it's fun.

Jonathan Shieber: I have to thank you for whoever provided the original visual that was up on the screen. That was amazing. That was really incredible and I would have loved to have gotten a picture of that for myself. I'm sorry I didn't bring any demos with jangly guitars, because I know apparently that's the thing. You all love the jangly guitars in your little animated videos. Think you could go for something a little more aggressive, considering how forward thinking this conference is, but whatever. I'm not going to complain. Getting back to the point at hand (sorry for my digression), when you think about the entire supply chain, or how this gets integrated, you all have talked a lot on your blogs, respectively, about the notion of fashion and retail moving from mass production to mass customization. When you think about the enabling technologies that are going to help move that along, obviously the visualization piece and the scanning piece is one part of it. How do you see that integrating with a broader supply chain to create a new kind of retail market?

George Borshukov: Right. All the apparel companies will love that model. It will be much more efficient. It will be much more efficient for the environment. It will be much more efficient for all of us, but it's a very slow process. Right now the way the apparel industry works is this 30-30-30 rule. 30% of the clothing is sold at full price, 30% is sold at the deep discount, and 30% ends up in the landfill. That's the business model. If you bring in mass customization, all of a sudden you're talking about having stations that only produce clothing on demand. Therefore you have no waste, essentially, especially if you have good visualization. So it could be a revolution.

Jonathan Shieber: You talk about a revolution and yet the stores that have been able to implement this most effectively that I've seen—and we can talk a little bit more about who you all think have been really good at this—have been bespoke stores that are doing tailoring anyway but are now doing really high end shit. Frankly, I'm a reporter; I'm too poor to afford that. I'd like to be able to afford it at some point, so what do I do? How does that happen for me? When does it come to my level where I can spend 20 bucks and get a nice shirt?

Alper Aydemir: I think we are still quite some time away from that when it comes to getting something customized, or entirely customized. Part of our team spends a lot of their time in manufacturing ateliers and factories and so on, trying to understand the craft of it. A lot of it is historical, and you're trying to digitize something, an industry, that is maybe a thousand years old—like shoe making, let's say. The thing is that we can't just come up with computer vision algorithms and hope that it will work immediately out of the box and they will be able to integrate that into their existing system. They won't be able to.

Jonathan Shieber: But if you look at something like Soles, or you look at something like, what's that earbud company? Normal, here in New York. They're both here in New York, and both amazing.

Alper Aydemir: But then again it's coming back to the thing that you said: it's quite expensive. It's bespoke, it's not like a 50 bucks insole that you can buy. You need to actually go somewhere, scan yourself.

George Borshukov: There's a special machine on the side. That's the one thing we know. There's the mass production area in the factory, and there's a couple of actual machines that are set on the side for the custom stuff. Those require more work, but they're more profitable, so how do you get the volume up on that stuff?

Jonathan Shieber: At some point do y'all envision a world where there are 3D printers, in-home 3D printers that will be able to take these scans, someone will customize the clothes, and they'll just print it out? Is that a future that you can envision, or is that a pipe dream? Is that science fiction?

Alper Aydemir: It's not a pipe dream, but it's quite Jetsons I would say.

Jonathan Shieber: Are people too lazy for that shit?

Alper Aydemir: It just doesn't exist. You can't really print a shoe at home. But then you can still deliver a ton of value. You don't have to go all the way Jetsons.

Jonathan Shieber: Who are the companies that you all are working with or that you all are looking to who are doing this successfully and doing it well? I mentioned Soles, I mentioned Normal, mainly because those are the two companies I can think of right now.

George Borshukov: For me I see it from a different angle. You're talking about startups doing the ninja stuff, really cutting edge stuff that's very small still. Maybe it'll make it big. My experience comes more from the fact that we have approached all the biggest brands out there. We've approached them first with the virtual fit idea, and trying to digitize their existing lines of clothing, and offer a virtual fitting room online. That idea is very hard for them to understand and fund and take. So we ended up pivoting to the customization piece. It's especially hard for the more fashion-oriented brands. Where we've found our best business ends up being with companies like Nike and Adidas, who are on the sports apparel side. They're very big companies, very technologically savvy. It turns out that they really do have a technological edge to them, so we've found them to be the best adopters of technology.

Jonathan Shieber: That was your previous business model, right? That's not the business that you're in currently.

George Borshukov: Right, so I think on the next one you go full on, you go direct-to-consumer. You don't go B2B with the brand. You just create a destination where, if you want to buy clothes online, you go to there, if you want to make sure they fit. Then maybe you get redirected to Amazon or wherever to make the purchase. But in order to make sure that it's going to fit, you go through that intermediary. Integration into each of the brands on an eCommerce operation is not going to work.

Jonathan Shieber: Direct to consumer is, in your opinion, is the future?

George Borshukov: It's the only way for the fit side of things.

Jonathan Shieber: How do you feel about that?

Alper Aydemir: Okay, for the fit side of things? Do you also mean the making-the-product side of things?

George Borshukov: No, no, no. I mean only for the fit. The product then is made-to-wear, basically.

Alper Aydemir: Right, exactly. I think similarly, and another strategy that we can capture is really the long tail and that's all these bespoke or smaller companies. There are a lot of them that you can have really quick roll-outs, get some really quick early adopter user base, and also revenues from it, and learn from that to catapult yourself to the really, really mass customized future while all these big giant companies are busy moving their various parts.

Jonathan Shieber: Right.

Alper Aydemir: That's one thing we are doing. We are working with bespoke manufacturers, for example in Italy, in Germany, in Denmark, all these places. Also, as of this week, in headwear, also in eyewear. 3D printing is much more doable when it comes to eyewear, for example, than shoes.

George Borshukov: Because it's hard surfaces.

Alper Aydemir: Right. When it comes to making the actual product, that's always something that is hard and you will be stuck with one kind of product, whereas the fit solution offers a lot of different opportunities.

Jonathan Shieber: I see. I'm feeling you. Let's talk some hard numbers now. How much does it cost for a company to implement y'all's solutions, and what's their ROI on stuff like that? How many have you done? You say you're working with a number of small companies. What's the time, what's the payback for these guys?

Alper Aydemir: Right. For example, I can give one example that is public.

Jonathan Shieber: Give us some private ones. Just a few.

Alper Aydemir: I would but I think I would not be able to afford to travel here next year, or be allowed to.

Jonathan Shieber: Killjoy.

Alper Aydemir: One example is a company that has about 20 stores in Europe and they have their own shoe masters. These are people who are the shoe nerds of this world. There are a few of them. What happens today is actually quite interesting. This person goes to 20 stores, one by one, and measures people's feet. They spend their life on a plane, basically. Obviously they can't scale. They are doing pretty well, but they can't scale. And so what if you could replace this person with a scanner and algorithms and stuff. We just did a man versus machine competition for that. We had the same 150 people, 150 shoes.

Jonathan Shieber: Did John Henry win?

Alper Aydemir: Hmm?

Jonathan Shieber: John Henry. The guy with the hammer and the railroad and the locomotive. Did the guy beat the machine?

Alper Aydemir: No, the machine beats them.

Jonathan Shieber: Damn it!

Alper Aydemir: They are quite comparable in performance and so on and when it comes to fit, but obviously it's a lot more scalable, it's faster, and so on and so forth. That's one thing that we are doing. To install that into the store is in the low or mid hundreds of dollars.

Jonathan Shieber: Per device?

Alper Aydemir: Per store.

Jonathan Shieber: Okay, per store.

Alper Aydemir: Per device.

Jonathan Shieber: All right, one device per store?

Alper Aydemir: Yeah. The capture time is about a second. So that's the thing. People shouldn't be even aware that they are being scanned by this Star-Trek-like device. You just walk in and say, "Yeah, I want to buy some shoes," and then the salesperson there assists you. "What kind of shoes are you interested in, please sit there" or whatever, and then in that instant you are done. The tech part is done.

Jonathan Shieber: Right.

Alper Aydemir: I think that's what we want to build to make this more like a utility. It's not visible, it's just like, you turn on electricity and it starts.

Jonathan Shieber: Does this become a software program? Do you see this as something that people have the option to download for shopping at home? Is this just in store? Is this a physical retail product?

Alper Aydemir: Obviously once that is done you can take that data and use it to see if eCommerce shoes on Amazon will fit you well or not. So you can take that data anywhere with you. Our system, we've built the system to allow for that.

Jonathan Shieber: George, how does that jibe with what your thesis is around Intervisual, now that you've launched out on this new venture and you've got this new thing going on?

George Borshukov: The Intervisual is to really restart the virtual fit idea and customization remains with Embodee. Intervisual is more about restarting the original idea of the virtual fit, in a better environment. When we started Embodee it was 2009, it was a very bad year. Now a lot has changed. Now we have 3D cameras coming around the corner that will be on every tablet, on every mobile phone. That means we're going to be able to capture our bodies… [Beers delivered to stage.] Ah, look at that! That's delicious!

Jonathan Shieber: What a wonderful guy.

George Borshukov: Thank you so much.

Jonathan Shieber: Cheers.

George Borshukov: That means we're going to be able to....

Jonathan Shieber: We're going to get rowdy in a second, we're going to get real rowdy. I'm going to take this opportunity as we're dropping cups everywhere to say that if you all want to ask questions, I'm sure someone will be coming around with a mic in the next five minutes or so, but y'all can also tweet at me and I will pose your question based on the tweets. You can tweet at me at @jshieber, that's j-s-h-i-e-b-e-r. Also this ups my follower account, which also looks good. So thanks for that. Cheers, guys. I'll get them both in there.

George Borshukov: Let me finish my thoughts. Basically, I think that it's a moment where it's probably still worth trying the idea. We're seeing what's happening with augmented reality and virtual reality. Something like the HoloLens could be an amazing device to implement the virtual mirror. The virtual mirror will require some of the same ideas. To digitize the clothing, you're going to need real-time cloth simulation. You're going to need to superimpose it over a real image feed of the user. I think it's an exciting time. There's so much buzz around these new technologies and this could be one of the killer apps.

Jonathan Shieber: I've seen a company actually. I was at South by Southwest—and I apologize to that company if this is a live feed and they're watching me, I forgot their name—but they were doing this virtual mirror. I guess my question is, what's the friction around that? You run into that same problem at the store where you're dealing with taking off your clothes, getting scanned, and it seems like there's a lot of friction there. Is there a way to reduce that process? Do you have the virtual mirror in the fitting room and you say, "Just take your kit off once and then we'll try on all these clothes?"

George Borshukov: We'll have to figure it out. I think that's what's exciting about it, but I think now there's all these new devices and new technologies coming on the market that, six years ago, there was nothing. It was the Dark Ages.

Jonathan Shieber: Do you see a supply chain for that stuff now when you think about the companies that you would partner with to make this happen?

George Borshukov: Totally. All the big technology companies I think love the concept of this idea. We all understand it. There is a problem with shopping online. I don't shop online. There's a reason why I don't shop online. It's because everything I order doesn't fit me when it comes home, because there's no universal sizing. That problem has not been solved. There's been a hundred startups that have crashed and burned in the last 15 years. We tried it. We had a trial for two years with Hurley. It was super successful from as far as the data, but the business model didn't work. This was four years ago. I think now there's technology companies out there, the big Top 10 names, that are starting to realize that what it's going to take to put this to the consumer. Because the consumer wants it. The apparel companies don't care, so we have to get a technology company to fund and seed that to really make it happen.

Jonathan Shieber: Who's pushing the hardest on that front, from a hardware and enabling technology perspective?

George Borshukov: I can't say. [laughter]

Jonathan Shieber: You must have thoughts, right?

Alper Aydemir: I do, of course.

Jonathan Shieber: Come on, just one.

George Borshukov: Magic Leap doesn't say anything so I get…

Jonathan Shieber: If you were to mention some companies that don't talk about what they're doing and are pretty impressive when it comes to this stuff. No, I'm sorry.

George Borshukov: Magic Leap will be one.

Jonathan Shieber: The HoloLens looks pretty impressive, too.

Alper Aydemir: Yeah, that's pretty cool stuff.

Jonathan Shieber: Do you see this integrating with AR or VR?

George Borshukov: I think it's more AR. For me it's more AR because it's more flexible. VR is really about experiencing a presence. I think AR is about mixing things in the real world, obviously. We're talking about seeing something on your body, and trying to decide whether you want to purchase, and when experiencing a product virtually in a way that sales will bridge the Sensorial Chasm. One of the reasons you don't shop online is because you're not quite sure. All you're facing is a couple of rollover pictures and that's not always enough to make a decision. But if you have an experience...

Jonathan Shieber: How does that play into Volumental's role in this?

Alper Aydemir: The thing about this is that it's really about building trust and if it doesn't work I think consumers are really quick to say "Oh, this is some gimmicky thing. I tried it once. It didn't really help, so I will never try this again—maybe in the next 10 years or five years." I think we are trying to be super careful in bringing this to the public. It shouldn't be jittery or it shouldn't be laggy. It shouldn't be all these things. If your dog runs in front of you, it shouldn't just crap out. All these things. We turn down a lot of offers from handset manufacturers who are building these 3D cameras into the tablets for fit. They came to us and they said, "We want you to build a fit app for this." You can, and you will probably make some money from that, but it would...

George Borshukov: It would be horrible.

Alper Aydemir: Exactly.

George Borshukov: Because that's just the first step, the scanning of the body is one of the three important steps. You need the content, which is the garment, and then you need the visualization, which is the garment on someone's body. Those are difficult problems. The body scanning is the easiest.

Alper Aydemir: Exactly and the same goes for algorithms and what we heard today previously. It's almost like that, I said this before, it's almost like that scene from that movie *Full Metal Jacket*. It's like, "This is my algorithm."

Jonathan Shieber: Did you just refer to *Full Metal Jacket* at a visualization conference?

Alper Aydemir: Yeah, it's like, there are many others like it but this one is mine. This is my deep learning. There are many others like it but this one is mine.

Jonathan Shieber: That's an impressive reference there, an impressive reference. We're about five minutes out from the end of the talk. A few people have followed me, which I appreciate, but no one's tweeting at me, and I can't believe I've asked all of the questions that I'm supposed to ask. There's one up front. One in the back and then we'll go down front. You all use that technology. It's there for a reason. It makes life simpler. Better. Easier. Happier. More productive. Sir?

Audience member: My question is, when you were replicating that shoemaker, did you actually replicate it independently, or did you have input for this talent, and then what happened to him actually eventually?

Jonathan Shieber: Are robots taking over? Did the shoemaker get fired because he wasn't as good as the visualization tech?

Alper Aydemir: No, he wasn't getting fired. Because simply he can't be in two places at the same time. We probably prevented there being more shoemakers, if that makes sense, but there wasn't anyways, so these guys were not able to scale because of that. What we did is that we sent a couple of people down there and spent some quality time with these people and understand how they are. This is what I mean. There is no institution in the world that says, "This is now .01 millimeter accuracy fit. Get your stamp." There is no such thing. It's also about the psychology of it, that you are getting this premium experience. You have this trust that it's going to work and you don't really care if it's computer vision or you are doing some freaky x-ray thing. It's just that it works.

Jonathan Shieber: Okay, let's move on.

Evan Nisselson: We've got a question here. Is the mic on?

Jonathan Shieber: There we go.

Audience member: Psychology is the perfect word for my question, which has to do with the virtual mirror. When I've talked to online retailers about, "Wouldn't you like to take a 3D model of the customer and show them in the clothing?" they're all very, very clear: "No."

George Borshukov: They don't like it.

Audience member: They don't like it, and furthermore, some of the smarter ones I've talked to, they've done A/B testing. They don't necessarily have the 3D model, but they say, "Look, we show a model in the best possible conditions, with beautiful lighting. They look gorgeous." Most of this is women's clothing. The psychology is you want to give them the sense that they're going to look really great.

Alper Aydemir: This is great, yes. Most people don't want to see themselves in 3D. It's disgusting to see yourself in 3D if you don't have the body of Cristiano Ronaldo.

Jonathan Shieber: We're all just wasting our time.

Speaker 3: I want to see myself in clothing before I buy it but I don't think my wife does.

Alper Aydemir: In some, not in all cases. I want to verify that, but you get my point.

George Borshukov: No, but this is my goal exactly. I know that, because I've lived this for six years and you're absolutely right about it. The brands don't want that. They want to glamorize. They want the glamour shots. That's what generates the sales for them. But we have done enough research with consumers and consumers ultimately want it. It's going to be a utility. This is why I said it's never going to work with a brand. They're never going to like it. But I think ultimately we as consumers have to take charge and it's going take the technology companies, the Microsofts, the Intels, the Googles of this world, to enable this technology to reach critical mass. Then it's all going to be good. But the brands don't want it. First of all they don't have the funds to pay for it and they're too conservative.

Jonathan Shieber: Is there another question out in the audience? Anyone? Oh, there's one over there in the back.

Evan Nisselson: We've got time for... Actually, there's only 40 seconds left.

Jonathan: Oh no! I'm going to do my follow up. Sorry, sir. What is the retailer then, or the brand company, that will come in and disaggregate from the bigger brands and say, "Look, we want to do this"? If it's hardware on one side, what's the...

George Borshukov: I think it's about aggregation. You basically drive a truck into any warehouse facility and scan everything en masse. You need a scalable scanning solution. You don't ask anybody. Just go and scan the stuff and build a destination that's brand agnostic, if you wish. To enable that you need a Google or an Intel to put up the money just to get it going.

Jonathan Shieber: With that we are actually out of time. Thank you all so much for paying attention. The panelists were great. I was hopefully more than mediocre. Thank you so much. You all have a good one.

George Borshukov: Thank you.

Alper Aydemir: Thanks.

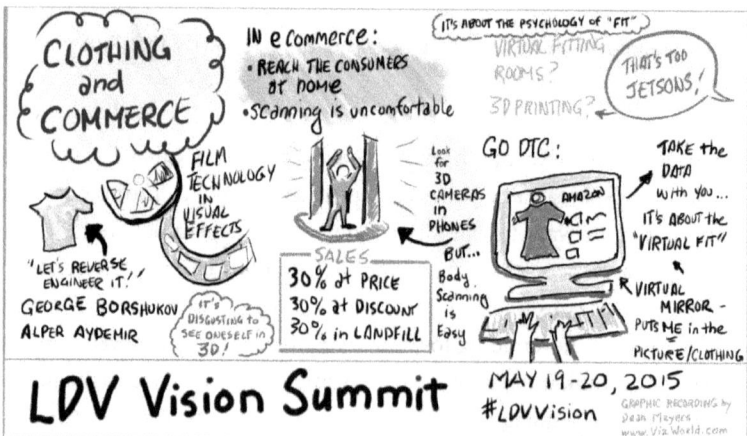

©VizWorld

17

HOW CAN DATA SCIENCE EVALUATE WHY SOME ADVERTISING CREATIVE CONTENT RESONATES MORE THAN OTHERS?

Claudia Perlich, Chief Scientist, Dstillery

WHEN EVAN INVITED ME TO COME AND TALK AT A VISION EVENT, MY INITIAL response was: "I am not sure. I actually don't do vision. I typically don't even try to visualize my data." But I thought about it and there's some very interesting development that's going on in digital advertising, in our company right now. I wanted to share the premise and maybe some of the promise of this work. It all starts with what moves us. Ultimately you would argue that the whole point of advertising is to affect people, to touch them emotionally in some way, maybe bring them closer or at least generate some interest in your product.

What I'm going to do is I'm going to show you a couple of images. And I'm asking you: What moves you?

This is a pretty well known campaign. I'm sure you have seen many of those before, maybe not all of them. I will not embarrass you and ask you to raise your hand on which of those you felt most touched or affected.

But chances are that we all had very different reactions to these images. I am not going to tell you which one my favorite is, but I wanted to take the opportunity to tell you a

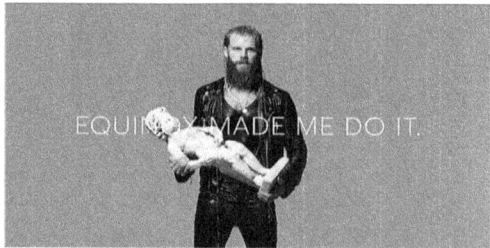

little bit of a story that few people know about me.

I grew up in East Germany and that in particular means—and that's the irony of my life—until age 15 I had never seen an ad because in East Germany there was no such thing as advertising. There was nothing really to sell anyways so why the hell would you want to advertise it? When the wall came down I became fascinated by ads. Not because of the product, but what I discovered was photography: beautiful images of things and nature. And where I discovered it was in cheap magazines that had incredibly well-printed and produced pictures (by my East German standards at least).

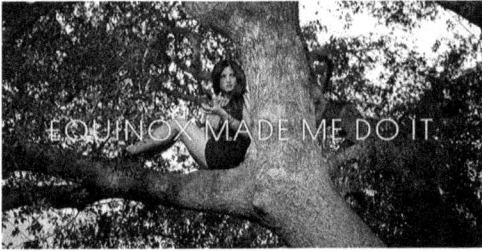

The dirty little secret is I spent my time as a teenager collecting those. It was truly because I was just incredibly amazed and touched by some of that photography. Not that I ever bought any of these things, but still.

Images have the ability to connect to us, to touch us. I want to challenge you right now: do you think you're able to express why a given image had such an impression on you? I'm sure you can tell me which one it was and I will recognize it, but do you even know why? I personally feel that we are very restricted by language when we try to explain things, when we annotate images, when we bucket them: "This is a happy cat and this is a grumpy cat." Ultimately, language is limiting in the ability to express our emotions.

Whether or not the tale is true, that northern tribes have an exceeding number of words to describe the various types of snow, just consider that the average person's vocabulary is estimated to be only around 17K. While this may sound like a lot in a specific context, that is all there is for all the things we may ever wish to say. Anyone ever trying to describe the subtle details of an image will soon realize how limiting already our ability to characterize colors are. How many words for different colors can you come up with? I have seen a list of about 150 and they contained a lot of analogies like "forest green." Do you want to guess how many colors your Internet browser uses? HTML allows for 16 million. That's one of the challenges that I feel we face in machine learning when we try to explain, characterize, categorize, and annotate things. We're stuck with language and in that process, a lot of the magic and subtlety gets lost.

As I said, I work in advertising so I want to show you some of the alternatives where we're trying to avoid having to characterize what it is about

Claudia Perlich, Chief
Scientist, Dstillery

an image that touches you. I don't actually have results for Equinox, so what I'm going to show you are results from a food brand. We ran an experiment where we show digital ads with a set of creatives that vary both in image and message. The interesting question is not so much "which of these variations is 'best'" but rather to understand how different people react differently to these variations. The methodology is a combination of a random experiment alongside with machine learning. I will start initially looking at the impact of just different images.

What is this graph here? It's showing the impact of the image as a factor for different groups of people. Unfortunately I cannot show you the exact six images, and now I have to eat my words and describe them a little but in terms of what they looked like: family oriented (probably a picture with the family), individual, lifestyle, just showing the logo, some variety of the product, or just a very close up shot of the product itself. And in order to give you a talk, I also have to describe the characteristics of different groups of people to you rather than the actual millions of details that our machine-learning approach is actually processing. Specifically, I have grouped people by where they physically go—fast food restaurants in this first comparison and gyms in the next. (We obtain the information about device locations from mobile advertising bid requests.)

The first thing you observe is that people who go to Chick-fil-A are really, really hard to sway to buy this product. No matter what it is that you show them, they're kind of happy with where they are, thank you very much. The next observation is that the "lifestyle" image is sometimes having no effect at all whereas the product image was overall the most effective across groups.

But you also see a lot of variance between these subgroups—they react very, very differently to some of the differences in the images. Right now, this is just a very high-level picture characterizing people by the fast foods they like. But you see clear differentiation and you can think about how would you use that information to schedule or to choose separate images for sub-population if you wanted to actually specifically reach out to any of these groups.

Claudia Perlich, Chief Scientist, Dstillery

Let's take a look at how the messages themselves fare. Some related more to the natural process of the production. Others talked more about taking a snack at a certain time of the day. Some tried to tell you that they're really healthy and good for you, talking about the benefit of this product. What you see here is now broken up by people who go to certain gyms.

In general here you see that there's overarching effect on the emphasis of benefit: it's good for me. I mean yes, people who go to the gym probably care about the benefit and this is very consistent. But what is fascinating to see are the implicit groupings of the gyms: Equinox and Crunch are similar, and so are YMCA and LA Fitness. The populations are similar to each other in how they are affected by the message of the creative. In the case of the YMCA and the LA Fitness, the descriptive message "This is a great product, it will taste perfect" is very effective. This is the sensory stimulation that is not just limited to taste but also includes texture and how it will make you feel beyond taste.

I wanted to use this analysis as an example to really challenge our industry as we move on. What's important here is, that I wasn't trying to characterize the images in any way, but just let a machine-learning algorithm estimate how different people are affected by the message and the image. If you think this forward, maybe in the future I actually don't need to ask you whether it was the guy carrying the little statue. I might be able to predict from observed data alone which of those creatives will most likely speak to you as an individual.

Thank you very much.

18

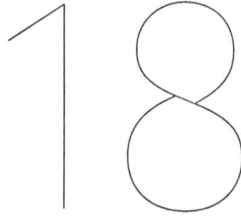

NEW APPROACHES TO
IMAGE SEARCH METADATA

Andy Parsons, CTO, Kontor

KONTOR IS A NEW COMPANY AT THE INTERSECTION OF TECHNOLOGY AND design, launching later this year. We had some fantastic TechCrunch press last week, highlighting the very exciting space we inhabit: workplace architecture and design.

Because I only have five minutes, I want to quickly give you an idea of what we are working on. You can think of it as Houzz for commercial design. Anyone know what Houzz is? Home renovators, home owners? It's a young, successful company in the home renovation space, serving a consumer audience obsessed with great home design. We're doing the same and more for workplace designers.

With office space being an important recruiting tool and retention tool, and things like Facebook office space design being on the cover of every magazine, it's a very interesting space to inhabit at a time when there is a lot of attention being paid to workplace happiness. Because our primary audience at this stage of the product is architects and designers themselves, we have a very high bar for the design of the site, the responsiveness of the site, but also for the visual search and storytelling through images. It would be awesome if we had technology available to us now that would enable us to determine what objects and styles were contained in a given image of a workplace. It could be furnishings, fixtures, flooring, wall material—it's amazing all the esoteric things that comprise interior design for professionals. It could be a particular kind of duct work, of course it could be furniture made by Herman Miller or a little known designer in Williamsburg, Brooklyn. We'd like to identify all those things, make them searchable, and link people and objects to the great interior designs of our time. We anticipate a time in the life of our business, which I hope will be many years and decades, where computer vision will help us scale to a point where we can ingest all the images of commercial space, eventually expanding even to restaurant/ bar and retail space, and identify those things at scale because we can't do it with humans alone. Our images contain all sorts of objective and subjective elements as well. Subjective elements are the ones that define the style of a particular space and embody the knowledge of professional architects.

We've heard a little bit about image similarity and the way certain characteristics of images make you feel. For us there's a joke in my office because I don't come from an architectural background. I've given talks where I use the term "neo-modern." It was pointed out to me that "neo-modern" is actually not a style. It doesn't make any sense since a design can't be neo and modern. But, if you did have a space that was neo-modern and you wanted to see similar spaces, that term may mean one thing to you and something altogether different for me or other people looking at it. Moreover, lay people who are CEOs or founders building workplaces for their teams don't have access to that vocabulary. We'd like to show them spaces similar to ones that inspire them, and through a visual experience, help them find the perfect designers and objects.

This view is taking a page out of Google's handbook. This is what the private beta site looks like now when you're logged in. It's just a search bar. Effectively what we're doing is taking an approach to image search that applies metadata in multiple layers to workplaces and their various spaces. You can navigate through projects, "scenes" as we call them, or spaces, individual components of spaces, and all of these items are tagged and stored in our database in a way that anticipates a time that machines will be able to do this highly accurate classification for us.

Andy Parsons, CTO, Kontor

I've got only one minute left. I'm going to show a couple of quick slides so you can see where computer vision will apply for Kontor. We do some algorithmic cropping once we locate something in an image. We do an algorithmic crop to bring it to the fore and make it searchable. This is an image that shows the iconic egg chair. If you try to do this on reverse image search on Google or Pinterest or other tools that people are using nowadays for this kind of thing, you get all sorts of things that look like egg chairs that are not the authentic egg chair. They may be cheap knock-offs or they may be similars. Kontor will be able to find the exact item across different contexts, different designs, different architects, and designers.

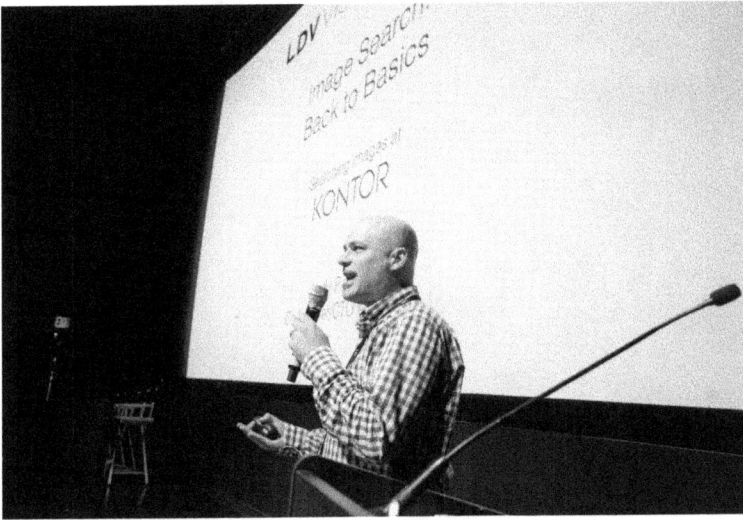

Andy Parsons, CTO, Kontor

Conversely, here's what you get when you search for "open plan lounge chairs." This is one of our famous search tests around the office. We want to show you a mix of lounge chairs in open plans, but also open plans that contain lounge chairs. And some of the intelligence that we use to do that involves anticipating a time when we'll be able to run images through computer vision algorithms that can find all the items, all the materials, and eventually all the designers and producers of materials.

Here's what a hot spot or an element looks like once we find it in the image. We highlight it there in the center and we show that little algorithmic crop over there on the left. You can imagine how this business works as lead gen or in any number of commerce opportunities.

We call this linkage of designs, objects, and style the "design graph," which is the market-y term that we use when we talk to the press. The technology behind this is our key driver for engagement and utility. What the design graph actually will become is a graph-based search that draws inferences about images and projects, threading design concepts and stylistic terms through spaces and the products they depict. Again, this is human-powered now but we're looking towards Serge and compatriots to figure this out for us so we can use fewer human architectural experts. We do have some early work that does a first pass using some computer vision stuff from a couple of companies, actually, that will give us fodder for our architectural experts to look up. The future looks bright. And that's it, with 50 seconds left to spare!

Thank you.

19

PANEL: VISUALLY MAPPING THE WORLD CAN IMPROVE OUR DAILY ACTIVITIES

MODERATOR

Erick Schonfeld, Co-Founder, TouchCast

PANELISTS

Jan Erik Solem, CEO & Founder, Mapillary
Luc Vincent, Director, Engineering, Google
Alyssa Wright, VP Partnerships & Business Development, Mapzen

Evan Nisselson: I'm excited to bring up again Erick Schonfeld, co-founder of TouchCast, who's going to be our moderator. And he's going to talk about how visually mapping the world can improve our daily activities and probably many other subjects that they're going to talk about. Thank you, Erick.

Erick Schonfeld: Thank you, Evan. Can we give a big round of applause to Evan for putting this together? Really great sessions. I'm learning a lot. So we have a really great panel here. We've got Luc Vincent from Google.

Moderator: Erick Schonfeld, Co-Founder, TouchCast, co-founder, Demo, executive producer. Panelists: Luc Vincent, director, engineering (Lead Geo Imagery Efforts, Street View, Aerial, Satellite), Google; Alyssa Wright, VP partnerships & business development, Mapzen; Jan Erik Solem, CEO & founder, Mapillary [L-R]

He runs many of the geoimaging parts of Google maps and other things he's going to tell us about. We have Alyssa Wright from Mapzen and Jan Erik Solem from Mapillary.

Erick Schonfeld (cont'd): It's the end of the day and I really want to give people something to talk about during drinks afterward. We're at this point where we all use maps every day. I know I do. It's probably after email and maybe after Twitter, it's my third most used app. I'm opening it up at least three or four times a day. And we all sort of think, *Okay, well, this is pretty good. It's pretty accurate. I don't need a GPS in my car anymore. I've got my iPhone or my iPad and it gets me there even better than when I have a GPS in my car.* So what else is there left to do? Aren't we done? Yes? We're done? We're not done?

Jan Erik Solem: We're nowhere near.

Erick Schonfeld: No?

Alyssa Wright: Yeah, hi. I'll jump in.

Jan Erik Solem: Go ahead, jump in.

Alyssa Wright: First of all, there will always be blank spots on the map. Whether that means that things are changing and so we are mapping movement or more detail or a building goes up... But I think also another question is even though it's mapped, I think it's also important to look at who owns that data, and who has access to it. It's not just like the data exists, but what's the community around it?

Erick Schonfeld: So why don't you tell us a little bit about where you feel we are with the mapping? Where are we in the evolution of digital mapping and where we are in this?

Luc Vincent: I think we've come a super long way. I think if you look back at Google Maps, for example, 10 years ago, we had the US only. When we launched Street View in 2007, we had like five cities. I think we have, at least at Google and elsewhere obviously, we have come a long way. We have the whole world. We have a lot of detail. But I think two things are key here... That first of all, user expectation keeps getting higher and higher. I think people expect now everything to work all the time, to be super accurate, for a system to know where you are and to give you advice depending on where you are. A few years ago that was not the case. In addition to Alyssa's point, I think the world just keeps changing. So whatever we do, we're just never done, right? So we have to keep up with the world as it evolves and changes and that's a challenge we'll never be done with.

Erick Schonfeld: Okay. Let's talk about the different types of data that we place on maps. We started with lines and a 2D depth representation of the world. Then that became satellite imagery and photographs. Then we started taking photographs. And whether that's street view or user-generated, uploaded images, that started to add another layer to that. Can you just talk a little bit about: What are the different layers? As you think about making maps, what is that data stack? And from the most obvious to the ones where we're just starting to place on the map.

Luc Vincent: Yeah, I can jump in on that one. In terms of visually mapping the world, there are several layers there starting from satellite going down. If you're looking at it from where we are now, the cars have covered the main roads in the most important countries. So we've done the easy part and we're now getting into, *Okay, how do we get the rest of the world, how do we get the details, and how do we get that updated at a rate that people want to see?*

And so we're in the early days, I still think that, even after Google has been leading this for a decade, I still think we're in early days here.

Jan Erik Solem: We used to think of the imagery as being there to enhance the experience—to help you understand a place, to help you get a good sense for your travel destination, to help you every day. It used to be a layer on top of the map in some sense, and now I think it's maybe changing. I think now more and more we can derive data from the imagery itself and use it to make the map.

Erick Schonfeld: So we're all talking about data here and how this real-time map is what we're trying to aim towards, right? Which, to me is really fascinating. Because if we look back at the history of maps, we had maps that stood the test of time and were the maps people used for hundreds of years, and then decades. And then even when I was going to school, we had maps in our textbooks that were probably created decades or at least years before. And now these maps are just being updated every day, every hour. And we kind of take it for granted. But even that whole real-time aspect of it is something fundamentally different that we just take for granted. We're updating this virtual representation of the world as close to real time as possible. Why are we doing that?

Alyssa Wright: Can I answer a data question real fast?

Erick Schonfeld: Can you answer the question that I asked?

Alyssa Wright: Yeah. The goal of mapping, at least the... So I'm the kind of person who needs a map just to walk around the block. I grew up with a friend actually who I would consider to be a mapper. She would read an atlas and it actually seemed to have words on it to her and made sense. And then she would get in the car and just drive around. I thought that was a person who liked maps and I'm the kind of person that gets lost walking around the block. But I ended up finding my way to mapping because I realized that it's a storytelling tool. It's a way for us to make sense of our world, our experience, our relationships. And it's an abstract representation. I think it's important to have many different kinds of stories and have many different authors. And so I think that... Wait, what was your question exactly?

Erick Schonfeld: Well, forget about my last question.

Alyssa Wright: Yeah.

Erick Schonfeld: So are you suggesting everyone should have their own map?

Alyssa Wright: I mean...

Erick Schonfeld: Just like everyone should have their own story?

Alyssa Wright: I mean, I do believe that. Everyone should be able to make their own map, whatever that looks like. Whether that's a napkin that you're drawing to tell somebody how to get to a direction or something that's like a more...

Erick Schonfeld: Shouldn't there be some kind of standard?

Alyssa Wright: Sure, and there are standards bodies that work on that.

Erick Schonfeld: Otherwise we'd be bumping into each other.

Alyssa Wright: Say it again?

Erick Schonfeld: But what does that mean? When you say everyone should have their own map. A map is supposed to get me from one place to another, right?

Alyssa Wright: I think that's one purpose of a map. That's not necessarily the entire reasoning for maps. I mean, there could be many different purposes for a map. I think that we're familiar with, especially with Google, getting from point A to point B. We work entirely in the open source, open data field, and I'm on the board of the OpenStreetMap US. So in the OpenStreetMap community, which is kind of the Wikipedia of mapping, they will respond to a humanitarian crisis with Cloud-source response. So that's not getting from point A to point B. That's trying to understand what happened after an earthquake or a hurricane and look at "before" pictures and "after" pictures to help with logistics.

Erick Schonfeld: Right, and this is probably a good time for everyone to just give a little brief synopsis of what you're doing with Mapillary, with Mapzen, and maybe you can tell us something surprising about what you're doing with Google.

Luc Vincent: Can I answer your question first? Why are people doing it? Because I don't think that people realize the maps that we have now are mainly created manually by humans. So, OpenStreetMap is an open and obvious example of that, but all of the major mapping players, they have lots and lots of employees that sit in front of screens and edit maps all day. It's a manual process. One of the most important pieces of data for that process is the type of imagery that we collect. And so we shouldn't forget... Why are people doing it? They're doing it because we need newer maps, we need updated maps in Nepal when something happens. We need better road infrastructure so that people can get in there with their trucks or whatever it is. It's a manual process fed by imagery and data that we can extract from that imagery. But it's far from an automated process.

Jan Erik Solem: So it's manual, but less and less so.

Luc Vincent: Yeah, exactly.

Jan Erik Solem: As the learning gets better and better, I think the manual piece is going to shrink and it's shrinking right now.

Alyssa Wright: Have you noticed that with your data teams at Google?

Luc Vincent: Absolutely. So for example, we have published some papers... In fact, Andrea, who spoke earlier, was key to some of this early work on this. From Street View imagery and other sources we can extract house numbers. And that may seem simple from a computer vision perspective, but it's doing it at scale, at super-high accuracy, that's really hard. And from these you can then help users find a place. This is like geocoding. You type an address somewhere on a map and you take the user right there where you saw that address. That's one example. More and more is happening on this space.

Jan Erik Solem: That's a great example and you shouldn't forget that what initially fed that work, at least for looking at it from the outside, was again humans. Because you put it into the recapture system and you had humans proving that they're not bots to train the initial algorithms that now make better navigation for maps.

Erick Schonfeld: If each of you could give us a little data point about how much of the world you've mapped, or your respective projects or companies have mapped, and sort of how far we have to go. Just maybe some data points that give some context to what we're talking about.

Jan Erik Solem: All right. So at Mapillary, we provide a service that lets anyone crowdsource street-level imagery for mapping and for other purposes. And right now we have 16- million-something photos we launched last spring and about 470,000 kilometers of roads mapped.

Erick Schonfeld: And this is a crowdsourced effort?

Jan Erik Solem: Yes. Most of them use our app for iOS or Android and then they drive around, walk around, bike, collect imagery, and then we combine all of that together.

Erick Schonfeld: Okay.

Alyssa Wright: Hi. I'm coming from Mapzen, which is a startup that operates like a research lab. We're about two years old, or a year and a half old, and everything that we do is based on open source and open data. I'm simultaneously representing OpenStreetMap as well and some of their initiatives, which is, again, the Wikipedia of mapping. There are two million editors registered on OpenStreetMap, which is used for a lot of international aid organizations. It's mapped a lot of the world, especially a lot of the world that's in developing countries that may not necessarily be the market for other mapping providers. So we've seen a lot of movement in those areas as well.

Luc Vincent: So Google Maps is obviously everywhere, but with varying degrees of quality and detail depending on the country and the area. My area of responsibility is the imagery, which is sort of key to helping make the map. There we think of ourselves as being the crawler and the indexer for the physical world. We sort of gather imagery from a variety of sources, from Street View cars, from users, from crawling the web, from airplanes, and now from satellites. It's imagery we either acquire or license or gather ourselves from Skybox. We organize all of this imagery, and make it useful for users in the context of a map for their daily activities. It's also about using the imagery to make sense and create the map. And so in total scale we have... Street View, which is the key to that effort, has now 66 countries on the order of over 7 million miles. But also multiple time layers for these miles so probably more than 20 million miles if you include all this. In terms of high-res imagery of the world, we have probably 90% of the population covered. Now more and more we have also 3D of key metro areas in super high-res.

Alyssa Wright: Yeah.

Erick Schonfeld: What do you define as "high-res"?

Luc Vincent: I think it's in the order of less than 30 centimeters. 15 maybe. I forgot the specs.

Alyssa Wright: When you said "the data that we put on the maps," I feel like the 3D aspects of where mapping is going is a really exciting piece. You even said "data layers." I feel like modern mapping has been a two-dimensional concept. You just put layers on top of each other. And what would it be like if we had a much more immersive mapping experience? Something maybe akin to video games or animation or movies? And so I think those kind of influences on our mapping experience is really an exciting place to be in.

Erick Schonfeld: So can we talk about that a little bit? Let's start with the photos and the street- view-like experience. Just explain to the audience what the value of that imagery stitched together does for making more of an immersive map. Because from my perspective as a user, it's, *Well, this is cool. I can zoom in. It's sort of immersive.* It's kind of like a panoramic sort of experience. It's not like VR, but it's as close as possible. I feel like it's accurate in an amazing way. But I think of it as, again, this layer that I zoomed in on. I don't think of it as something that is making the actual contours of the map any better. Am I wrong about that? Are you actually taking the visual information, from photos or from maybe video, and is that giving you a more accurate map?

Luc Vincent: Well, not directly. What can help you from street- view-like data is you also have other metadata. For example, the location where the image was taken, which itself tells you where the roads are. You can use various vision techniques to extract 3D and understand how far buildings are. From this and other sources, you can create these building footprints we have on Google Maps. So that really helps you create the more accurate map. It's not the end of everything, right? I think that what you said is actually quite right. It's kind of a precursor to... Or poor man's VR. But it's going a step in this direction. It's giving you a virtual experience. You can preview and understand a place, which is often very useful.

Erick Schonfeld: Right. The question I want to ask you is: I understand why Google does Street View. Why do people do this work on behalf of the crowd? What's in it for them?

Jan Erik Solem: So, what they get is they get a way to capture the imagery themselves. For example, one of the things when I started this was my hometown doesn't have a street view. I live in Sweden. Well actually, this goes for a lot of places in Europe. Smaller towns. It's a town of 10,000 people, just outside of the third largest city in Sweden. We don't have street view and we probably never will have from the bigger providers, because it's not cost-effective to go in and drive up and down all the residential roads and map it on any frequency at all. But I can do it myself in two weekends with my bike. And so there's now a tool and a process that lets anyone who has that frustration to go out and get the imagery they want. There's a lot of people doing it for that reason, and then they get credit as the photographer. But then there's also a large crowd doing it for open-mapping purposes. So we have strong ties to OpenStreetMap, where a lot of our power users do it because they want to help create a better open map. For example, the Red Cross and a couple of other initiatives are out in developing countries taking imagery for Mapillary. So it goes into the OpenStreetMap editing tools so that people can sit there and fix roads and things like that. There's anything from the frustrated local that doesn't have street view to people who are out recreationally mapping on their vacation to people who are doing it for the work of creating better maps.

Alyssa Wright: I think there's also an urban planning aspect or potential. I was just a guest critic at a university finals. And they had put a kind of street view on the back of a car in order to record potholes. And so this would be something that they had recommended for city vehicles to have so that they could identify if there really was a pothole, what lane it was in, would it be fixed, etc., etc. And so I think potentially street views can help with city government planning.

Erick Schonfeld: What about indoors? I'm the producer of the Demo conference and one of the more interesting companies that has launched in the past few years is a company called Indoor Atlas that basically creates the magnetic footprint of an indoor space. And there are other companies that do other techniques. But to me the inside of buildings is the next level. When we're talking about immersive experiences, it's not just the shapes and the outdoor part of it, but it's also the inside, too. So where are we in mapping the indoors?

Luc Vincent: I'd say we're less advanced at this point, but I think there's a whole bunch of activity all around. In terms of what's happening at Google, we've been trying a few things, or ramping up a few programs. One is going into museums, in particular, in large spaces, where we can replicate the Street View car on some sort of push cart. Essentially put the whole set-up on something you can wheel around and take pictures. That gets you a pretty good experience, really a VR for a museum. Now we have, I think, a few hundred museums that are live-in maps as part of the art project. That approach doesn't really work for small spaces. For a restaurant or even for a mall, it might not work. So really what you can do there is use either an off-the-shelf 360 camera or use an SLR with a fisheye lens. The process where you just go in and take a sequence of photos and create a kind of virtual tour. It's happening, but at a slower pace.

Erick Schonfeld: So if we shrunk down the 360-degree camera into something that was more like a GoPro device, then all of your users could go into the most interesting buildings and that could be uploaded to the map as well?

Jan Erik Solem: Yeah, absolutely. And there are a lot of those devices in the $500 range coming out on the market now. So I think we will see an explosion in that area. Absolutely.

Erick Schonfeld: So if people have questions, just raise your hand and somebody will come around and give you a microphone. We'll continue the conversation up here. So I really want to push you guys on this whole idea of sort of indoor mapping, the immersive experience, the 3D experience. So Google has SketchUp, right?

Luc Vincent: Well, we *had* SketchUp.

Erick Schonfeld: Oh, you *had* SketchUp?

Luc Vincent: We don't have SketchUp anymore.

Erick Schonfeld: What happened? Was that closed?

Luc Vincent: We sold it.

Erick Schonfeld: Who bought it?

Luc Vincent: Trimble Navigation bought it. So what was SketchUp? I think in 2006 or 2007, I forget exactly when, the notion was that it was going to be a crowdsource of 3D buildings to enhance maps and Earth. But I think we realized after a while that we couldn't really get the scale we wanted from this approach. Just too hard to do. Perhaps it was just too hard to do. And in the meantime, I think algorithms have gotten better and better and the ability to do this from automated systems has now surpassed what we were able to do with SketchUp.

Erick Schonfeld: I think you mentioned that you've got a certain critical mass of cities or parts of cities that are now in 3D and you can see that on Google Maps sometimes, right? That is based on what technology?

Luc Vincent: Based on the imagery captured from planes at a 45-degree angle. Microsoft has something similar, too. I mean essentially, you take this imagery and using stereo-type techniques, multi-view stereo, you can reconstruct a 3D mesh and then you can texture map the mesh. And you've got the 3D map.

Erick Schonfeld: How does the advent of drones and the ability to put pretty sophisticated imaging cameras on drones affect this? Do you use drones?

Luc Vincent: We don't use drones. We have played with drones, but we haven't used drones for any kind of commercial application at this point. However, there are a bunch of companies doing this—a number of them in Europe. I think it's great for applications like… I think they have mining applications where they send a drone and the drone can very quickly sort of map the mine and create a 3D rendering and you understand how much is left of the mine. Lots of application around capturing a landmark building, like a castle. It's hard I think to take this approach and scale that to a whole city. But it might come one day.

Erick Schonfeld: Are any of your projects using drones for capture?

Jan Erik Solem: We have one or two users that played with it. Low-flying drones. It doesn't scale right now unless there's some kind of battery development out there. It doesn't really scale. Most of them can fly for 15, 20 minutes. You can cover a castle or a city block or something like that, but if you want to do large-scale mapping it doesn't scale right now.

Alyssa Wright: There are some initiatives within the OpenStreetMap community to start an equivalent drone map. It's called "Open Drone Map." It's one of the tools to stitch together imagery. And then there's another initiative that just restarted called the "Open Aerial Map," which is like a close sister to Open Drone Map. These I know very little about. I don't know much about drones. But if you're interested, I can connect you.

Erick Schonfeld: Do we have any questions or suggestions? Feature requests? We're taking feature requests up here. What do you guys want to see on maps in the next two to three years?

Audience member: I'm thinking a little bit long term. And I'm wondering if what we now think of maps is a kind of temporary stand-in and in X number of years, 10-20, we'll just have real-time video imagery of every point in the world? And then maps will just be images of some labels? Or do you think maps would be made for another hundred years as a separate type of representation, which has something which video does not have and will never have?

Alyssa Wright: Are you asking will maps go away?

Audience member (cont'd): I mean, maps… My perspective, maybe that they haven't really changed in a hundred years or more. Maybe it's time for a change. Maybe you don't need any more *better* representation because you have a phone with you at all times, which has all the information. Maybe it can give you some other way of showing what you want to know. If you're trying to go from point A to point B, do we need to show you a map? Maybe not. We need to show you pictures and maybe some arrows and that's it. I don't know.

Alyssa Wright: Yeah, I do think it's a really exciting time for mapping because it's so much a part of our experience and we do have a map in our pocket all the time. Or multiple maps. So I think it's changing. In my opinion, I think we will always have something, some sort of map. Some abstract way of understanding the world that's not necessarily video. Maybe it's interactive, some simplified way of understanding what's happening.

Luc Vincent: Yeah, I agree. I think that the flat yellow map that we see today is not the future of the map, but it's also going to be much more visual than we see now. But probably not real-time video of every location.

Erick Schonfeld: Are we going to continue to be the center of the map?

Alyssa Wright: When I first started to get into mapping, somebody said to me that they didn't like what was happening in digital mapping, because when you had a paper map you had to locate yourself in a space, as opposed to the way our maps are designed right now. It's like you're in the center and it's almost like an egotistical view of the world. The world kind of revolves around you. This has been a metaphor that has stuck with me. I believe in being authors of maps, but who we are on maps and how we represent ourselves and the people around us will also be a way that we can shift.

Erick Schonfeld: One last question over here.

Audience member: So it will be a political question. Some maps cause borders, or maps identify borders simply for the regular people. And countries have gotten in trouble for using Google Maps for wars. Border wars. Alyssa, you said something about who owns the data on the map. Is there a possibility in the world that just regular people deciding borders... For example, anarchists border mapping where the borders of locations are actually changing based on who controls the map. And if the map is based with the community, then the community decides what the border of the country is. Or the border of that anarchic state.

©Anne Gibbons

Luc Vincent: I think that's what treaties are for, right? Treaties and international law define boundaries and that's why maps are important.

Jan Erik Solem: True, but I think he's saying, "Let's put some limit on what you can do with crowdsourcing." That's dangerous to rely entirely on crowdsourcing to decide on borders.

Erick Schonfeld: Which is back to my original question. There are certain standards—I expect a border on a map to be the real border. Or I expect Times Square to be where Times Square actually is. Which wasn't, I don't think, what you were saying about different maps for different stories. Don't want to put words in your mouth, but maybe a map for me would show different layers, right?

Alyssa Wright: Sure. I mean, I think there are many different sources of authoritative data, even for a border. You could look at a U.N. border as opposed to what individual countries are saying is a border. It's a really interesting question. Where exactly is a border? Does it follow a geological formation? What happens if that's moving? So part of my dream—if we're going to talk about if we will live in a world without maps—is that there isn't one map, but potentially a series of looking at the world and lots of different maps. So that there doesn't have to be one border, but potentially like multiple borders.

Luc Vincent: That's already the case.

Alyssa Wright: Yeah, yeah.

Luc Vincent: Depending on how you look at Google Maps, for a different country, you see a different border.

Alyssa Wright: Right, yeah. True

Luc Vincent: Or disputed areas.

Erick Schonfeld: So we'll just leave it at "you really can't trust anything you see."

Alyssa Wright: Yeah.

Erick Schonfeld: Thank you very much for taking us through this journey through the world of maps. I'm now thinking about it in a very different way. I hope you all are, too. So, please, big round of applause for our panelists and I'll bring Evan back up to close out the day.

Evan Nisselson: Thank you, Erick. Thank you, panelists, very much.

©VizWorld

20

VISUAL CONTENT WILL INSPIRE, MOVE FINANCIAL MARKETS, ADVERTISERS KNOW WHEN WE ARE

Evan Nisselson, LDV Capital

VISUAL TECHNOLOGIES WILL MOVE FINANCIAL MARKETS. I WROTE AN article in *The Wall Street Journal* on Friday that talked about this and I wanted to share it with you personally. Advertisers know when we are happy or sad and technology is going to help us improve our health.

A picture is worth a thousand words and I want to take a second here to comment about something that is very important to me. Yesterday we had Serge bring up his kid, about a four- or five-week-old kid, to the stage when he was Master of Ceremonies for several keynotes. There was a little bit of noise coming from the kid throughout the day, but it's a family affair. We're all in this together. On that note I think we should all celebrate our parents, and I think we should all give a round of applause to mine who are sitting over there because this thing wouldn't happen without them. [Applause]

A video is worth a billion words. Humans can review visual content but not in millions. Computers can quickly review billions and trillions of images and video. This is a huge opportunity. The unique piece of the computer vision and AI part of this Summit is how we read trends in content. How do we see content in a different way? It's not just a picture, it's not just a video.

All of this potential is a huge impact to society. But why now? I've been a photographer since 13 years old—and these photos were photographed on 3200-speed film which was pushed to 6400- speed film.

I processed them in the dark room with the fixer smell on my fingers, which I miss. Real time access and computer vision, artificial intelligence, and Cloud infrastructure is what's delivering this huge transition and opportunity. Let's talk first about satellite imaging, which is fascinating on many levels.

Financial investors can soon see real time economic trends from satellite images rather than waiting months to get research data. Recently a couple of companies that are working on this space have raised capital or sold. Skybox to Google, Planet Labs. Orbital spoke yesterday and Skybox is speaking today.

Let's talk a little bit about this. Now they can capture images of a shipping port from the sky and understand how many cranes are in use, how many ships are there, and they can track that on an hourly, daily basis. Sure, we need more satellites in the air and we're not there yet, but this is the trend that I think is unbelievable. You can check loading docks and know how many cars, new cars, came from different countries like Japan. Are they moving quickly

©*Skybox Imaging*

or are they sitting there? Direct information to economic trends that are going to be in real time versus months old. How about oil reserves? Knowing how many tanks have how much oil reserved is unbelievable.

We can understand the shadows of the ceiling of these tankers that move up and down depending on how much reserves are in the tank in real time. Not only hedge funds but governments and others are going to have tremendous value in this data. Agriculture—we can know if crops are healthy and estimate yield faster and more accurately than manually. Deforestation—huge human benefit from data and from visual content.

My next camera hopefully will be the death to the selfie stick, which I can't stand—a satellite selfie. I can't wait to click a button on my phone, grab my friend, and say "smile." That is going to happen in our lifetimes, guarantee it. We'll have a discussion about the future of cameras very soon.

Can anybody see me?

©Brandon Carter

Medical imaging is another opportunity to improve our health. Several companies have raised capital. CaptureProof is speaking later and so is Zebra Imaging. People go and get x-rayed for different health issues. We go and get an x-ray and the doctors look at them.

Maybe they look at a past x-ray or two. What if they could look at x-rays from multiple years of millions of people? What if you could track mammograms over time, which they don't do? What if you could track sclerosis, not only for yourself, but for millions of other people by giving data to identify trends in sclerosis or other diseases like cancer to predict it earlier? This is happening now and the opportunities are unbelievable.

Evan Nisselson, LDV Capital

Brain scans. Why look at just one person's scans? Sure, there are people looking at them and leveraging their expertise, but why not have those experts look at trends from millions of scans? Doctor follow-ups... Frequently, you go in to the doctor, wait for hours, you get in there for five minutes, they look at you and say "You're doing great. Come back in three weeks." Why don't we just send pictures with our mobile phone? We've all heard, "Take two aspirin and call me in the morning." It should be, "Take two photos, share them with me, and I'll text you in the morning."

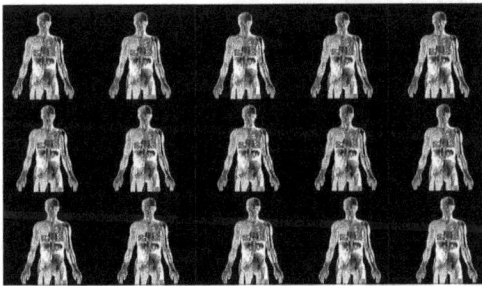

Here's a fascinating company that wants to help you grow your own bones. Unfortunately, Nina, the CEO, was going to speak but could not. EpiBone uses 3D to customize bone transplants via scans and 3D modeling. This has huge opportunities and a huge impact on society.

Visual sentiment analysis can empower advertising and other business sectors. Yesterday we heard from Marni from Emotient and today we will hear from Affectiva. Understanding our emotions when we're watching TV is possible. When we're looking at advertisements. Maybe we can tell right now if everybody is happy in Times Square. How about late night TV: understanding where advertisements should go depending on whether the people watching are happy, perplexed, surprised or sad? This opportunity of seeing data, seeing trends in millions of videos in seconds, has never happened to the scale it can happen now.

Commerce is another business sector that will be impacted from visual technologies. Volumental was speaking yesterday and Souls is speaking today.

I love Italian shoes but I don't go to Italy often enough to buy my next shoes. I'd rather have a photograph of my shoes, send it off to the manufacturer, choose the colors, and have it fit perfectly. Is that going to happen? The trend of more customized orthotics and clothes is a huge opportunity to figure out for those of us like me who don't like to shop and for others who want clothes that fit better.

Visual content is extremely valuable. Visual trends are exponentially more valuable. It's just the beginning. Think big, have fun, and carpe diem, which is obviously the trend here because it's also the password of the wifi. I don't have all the answers, as I've said before, and I believe that. I'm in awe of the people who are here and will be speaking throughout the day. We can hopefully learn from them and hope you enjoy the rest of the day.

Just the beginning... ©Evan Nisselson

#carpediem
@Evan Nisselson

21

JPMORGAN ANALYST TO WORLD TRAVELING PHOTOGRAPHER WITH OVER 273,000 INSTAGRAM FOLLOWERS

Natalie Amrossi, Photographer, Misshattan

GOOD MORNING, MY NAME IS NATALIE AMROSSI. I CAME FROM A FINANCE background and am now a freelance photographer using social media to market my work. I was working at JP Morgan about six months ago and one morning I decided that it wasn't for me anymore. I knew I wanted to pursue photography. And thanks to a lot of you guys and technology, I was able to do that using the platform of Instagram and showcasing my work to the world.

Back when I was at my corporate job, I used to carry my camera everywhere. There was never a moment I would leave my camera behind. It was great being able to take photos on my way to and from work. I took this shot right after leaving my office building. Every time I look at this image, I'm reminded of how much I've grown during this short freelance time. Photography helped shape and give my life purpose.

©Misshattan

With photography, I'm able to show the world my perspective visually. I hope to bring out any emotion when people look at my images.

I love Instagram's platform because of its community. I'm able to reach people and receive instant feedback on my work. I love hearing thoughts and interacting with my audience.

I downloaded Instagram in 2012 but it wasn't until two years after that I would heavily use the platform. The more active I was on it, the bigger my network would grow.

Aerial photography is one of my favorite ways of shooting. Here's one after the snowstorm Juno that got a great reaction and a lot of reposts from my viewers. With using social media, I've been able to book jobs with various editorial clients such as *The Wall Street Journal*, Thrillist, and NASDAQ.

©*Misshattan*

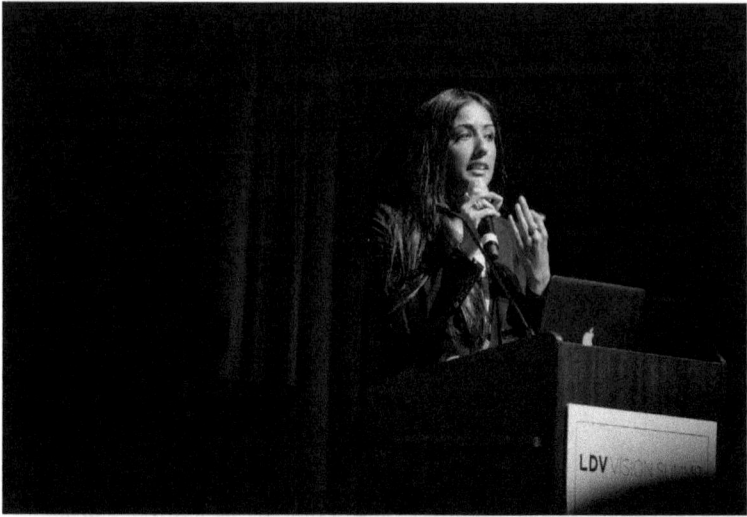

Natalie Amrossi, Photographer, Misshattan

Here's a favorite image of mine. I love the perspective and colors in this photo. It was exciting posting this photo and reading the comments that were coming in. It's the main reason why I love Instagram's platform.

©*Misshattan*

©Misshattan

I've worked with many different brands to promote their work by incorporating my photography perspective and their product and services. One brand I've helped is an aerial company based in New York. I believe their business grew about 400% by using Instagram's platform and having people like me promote their company. I love my job and being able to capture images of everything from people to iconic architecture. It enables me to challenge myself and find different creative ways to photograph my subject.

One of my favorite clients would have to be Jaguar, and it's not just because I get to ride around in an F-TYPE. After I was introduced to the brand, I instantly believed in how great their products are. It made me want to push my creative side and feature their products on my feed in an organic yet fun and creative way. I think my audience can notice the difference between an organic post and a straight sponsored post. I love that Jaguar lets me be the creative director of my sponsored shots as I know what my network likes to see.

By being my own creative director, sponsored posts seem much more organic to my audience. When a post goes live, seeing all of the engagement truly makes all the hard work pay off. Brands are instantly able to get feedback through Instagram. Great results on Instagram lead to many other opportunities to work with brands on things like content for their websites, billboards, emails, blogs, and advertising.

©Misshattan

    Instagram is the most efficient way for me to put my work out there for the world to see. It helped me book many jobs in and out of the platform.

    Some of the other brands that I've worked with are AT&T, Jordan/Nike, and Budweiser. Collaborating with brands is always fun because it gives you an opportunity to use multiple channels to tell a story. For instance, posting an image on your feed and continuing to tell the story by posting on the brand's feed. This is really a powerful way to engage with your audience and have them follow along on the brand's social networks.

Here's a fun color-splash image for a project I did with Coach to feature their new bright-colored purses that were soon-to-be released. ©Misshattan

This past winter, I was able to work with Rockefeller Center. We collaborated for the tree lighting ceremony. I was able to capture exclusive perspectives and then release it on my feed during real time. It was awesome to receive instant feedback and have people tune in by either watching it on TV or following along using social media.

Where I see technology going: technology will forever evolve. I definitely see more brands using images and videos on different social platforms to promote their products and/or services. We are naturally social people who like to simply tell a story by using social networks. The easier the platform is to use, the more active users there will be. What I'd love to see is a way to easily share videos without it lagging or having the sounds cause attention in public places.

Thank you.

©Misshattan

$$22$$

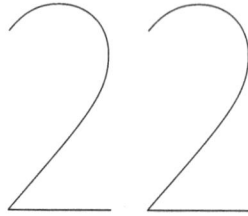

PANEL: FUTURE OF STILL AND VIDEO CAMERAS

MODERATOR
Evan Nisselson, LDV Capital

PANELISTS
Hans Peter Brøndmo, Serial Entrepreneur
Julian Green, Group Product Manager, Google

Evan Nisselson: I'd like to invite up my fantastic panelists, Julian from Google and Hans Peter, a serial entrepreneur. Come on up here. A round of applause for these guys. What have you got back there, Hans Peter? You're causing trouble. You brought jewelry. We're going to talk about why I call some cameras "jewelry" in a second. Hi, guys. Oh, you haven't met yet. Why don't we do the intros really quickly? Two minutes from each of you. What the hell are you doing? What do you love about photography?

Julian Green: Hi. I'm Julian Green. I have been living in San Francisco since 1998 at the rock face of technology.

Evan Nisselson: The rock face of technology?

Julian Green: The rock face. Four startups. Six years at eBay. A startup selling books online in '98. Selling bulldozers online from 2000—I think I auctioned the first bulldozer on the Internet, which was... They said it couldn't be done, but it could be done.

Evan Nisselson: You did it. It's a classic entrepreneur story.

Julian Green: Six years at eBay, helping sellers sell on the platform and then scalping tickets, an important contribution to society. Then started a company called Houzz, which is a home design inspiration website, and iPhone, iPad, Android apps. Started Jetpac for travel inspiration. Jetpac got bought by Google, and now I'm at Google. I'm the product manager for the mobile vision products, so putting all the computer vision technology onto phones and drones and robots and satellites.

Hans Peter Brøndmo: Hans Peter Brøndmo. I am, as you introduced me, a serial entrepreneur, currently mostly skiing.

Evan Nisselson: That's entrepreneurial, isn't it? I know how to do that, too.

Hans Peter Brøndmo: It can be quite entrepreneurial.

Evan Nisselson: Yeah. Depends on the jumps and...

Hans Peter Brøndmo: Especially with the snow conditions we've had.

Evan Nisselson: Depends on what kind of cameras you're wearing.

Hans Peter Brøndmo: I've been a photographer since I was a kid. Started in the digital imaging space based on research I was doing at the media lab [at MIT]. Co-founded a company in 1990 called DiVA—it stood for "digital video applications"—and we sold it to Avid Technology in '93. And it was very early on, somebody was talking about, I think one of your earlier colleagues here was talking about video starting in 1998, and we were doing digital video back in 1990. I'm dating myself here. I've been tracking that space and have done several other companies since then, but imaging— storytelling, more importantly—has really been a core thread through my path to where I am today. I sold my last company [Plum, a private group social networking play] in 2009 to Nokia. Stayed there for a while. Worked on some very exciting stuff in the imaging space that I can't talk about, and left about six months ago in December 2014.

Evan Nisselson: We'll try to get some wisdom out that relates to it potentially, but thank you very much, guys. So, guys, tell me what cameras did you guys use—or video, imaging, photo—cameras that you guys used 15 years ago? What was your go-to camera 15 years ago? What's your go-to camera now? And why?

Hans Peter Brøndmo: I'll start. I think the answer is very simple. It depends.

Evan Nisselson: Okay.

Hans Peter Brøndmo: I think there's the classic hierarchy of why, what, and how. I think the first question I always ask myself is: Why do I need a camera? And then: What camera do I choose? And then: How do I use it and how am I capturing the story or images I want to capture? I think we tend to get caught up in the discussion about, is it the cellphone or is it the DSLR, what is the technology itself? But I think the more important part is to stick with the why—why we are trying to capture images, why do we have cameras in the first place? We have cameras to tell stories. That's always the starting point for me. So specifically, to answer your question, it was analog cameras back 15-20 years ago.

Evan Nisselson: Which ones?

Hans Peter Brøndmo: I had a Contax. That was my favorite. I originally bought cameras on the back of cereal boxes when I was a kid, little Kodak Instamatics.

Evan Nisselson: Now you're really dating yourself.

Hans Peter Brøndmo: Yes. And I think now, today, it really depends. I mean, I just did a great trip in the mountains a few weeks ago. I carried a Leica on my backpack and I carried an iPhone in my jacket. The iPhone was the go-to if I quickly needed to get to something. If I saw something beautiful and I saw the right light and the right image, I would take my backpack off and reach for the Leica. So it really just had to do with what I saw and what I wanted to create.

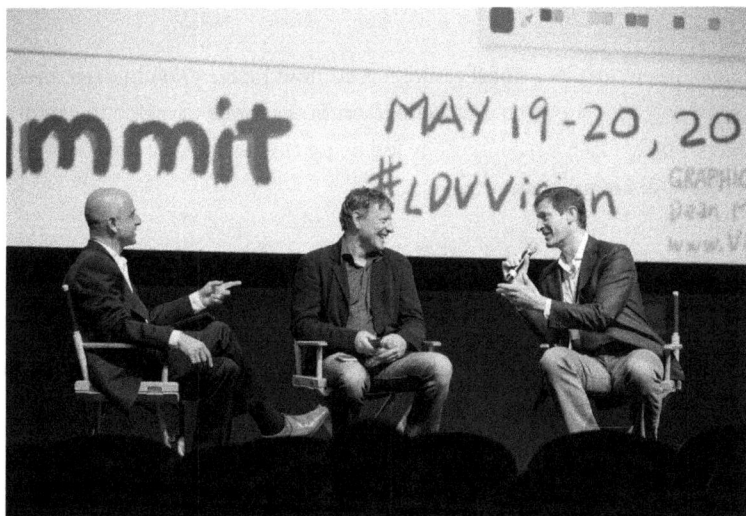

Julian Green: Yeah, I grew up on my dad's LM1, which was this beautiful, light—light-ish, by common standards—film camera. 15 years ago, I think just about, got the digital Rebel, the Canon 300D. Again, just a nice, light, easy camera. Although that competed with, you remember those pre-digital, sort of throw-away Kodaks?

Evan Nisselson: Sure. They used to put them at the tables of the weddings.

Julian Green: Spin the wheel, they even have a flash. Yeah, they'd be on the tables at the wedding.

Evan Nisselson: We used to do these great things at weddings, where they would put them on the tables... And some of these weddings would have multiple different weddings happening at the same venue. Our game was to take those, go to another one of the weddings, and see if we could make photos of us dancing with the bride at the other weddings on the camera from our friend's wedding.

Julian Green: The wedding photo bomb.

Evan Nisselson: Exactly. Before all that. It was fantastic.

Julian Green: Those were so great because in my lifestyle 15-20 years ago, if you lost the thing it didn't matter.

Evan Nisselson: Right.

Julian Green: It was a good camera. And now I have two children and a wife and I have a D7000 and I either am in the mode of trying to capture every pixel of every second of my kid being cute or I am... I love being on Instagram. I'm no Misshattan, but....

Evan Nisselson: How many followers do you have on Instagram?

Julian Green: A few thousand. Yeah.

Evan Nisselson: Misshattan, how many followers do you have?

Misshattan: 275,000.

Evan Nisselson: How many? Misshattan has 275,000.

Julian Green: She is 270,000 times as followed as I am.

Evan Nisselson: Somebody actually saw her before going into the speaker dinner last night. She was in one of those bicycle cabs, and they jumped out and said, "You're famous, you're famous! Can I take a photograph of you?" That's amazing.

Julian Green: That never happens to me.

Evan Nisselson: That doesn't happen to you?

Julian Green: That never happens to me. Although I did sit by the founders when they started it, and I was sort of in the dogpatch in San Francisco and I was like, "Why are these guys taking photos of Coke cans and giggling? What is this thing?" And I was working on a startup and I was like, "What are you guys doing?" And they're like, "We're playing with filters. It's so fun!" But they're really very smart, very cool people. So I managed to get Julian, @Julian on Instagram, because I was one of the first people on it. And now I get, this is a double-edged sword, now every day I get, "Can I have your username?"—which is the polite version.

Evan Nisselson: What's the not polite version?

Julian Green: "Can I have your [expletives noise] username?" and "You're too old to be on Instagram." I get that from my profile photo.

Evan Nisselson: Do you agree?

Julian Green: I'm probably older than the average, but I love it. Then the other thing you get is everyone, when they're commenting and putting in Julian-somebody-else's name, they always put a space and so that's @Julian. So I get a million comments every day that do not relate to me.

Evan Nisselson: But that should increase your followers, but I guess it's not working.

Julian Green: Could be the photos.

Evan Nisselson: So just as a perspective, I used to shoot with a Roloflex, a Nikon F, and a Nikon FM, and I totally agree with you that it depends on what you're going to shoot. The hardware is means to an end, and we'll talk more about that. Now I only photograph—since 2003—with my camera phone and my Narrative wearable clip, which I'm also an investor in. And some various head-cams when I'm doing sports.

Julian Green: And I think with the iPhone, which I shoot with, it's partly because the iPhone's a great camera and it's my primary phone. It is also because it is the fastest to get it to, in my case, Instagram or if I'm sharing with my wife.

Evan Nisselson: Is that the primary benefit of it?

Julian Green: Yeah, and I think somebody yesterday said photography is not just capturing, its capture, edit, share, view. And so you don't think just about the camera, you think about the camera and the sharing.

Evan Nisselson: So, my view is we only make images to visually communicate. Why do people make pictures? And like you said, it's a question about what story do I want to tell? So tell us a little bit, Hans Peter, about your view on what is the ideal—or not the ideal camera, let's choose one use case—maybe traveling.

Hans Peter Brøndmo: So if you step back and you look at the human visual system, we've got a few really important components. It starts with the cornea, the lens. Then it goes to the retina, which is the sensor. Then you've got the optical nerve, and there's a lot of processing in there, and then ultimately the brain. The brain is where most of the interesting stuff happens. You might think you see me right now, but especially if you look at him [points to XXX], you have one-degree field of vision, field of view, so you're not seeing me, your brain is seeing me. It remembers that you saw me, but you're actually not. In your peripheral vision, in case there's a saber tooth tiger attacking you from the side, is very attuned to basically directing your attention and all that to the potential threat. [Evan snaps a photo.] There we go. I saw that because there was movement.

Evan Nisselson: One more time.

Hans Peter Brøndmo: I saw that because—see, you didn't catch it.

Evan Nisselson: I didn't catch it.

Hans Peter Brøndmo: Because you were using an iPhone in full flow. But I caught it with that thing standing back there.

Evan Nisselson: Yeah, you're going to tell us about that.

Hans Peter Brøndmo: Yeah. So, the thing is we have a very well developed—and evolution is a great thing, right?—we have a very developed visual system. The problem with most of the traditional cameras today is you've got the lens, the cornea, you've got the sensor, the retina, and then dumb storage, and to [make pictures you] have to pull the [memory] card out, and stick it in [a computer]. So the iPhone is much better because it actually comes with a brain. So the starting point is to look at the architecture of [cameras], but then I start thinking, there are many new scenarios [we should also consider]. Say you're in the mountains and you want the camera to follow you down the hill, hovering a few feet above you. That's just starting to become real, right? There are other scenarios where you want a third-party view, you want some guy on the bottom catching you coming down [a hill on skis]. So in some ways, what you really want is a distributed set of eyeballs all connected using high bandwidth connectivity to a brain, and ultimately those brains constituting a hive mind. So I think what's interesting with the Cloud and with all the stuff that's happening [with cameras in smartphones], is that we kind of have two brains now.

Evan Nisselson: You only have two?

Julian Green: Steve Martin?

Hans Peter Brøndmo: Like in *Doctor Who*... Well, we each have two [brains]. We have our own brain and we have the collective brain, right? Let's not get into neurology here, but I think that the camera is similar. You're going to need a brain, a local brain, local processing, local capabilities. You're going to need an eyeball or multiple eyeballs. You've got one on your lapel—you've got two, actually: one right there in your lap and another one in your pocket. I've got a couple with me today, three actually. And those are the eyeballs. And then I've got some compute, a little bit right here [holds up iPhone], and a bunch up in the Cloud. And so [the interesting question is, what's the] architecture for how it all gets put together? And then the [capabilities of] the hive mind, the Cloud brain. What can it start extracting from the images? We heard a lot about that yesterday, some really cool stuff. [Imagine] if the weather wasn't quite what I'd expected and I wanted to blue up the sky a little bit when I was taking that picture of the Eiffel Tower on my family trip I'd saved up for for years. Well, why not just pull some images from the hive mind to kind of reconstruct the slightly better scene than what I actually happened to experience? Now you might argue that that's not traditional photography, but it's what we do [already] in Photoshop and with filters on Instagram. So that's where I think it gets really interesting, is when you combine all [of the eyeballs, the local computer, and the Cloud computer] and you start looking at it as an architecture, not just a "What does a single camera look like?"

Julian Green: Yeah, and I think we can take that further. I think we have some interesting questions as a society. People seem to have an unquenchable thirst to save their memories. What is it, *Blade Runner*? "All these moments will be lost like tears in rain." People want to keep every moment, and each moment, whether you're skiing or whether it's a portrait that you want, requires a different viewpoint to capture it the way that you want it. So if you assume that people want to be able to capture every moment, how do we set things up, now that cameras are so inexpensive, to essentially be able to capture everything all of the time? And I think we're going there. I think that if you look at countries like the UK that have closed blanket CCTV coverage of public areas, increasingly, whether or not you want to be recorded in public space, you will be recorded in public space. The choice that we have as a society is: "Do we want to be always recorded, video and audio, all the time in our private space?" Other people in your private space doing that will inevitably start recording you as well. So it's sort of a question that we want to ask ourselves, both from just a comfort and privacy question, so there's that part of the question. But then there's the, "Okay, how do you do that?" How do you make sure that you can go back and see every moment? Can you do framing and post processing? You can do focusing in post processing. How much of it is just sort of, grab all the photons and then just work it out later?

Hans Peter Brøndmo: Well, you know the image you showed of your satellite...

Evan Nisselson: Satellite selfies.

Hans Peter Brøndmo: Yes, satellite selfie. If you start with that idea and then you say, if you think about Cartier-Bresson, and the tradition of street photography, which is my favorite genre...

Evan Nisselson: Mine as well.

Hans Peter Brøndmo: Imagine a future of street photography where you're effectively sitting in front of your computer and looking at Google Street View and it's live streaming and you can just go wherever you want and you can just zoom in, hone in on what you're doing. [This could be] incredibly boring.

Evan Nisselson: Depends on what you're looking at.

Hans Peter Brøndmo: Depends on what you're looking at, exactly. There's some very interesting art...

Evan Nisselson: You're looking on the wrong streets.

Hans Peter Brøndmo: Yeah [laughs]. There is some very interesting art that has actually come out of Street View, people just going through until they find something interesting. But imagine for a moment it was live streaming. Imagine that you had high quality. And at some point, versions of that will happen, maybe not ubiquitously, but it will happen in a lot of different areas. So I think the whole idea of what it means to be a photographer, what it means to capture images... To your point, do you want everything to be live streamed? Why not? If you're at home, you could just talk to the camera and say, "Please shut off now." And you should be able to gesture to it and say, like...

Julian Green: Off the record.

Hans Peter Brøndmo: Off the record, exactly.

Evan Nisselson: I think that's actually going to happen. I mean, that satellite selfie, or others, it is happening. But the question is, what cameras are going to be where? What capture device, or whatever we call them, which we need to determine.

Julian Green: I've got a question about the satellite selfie. You've got latency issues? So do you set up at 10 a.m., I will be looking up?

Hans Peter Brøndmo: Are we going to try and solve this here and now?

Evan Nisselson: Well, I think those are great questions, but I'm looking at it more of... It will happen and there are a ton of technology questions that have to come into effect. There was actually one done by a group that was working with...

Julian Green: So you faked that one?

Evan Nisselson: That one was faked, yes. Sorry. You knew that, but that was a drone that did that. But I think that's going to be real. There was a group, I think an Israeli computer division group, that actually went outside and shot a photograph of them from a satellite, and planned it and set it up and obviously it cost a lot of money.

Julian Green: They were probably looking up for quite a long time.

Evan Nisselson: They could have been, but I don't... I think it's coming. We also talked last night in relation to what you guys both said. There's images everywhere. I would love to be able to see my trip this morning to this Summit or elsewhere... With all the security cameras I went through, they should be able to know that that's Evan's face.

Hans Peter Brøndmo: Careful what you ask for. Somebody already has that.

Evan Nisselson: Exactly. It's going to happen, and they may already have it, but they're not sharing it with me. I'd like them to share it with me.

Hans Peter Brøndmo: And that's an interesting thing, right? The whole transparency thing. It's like, why not just live stream all our stuff?

Evan Nisselson: Exactly. We heard earlier about capturing memories and talking about a lot of it is for capturing memories and there's other reasons we make pictures and share. So many of my memories from my childhood—I don't really think of them as my memories. It's a photograph that I've seen and I have created a memory. So if there is one more picture or image per month that we get that we never captured, that would help me in the future think of a moment, on stage here or outside talking to people, that I wasn't capturing. I'd rather be in the moment and have another way to get that visual moment or be inspired.

Julian Green: Yeah. I love photography, but I would love to have the photograph without spending this great moment squinting through a viewfinder.

Evan Nisselson: The other one I've been talking about for years is a retina cam. So "click," I just took a picture of Hans Peter. Two pictures...

Julian Green: You're scaring me, Evan.

Evan Nisselson: Okay. So let's take it to the next step. You've got a couple of cameras, and I referenced jewelry... There are many photographers, professional photographers, who frequently carry their Leica or fancy camera around. They're great cameras, and there's other brands, but it's just like a piece of clothing. They might be at an event. I used to go to a photography event with about 2,000 great professional photojournalists. There was no news happening there but they probably wanted to have their camera available like anyone else. Tell us what camera you have here, and what camera you have there.

Hans Peter Brøndmo: So do you have the big iPhone or the little iPhone?

Evan Nisselson: The little one, yeah. I'm not an ego guy, sorry.

Hans Peter Brøndmo: Mine's small, too.

Evan Nisselson: It's not the size.

Hans Peter Brøndmo: So I think, and he refers to this [holding up Leica camera] as jewelry. There are probably versions of this you can get with diamonds.

Evan Nisselson: And that one has them?

Hans Peter Brøndmo: No. This is a tool, right? And it's a tool for capturing images and telling stories. I think that from a tool perspective, as a craftsman or as a hobbyist or whatever you might be, you carry your tool with you. What I happen to like about this particular tool is that because it's a range-finder camera, it stops me, slows me down. When I take a picture with this camera, I have to think and compose differently than when I take a picture with other types of cameras. I use this tool in ways that I can't use other tools.

Evan Nisselson: Do you take pictures or make pictures?

Hans Peter Brøndmo: Well, I take them and then I make them. [Let me] take a picture of all you. I'm not sure I'm going to make a picture of this. I just recorded an image right here and then I'll use Lightroom [on my computer— my local brain—later to decide whether to make a picture that I consider worth holding onto].

Evan Nisselson: So but Cartier-Bresson used to make pictures in the camera.

Hans Peter Brøndmo: He used to frame it. And there's a whole school of thought which—probably a different panel—but there's a whole school of thought around what you capture is what you do. But whether you capture it by framing it right here or whether you frame it later in Lightroom, honestly I don't see the difference. But that's a personal, a philosophical debate. The thing behind me is a 360 camera? It's a very cool little design.

Evan Nisselson: Is it video or stills?

Hans Peter Brøndmo: Right now it's time lapsing.

Evan Nisselson: Is that what you're controlling on your phone?

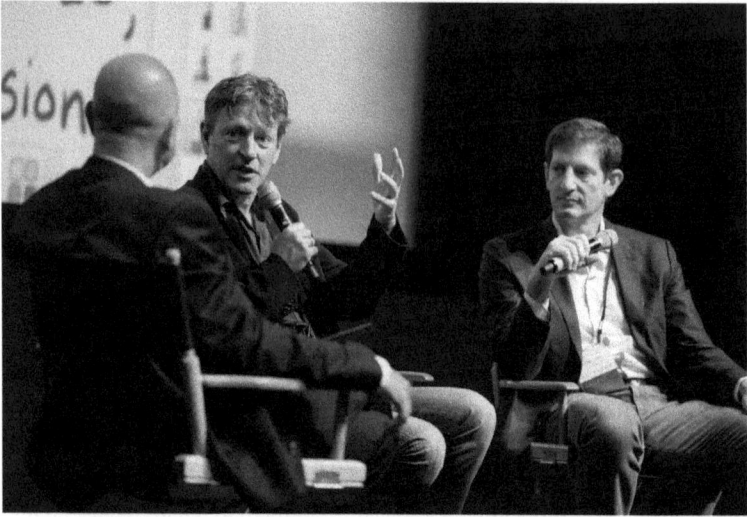

Hans Peter Brøndmo: Yeah, so you control it from your iPhone and it just time lapses. I should have put it up there because then it would have gotten us and you all in one.

Evan Nisselson: Well let's do that, let's move it.

Julian Green: It's actually my best angle back there, so...

Hans Peter Brøndmo: Is it? Mine too.

Evan Nisselson: Yeah, move it. Put it somewhere where you keep talking.

Julian Green: I think the camera, the perfect camera, the connected camera which shares very rapidly and does as much as you can of all the things you want to do, or on the phone, I think that's going to be my go-to camera. I think the one situation where I want a camera is for portraits and in low light conditions that sucks in as much light as possible and enables me to never use the flash, because that is so off-putting to people in most situations. Just a big lens and sucking in as much light as possible. That's the one camera I think I would have other than the... If money were no object.

Evan Nisselson: It's of no object—right, Google? So I totally agree. Let's talk a little bit about the actual experience of the visual content. We've got stills, we've got video, we've got a presentation yesterday by Fyusion—more interactive, almost 3D—we've got virtual reality and augmented reality. There's all kinds of different experiences and it's about capturing a visual story and communicating it, right? How much video capturing do you do on a percentage basis? Is it still or video? Do you do much?

Hans Peter Brøndmo: Very small. I actually consider videos and stills to be a very technology-limited artificial distinction. What is a video? It used to be called motion pictures. Why? Because you took a bunch of frames of film and recorded them and then you played them back. And low and behold, stuff moved, right? The camera, I would like to see, is a camera where I just click, I hold the button, I wait for a little bit, I release, and then at some point later, I can go back and say, "Well, that was the perfect moment. That was the image. That was the decisive moment." So I think that that distinction... And on the contrary, why shouldn't I just be able to just walk up to an image and say, "Wow, that's really cool. Give me context."? And that could be sound, it could be movement, it could be location, it could be a whole host of other things that I want to double-click on the image and go into it and learn more. Or I just want to stand back from it and enjoy the beauty of it and the expressiveness of the image itself.

Evan Nisselson: So hang on, Julian, one second. Questions. We've got eight minutes left or so, and I look forward to great questions from the audience.

Julian Green: So I take a lot of video because I have a two-and-a-half-year-old and a four-and-a-half-year-old, and boy are they cute in video. So at least I think so. So I shoot a lot of video. I think the other part of that is there are some interesting new formats. So the hyper lapse, whether it be Instagram or Microsoft, enables you to do some fun things and get some fun effects with motion and moving through spaces and getting an overview of an area. It just has a different feel to it and it can be beautiful, so I enjoy that. And there's a lot of different, new apps playing with motion and effects.

Evan Nisselson: And animations or animated gifs, so there's a bunch of those. So, on that note, there's a conversation again... Let's not talk about which use cases, but in general, do we think that most people would rather see a 3D image than a 2D image? If they could, forget the ease to capture and all that kind of stuff, if there was a 3D gram, rather than Instagram, would people want more of that, do you think?

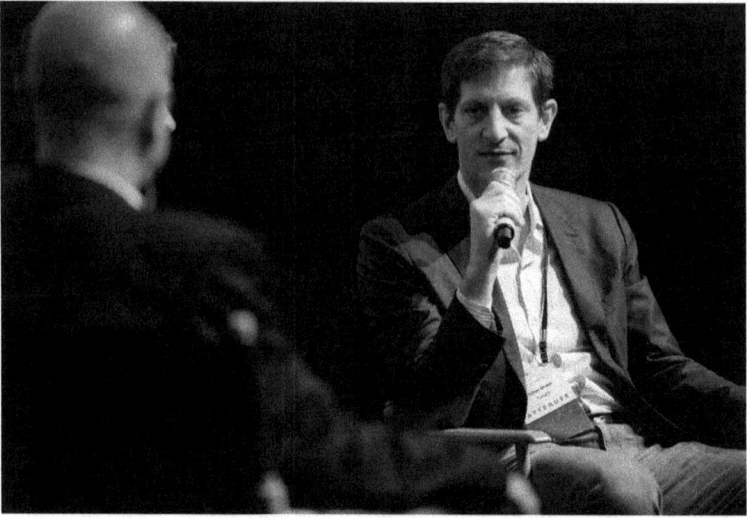

Hans Peter Brøndmo: Isn't that like asking, would they rather listen to music or watch the music video? Would they rather listen to radio or watch television?

Evan Nisselson: Yeah, so what do you think the answer is?

Hans Peter Brøndmo: Yes.

Evan Nisselson: Okay, I agree with you.

Hans Peter Brøndmo: I think you'll have beautiful art on the wall, and it will be still images. Maybe eventually you'll be able to walk up and touch it and interact with it and learn more. I think that still images, what I call the invisible moments, the moments that you never see but they're actually there in plain sight and that you can capture, I think invisible moments are unbelievable. It's a beautiful way to actually capture life. I also think that kids can be really cute, and if it's other people's kids, you can only watch so much cute video.

Evan Nisselson: Did you put that filter on Facebook where you can not see your friends' kids?

Hans Peter Brøndmo: Exactly, pets and kids. But I think that about 10 years from now, they'll be even cuter, right? So that ability to archive and remember [your children in video] is very desirable.

Julian Green: 3D is interesting because 3D TV and film has been underwhelming. And I think the brain somehow turns everything into 2D.

Evan Nisselson: Well, that's my question.

Julian Green: I mean, after the beginning of a 3D film... They always put the good stuff at the start, and the thing comes out of the screen and you're like "Whoa!" And then 15 minutes in, you're like, "Am I watching 3D or not? I don't really know." It looks maybe a bit more 3D. I think 3D is important when you can interact with it, so when you can move into it you can see a different perspective, you can move in space and time. Sports would be great on 3D TV. When you can change the angle, that becomes interesting. But just a 3D representation, where you're not interacting with it, I don't think has done much.

Evan Nisselson: Well, I mean what's the potential? Where's it going to go? We have the HoloLens.

Julian Green: So VR or AR, when you're able to interact with the visuals in time and space and then also have a whole bunch of augmented information and be able to interact with things, that becomes really interesting.

Evan Nisselson: Okay. Any questions in the audience? Right there, Myron. Stand up, Myron.

Audience member: So speaking of capturing more of the visible... You know Lytro did this whole light field thing and now you can unfortunately buy their cameras for $79 on Woot, and that doesn't seem to have taken off. So any thoughts on where that fits in the spectrum of things? Because it gives you a lot capabilities.

Hans Peter Brøndmo: I think that's a good question. I think Lytro is a feature. It's not a [stand-alone product, nor a platform onto itself].

Julian Green: I agree.

Hans Peter Brøndmo: The ability to focus can be done a bunch of ways and one of them could be post facto. In fact for most consumers, I think just a little bit of blur—with the center [in sharp focus] and with a little bit of computational, low-pass filtering applied to the rest of the image—gives you pretty much the same effect, the sense of a depth of field. So Lytro [aka light field] is a very expensive and complicated way [to do something that can be done much more easily, achieving comparable results, at a fraction of the cost and] with no new infrastructure required. By "infrastructure" I mean that Lytro images don't work in Lightroom, they don't work in Photoshop. In short, they don't work in all the tools people currently use. Light field is a pretty cool feature if it could be built in at the infrastructure level, but it's not a movement, it's not a new kind of exciting format in my view. Just one more small point there... What they're demonstrating is that computation can change the way we manage images. Photoshop has been demonstrating that for decades. But I think that part, the fact that computation is changing the way we manage and interact with, record, capture, replay images, that's the trend that we should be looking at.

Julian Green: Yeah, you know you've got to remember, people that actually know what a Leica is, and care about it... We're a tiny minority of people. If you look at visual images looked at by people in the world, it used to be driven much more by a small concentration of professional photographers who put all the images in magazines that everyone would look at. Now people spend much more time looking at their own camera roll, their Facebook friends' photos, Instagram, than they do looking at Cartier-Bresson magnum photographers. So increasingly, that concentration of image makers is becoming less relevant. And so, when you think about which cameras, and you think about what effects are required, not that many people want to change the focus of their photo. That's the problem with Lytro. It's a niche activity. Most people just want the photo to be in focus for whatever they're taking to be in focus, and maybe they want a little blur. That's it.

Evan Nisselson: Any other questions from the audience?

Audience member: Yeah, I've got one here. Andrew. So we're talking a lot about where things are going, so on and so forth. My question is, do you still put together physical photo albums?

Evan Nisselson: I do not.

Hans Peter Brøndmo: I do books occasionally. But probably much less than prior generations did and probably the next generations in turn, even less so. But books are, I think, a really powerful expression.

Julian Green: We've done baby photo books for the first years of our kids' lives and we do a calendar for the relatives.

Evan Nisselson: So even more so. Why? So is it for you, or is it for the relatives, or is it for something else? Is it like we did for the Summit? It lets a moment in time continue?

Julian Green: For the kids... We're holding on to the fact that we had a baby book, and we think therefore to be good parents, we have to give them a baby book, right? But for the calendar, it's a nice way to have something on the wall that is functional as well as fun.

Evan Nisselson: So the goal is phenomenal, it's memories. Just a couple of weeks ago I was with my family, looking at my parents' wedding album, my Bar Mitzvah, my sister's Bat Mitzvah. It's phenomenal, sharing visual experiences. We probably do that once every, unfortunately, 10 years. A couple of those photographs I'd never seen before, which was fantastic. But what I did was I photographed them, and I now have them on my phone, and now I share them with a lot of people. They're both valuable, but to your point, my gut is I want a way to visually communicate to many people wherever I am and whenever I am. Whether or not it's on the phone or a server.

Hans Peter Brøndmo: Yeah, you know again it gets back a little bit to use cases. These are memories and they're stories, and you have different ways to communicate and tell memories and stories. Yesterday there was a conversation about maps and I think part of what we're talking about is putting our life on the map, right, in different ways? And that includes actually, literally maps like "Where was I? What was I doing?" But it also includes the images and the history of my life. So what I want to see is a full-on service that lets me put my life on the map and go back and relive it and revisit it later.

Evan Nisselson: I agree. I think that makes a lot of sense. Question over here? Samson.

Audience member: Yeah. Depth. Do you think that a 2D image plus depth might be an interesting camera? It's something that the community could certainly build.

Hans Peter Brøndmo: Yeah, depth maps are going to be critical. You're going to have stereoscopic imaging [which is critical for advanced computer vision applications]. There's Pelican Images with their array cameras, Apple just made an acquisition in this space. So depth is going to be an important part and it allows you to sort of poor man's focus pull a poor man's post processing to get the focus to be the same effect as Lytro, but a little cheaper.

Julian Green: And we've talked a lot about consumer photography. There's obviously lots and lots of applications of depth for business and industrial applications. I think because cameras have gotten so cheap, I think industrial applications are things we haven't talked about that are going to be massive. So whether it's using vision to recognize parts or whether it's monitoring processes or whether it's testing, visually checking assemblies, there's going to be a ton of vision driving industrial processes.

Evan Nisselson: Okay, fantastic. So we are out of time. But what I'd love from both of you because you're very successful serial entrepreneurs, you know the ups and downs, and we had some interesting conversations yesterday about transparency and building businesses... In one sentence, what would be your advice to entrepreneurs or people that want to build a business or are building one?

Julian Green: It's as much work to do something small as to do something big, so find something that's going to make a dent in the universe that you love, and do it with people you love.

Evan Nisselson: Great.

Hans Peter Brøndmo: That was good.

Evan Nisselson: You're going to have to do another one. You can't take his!

Hans Peter Brøndmo: It's about the commitment. Don't worry too much about the idea. It's going to evolve and change. Make the commitment and go.

Evan Nisselson: Fantastic, guys. Thank you very much for this conversation. Let's give them a round of applause, guys.

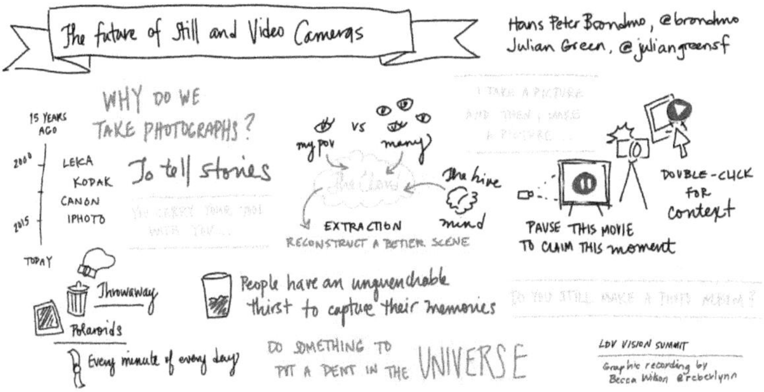

©Rebexlynn

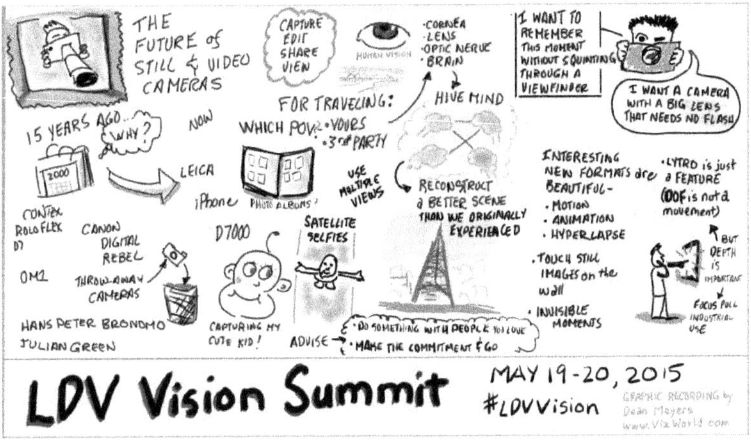

©VizWorld

23

AS AN INTRAPRANEUR, BRAD SHORT INVENTED A NEW PC AND CREATIVITY STATION

Brad Short, Immersive Computing Group
Chief Architect, Hewlett Packard

Serge Belongie: Brad Short is the inventor of HP Sprout, and I've known Brad for as long as I can remember. It turns out he's a family friend going back to the mid '80s. He's a Cornell-trained engineer and he's also extremely generous with his knowledge. I used to advise groups of undergrads who worked on assistive technology for the blind, and on top of all that he was doing to invent Sprout, what would become Sprout, at HP, which you'll learn about shortly, he helped me advise just as a volunteer. He mentored these teams of undergrads to help us create these cool technologies. I think what Brad has done at HP is an incredible example of "intrepreneurial" innovation. We're very fortunate to have him speaking today.

This is a real pleasure to share the story of Sprout with this audience. What I'm going to do is talk about Sprout and give a quick description, a little sales pitch, but also the story behind Sprout, which I think is actually just as important and fitting for this audience. I think you'll really appreciate it.

What is Sprout? There's no fundamental invention here. It's a combination of unique, existing technologies in a unique configuration. We have input methods, projectors, scanning systems using a high-res camera (fitting from the last panel discussion about the high-res camera), combined with a depth camera. That is essentially what Sprout is doing, all aiming down onto a touch screen.

Sprout is a dual-screen, immersive, new platform that integrates all of these existing components into a single device. In addition to this hardware platform that we've developed, we had to create a software framework. It's a software framework that we call an "apperating system" that's running on top of Windows. This gives us an SDK and APIs that enable this immersive playground for users.

Combining all of this stuff into one integrated system gives you many new use cases. It blends physical and digital, and I'll get into that. It's this intuitive interaction space right in front of you. You can capture 2D and 3D content effortlessly and seamlessly, and you can collaborate in real time remotely and simultaneously. You can create like an artist out of the box. Immediately.

It's really hard to just explain this in words. I'm going to now describe this with a home video that I put together of the journey of Sprout over the last five and a half years, with this in mind: that it's an intuitive creation station that makes creating as simple as the analog world of creation. You have physical things on your desktop, and the way artists work, that's really what we wanted Sprout to do. That allows people to go from thought to expression in an instant. I have an idea, I need to get it out and express it instantly—that's what Sprout does.

Okay. The home video. This is a personal journey of the making of Sprout. Five years ago or so, week two of the first concept idea, I presented to an internal HP innovation fair. This was a poster fair, informal. I had this idea a couple of weeks in advance. I was working on vertical printer scanner all-in-ones. This was 15 years into HP working on imaging and printing systems. I was using this vertical all-in-one to project down onto a surface in front of me, just to kind of save desktop space. I rigged up this little projector system, pico projector at the time, and I started showing what can happen when you integrate projection with the combination of projection and physical content.

That was the first moment there where I put a photo down and captured it, re-projected it one-to-one on top of the photo. That was an "aha" moment that was really amazing to colleagues as I shared it around. You can do things like, I'd scan the book and barcode on it. Now I know the book, so I can actually do a copy of the page of the book without even imaging the book. This is really kind of a breakthrough concept that allows scanning and digital content into the workflow.

This is what I worked on with Serge's team, to actually do a real computer vision prototype. All the stuff you see here is real. That allowed us to explore this idea further. In order to get it in front of executives and get some enthusiasm for the idea and funding, I created this prototype. This is off the shelf. I have a Wiimote to enable a stylus. Made [the projector] brighter. But I start to show this use case of WYSIWYG, kind of one-to-one editing, where I'm bringing physical content in, instantly transforming atoms to photons in an instant. (That's what I called it.) And all of this with no touchscreen. This is just a projector and a Wiimote stylus.

This actually, this idea, and sharing this concept, helped me get it green lighted. That's a big deal within HP. It took about a year to get it to this stage. At the green light, we started developing the experience. It's really the experience more than it is the technology. I had to reiterate that over and over to the investors.

Here we start to show what this experience on Sprout really looks like. You've got an immersive horizontal computing system. You're bringing in physical content and interacting with it seamlessly and we start to show some use cases: life management, bringing in receipts, checks, documents. This is coming out of the printing/scanning division at HP. That's kind of where we were.

Now, you can see augmented reality. This is direct augmented reality where I'm turning passive blocks into physical things that can interact with the digital content. Here, I've got these little plastic letters, and I'm creating a video using plastic lettering, so maybe stop-motion animation effects. This is our former CEO and VP, Louis Kim, who was kind of the mastermind behind developing out this concept from a business perspective. So now we had to get serious about developing hardware, building up a team within HP to about 75 engineers. We built a custom projector. We added touch—this is one pivot moment in the development of the hardware. Now you can start to see some more use cases. I'm writing on top of physical objects with a 3D stylus. Capturing [areas] before, giving it the areas I want to capture. Working with digital content the same way that you work with physical content. This is really the essence and key to all of Sprout. You don't need to learn how to interact with this computer because it's working the same way you've worked with physical content your whole life.

We start to look at augmented reality there where I projected detected text on top of that wedding invite and then it gave me contextual links. Here, this is like a wedding thank you note where everything is very seamless.

Here's a Fruit Loop counting game. Putting in real Fruit Loops for "edutainment" and kids' fun activities. This was our first look at 3D scanning that you could actually do 3D scanning with this hardware and have it to be a user-interactive experience.

Momentum started building. I'm presenting now to the next CEO. We're going through CEOs. So I had to build up the entire use case and explain the vision again. We got serious about the industrial design. We needed to redesign the entire projection system. This was another year and a lot of money to do these projection systems. We went to the left one, to the middle one to reduce the size and cost to get this more manufacturable and to make a product out of it. We got to lots of IP, lots of patent filings. A big team. This was the final ID that we were about ready to release. And bam! Two and a half years into this, the project was cancelled.

This was a big blow. Sure. I like this audience because you can probably relate to this. I wasn't as used to being exposed this way. But we hit the road. The whole team was disbanded, except me and a few other key people. We hit the road and went and tried to sell it around HP, VCs, external partners. Trip after trip. Thousands of miles. Setting up this elaborate equipment the night before, because it was very sensitive. We created fake *Wired* covers to create a vision to these investors as what they're going to get when this actually comes out.

I presented to all the GMs of the commercial groups. The work station team in Fort Collins, Houston, the PC group in Sunnyvale. Over and over, the response was always, "No. But this is awesome. Really cool. Great work. But can't fund it." We just kept doing this, over and over, for eight months. We were about to the last month. We gave ourselves about nine months. We said, "Okay. We've got to pivot again." So we added a second screen.

This went from a WebOS, kind of Android-type of experience, to a full-fledged PC experience since I was talking to the PC group. This really opened up and exploded the use cases. Now it's a no-compromise computing system, dual screen immersive. This is some of the stuff you can get. Immersive collaboration. The vertical screen has the forward-facing view. The horizontal screen is the content I'm sharing. Here, I shared a PowerPoint presentation really quick. Edited it. Said, "Okay. Thank you. Goodbye." That can replace an entire meeting in just a few seconds.

Then, we started doing focus groups, bringing in people to see this crude technology. This was an Apple fan boy, no offense to Apple, but he said this is the first time he would try Microsoft again—after seeing this. We're seeing that adding these use cases was more than just making a desktop slimmer and faster and more powerful. It actually created new use cases.

Now we had to go to the engineering side and actually develop this product. Up until this point, a lot of this was just show. We developed an ID. We got the second green light. It was about a year after the first cancellation. A new GM came into the consumer group and said, "Yeah. We have to do this." Now, the catch was we needed about two years, because a lot of this was just a video that we're showing. They gave us about 10 to 12 months. This was a double-edged sword. You get green lighted but you have to do it in half the time.

So we went to the team and the PC group in Sunnyvale and built a software team, because a lot of this is software now, and we had some hardware that had to be developed. Turned on the HP machine and supply chain to develop hardware quickly. Ramped up a software team. You can start to see that the software couldn't just run Windows as is. We actually had to create what we called an "apperating system" on top of Windows.

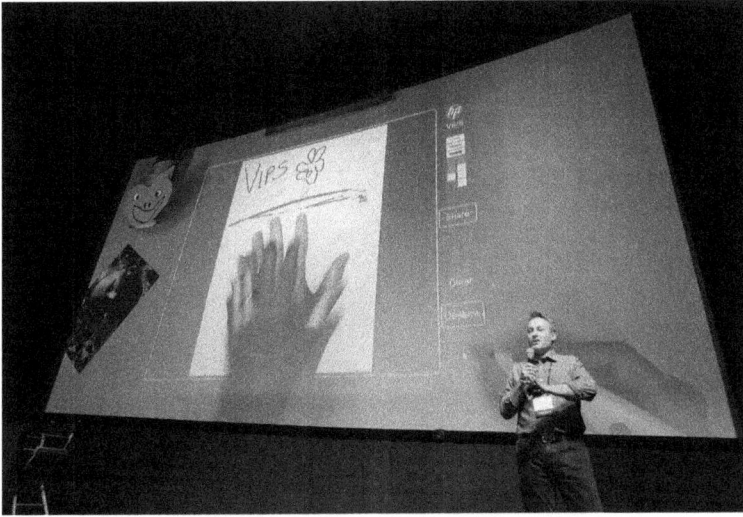

Brad Short, Immersive Computing Group Chief Architect, Hewlett Packard

This went into the way in which creatives work, where you're pulling in content visually from galleries. At the same time, HP was looking at 3D printing, and needed a tie-in to what we call "blended reality": going from physical to digital and back to physical. We, as Sprout was a fundamental hardware platform, were able to create 3D scanning in record time and actually introduce and launch that.

This was about five years after the initial concept. We actually launched HP last November for $1,899 into the market. Then we had to go to the difficult job of selling what this means. What is the vision of Sprout? It's a very hard concept to get unless you actually see it. We spent a lot of time traveling around, going to fairs. This was last weekend at the Maker Faire, which was a perfect audience. Makers really love this technology. They get it. Here we scanned a face, did some editing on it, then printed out the face onto a 3D printer. So five years, many successive iterations of this project, ups and downs, but we finally got into market.

Okay. Hope that gives you a sense of what "blended reality" is. 2D, 3D inputs—so scanning and high-res imaging—then, 2D, 3D output. Physical and digital can coexist into the same space. This idea of immersive 3D computing... Imaging and printing, we think, will democratize creativity. I believe it will democratize engineering to some extent. There's already a movement, called the "Maker Movement," which I think will be a renaissance, which brings art and engineering combined together and allows the masses to develop content.

Brad Short, Immersive Computing Group Chief Architect, Hewlett Packard

Today, computing is somewhat flat, even though [based on] some of the presentations I've seen yesterday, I think there's big advances already. It requires tool expertise to create. Photography on computing is more static, and video is kind of a single perspective. We think what is needed in the future is a hardware platform and interface that can keep up with this new technology, new content, as well as usher in a new way in which you can edit and work with this content. The first step will be seeing the content, consuming it, but we believe the Sprout and Sprout follow-ons is just a first step in actually making that content editable.

As art and technology converge, different types of people are going to want to create content without learning the tools and software. This is one of the ways in which Sprout, we believe, will contribute to this field. Sprout is just the first step. We're going to be expanding into different form factors. It's all about the experience. We'll be looking at mobile computing. This is what we call "immersive computing platform."

Thank you very much. We're, of course, looking for talented skills, especially in computer vision, to help build out this next generation of immersive computing.

Thank you very much.

24

VIDEO ADVERTISING CONTINUES TO EXPONENTIALLY GROW ONLINE

Erika Trautman, CEO & Co-Founder, Rapt Media

Rebecca Paoletti: I'm excited to hear from Erika. She's going to talk about millennials, she's going to talk about razors—this kind of razors [gestures to face], not the scooters—and some other things that are really hopefully going to shed some light on how we can all make more money with video through engagement.

I am excited to speak about interactive digital video technologies. What is true interactive engagement? And which brands are benefiting?

I'm going to start immediately with an example. So this was an international video experience that Philips created to promote the launch of a new electric razor. Their goal was to reach a generation of young men who maybe weren't familiar with electric razors and invite them to play with their brand story. We're going to go right into the example here. I've taken the audio out just to reduce points of failure here so I'm going to do his voice. So he just said, "What a night, huh?" And then he looks in the mirror and he sees: clearly it was a good night. He's covered in lipstick. Then he goes, "Help me remember. I know it started with shaving, but which style?" And you get six beard choices that come up below, so you can select the beard choice you like. We just clicked on Van Dyke, which was the goatee. And then it takes you back to the night before. You get the idea. As you make different beard choices, different adventures ensue the night before. I think five out of six of them wind up with a happy ending with a lady.

That's what we mean when we talk about interactive video, where a viewer clicking on the video changes the storyline of the video, changes the content they get; or a viewer clicking on the video can actually change the website or clicking on the website can change the video. This is about bringing web functionality into the video and integrating them seamlessly. These are the results that we saw: 65% of viewers were accessing this content on iPhone or Android. The average mobile viewing time exceeded four minutes. That's a really long time to spend with a brand message on a phone. The average viewer interacted three to four times.

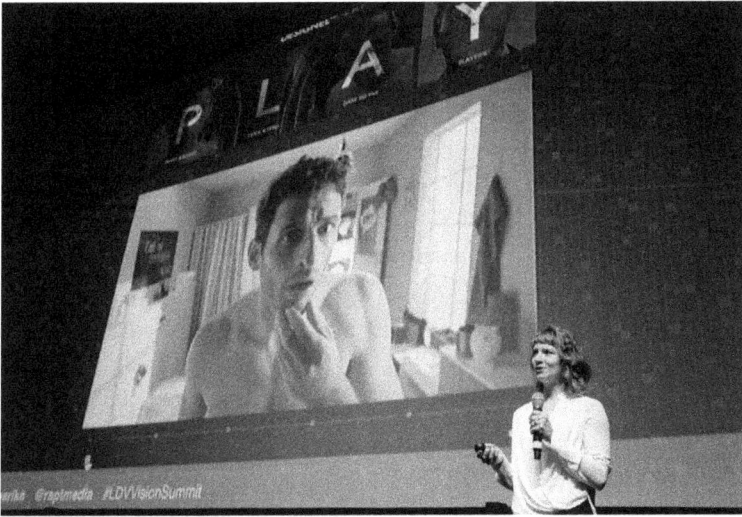

Erika Trautman, CEO & Co-Founder, Rapt Media

All of these are great indicators of engaged viewers. But so what? Did it impact Philips' business? Well, it turns out that it increased purchase consideration by 6% and they sold 16% more razors as a direct result of the campaign. Furthermore, with interactive video, each of those interactions requires the viewer to offer up a little bit of insight about what they like, what they don't like, what works, what doesn't work—so they also received really unprecedented data. We were able to share with Philips information like which beard choices were most popular by geography. (It was the full beard, by the way, that was most popular last year.) Which characters were most engaging? And the click path that was most likely to lead to a sale. It was also an award winner—Ogilvy - Dusseldorf did the creative on this—and so that was a great perk.

So why did this work? The reason Philips engaged in this—this is what Willem Schungel had to say: "Engaging and interacting with consumers is paramount when it comes to video and a big component of this interactivity is meeting the demands of an increasingly mobile consumer." So the three points here are video, mobile, and interactive. Those things are self-reinforcing trend. We actually think that there are two things going on in society that are driving the adoption of interactive video. The first is that we're in the midst of a new wave of storytelling. The second is that we're experiencing a societal shift driven by the rise of the millennials.

Technology's driving a new wave of storytelling. I think everyone here has seen that. This is nothing new. It started with paint on cave walls and it's continuing today. But if you think about what the Internet introduced to how we sort and process information—we had menus, we had links, we had search—that happened back in the late '90s, early 2000s and then with the mobile web it brought that functionality and put it into our hands and drove our expectation around intuitive user interface. Video has lagged in that until now. We're really starting to see video evolve to take advantage of that.

So why does this matter? We're in the midst of a shift driven by the rise of the millennial. The brand managers that I've talked to spend a lot of time thinking about this problem. Communicating with millennials, it turns out, is a different game than communicating with older generations.

So I'm going to talk about the profile of the millennial for just a second here. They account for 30% of the population. They're of course digital natives. They've grown up with infinite access to information at all times. In fact, they check their smartphones about 8,000 times per year. They're video gamers. There have been some really interesting studies that have come out that show the appeal of the video game is not about violence, it's not about tallying head shots, it's about exercising choice and experiencing consequence. It's about decision making. They're also incredibly skeptical, but very idealistic. They're not going to sit back and receive a brand message because you push it to them. On the other hand they're also seeking authentic communication. They want brands to reach out to them as individuals and speak to them as individual people. Clearly you can see how traditional video, which works really well in a mass media environment where you're pushing one idea to a lot of people, fails in this environment. Interactive video, on the other hand, is a great tool.

So to wrap up, what additional brands are out there that are using interactive video? We're seeing Condé Nast Entertainment, for example, innovating with interactive video. They're using it to create entirely new digital first video experiences—both for editorial and for advertising. Deloitte is using it to recruit young people. They saw a threefold increase in video engagement. These videos are actually out-performing their YouTube videos.

Anheuser-Busch: using it to tell complex brand stories. This was a really interesting use case. The average viewer clicked over 20 times within that video experience.

I'm happy to talk further—if you've got additional questions, come find me afterwards.

Thank you so much.

25

ENTERTAINMENT IN THE AGE OF PARTICIPATORY LIVE STREAMING MEDIA

Adi Sideman, CEO, YouNow

I'M EXCITED TO TALK TO YOU ABOUT INTERACTIVE SOCIAL NETWORKS using live video. A little bit about me: I've been doing user-generated videos since before the Internet; I co-founded the world's first online karaoke, which was sold to Fox Myspace at the height of Myspace; did the world's first audio ad network with CBS RADIO, focusing on anything media consumers can create and share.

I've been in the space since the beginning of user generated and have seen the progression from text to images to video to real-time text, real-time images, and now real-time video.

You've all, I'm sure, heard about Meerkat, Periscope, and other such technologies, which are basically live video social networks in real-time. It's curious to think of what's coming up next. Facebook paid billions for Oculus, a VR company, last year. We're going to see more and more wearable computing. We're going to have social network work through tactile means. Anything that makes interaction and simulation easier, simpler, and more "real." Sensory vests for gaming are already commercially sold, and the ability to talk in any language to anyone and have it simul-translated exists. All of these trends are happening right now and they will make social networks much more tactile, much more seamless, and much more immediate. In our case, we're seeing teens, who have more time than other demographics, use more of these technologies all of the time.

We're seeing youth broadcasting themselves while they sleep to wake up and see how many likes they got.

With that, I'm going to take a look at one service, which is called YouNow, and hopefully we'll find something interesting on, right now, and do a quick, fast demo.

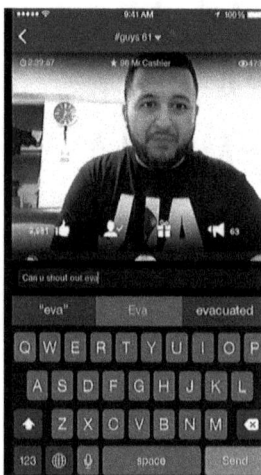

[Live feed video of YouNow user Tayser Abu-hamdeh/Mr. Cashier up on screen]

Tayser Abuhamdeh: But does it mean a lot? Does it mean something special?

Adi Sideman: Okay, so let's ask this guy [typing]: "Can you say hi to my friends?"

Tayser Abuhamdeh: Hello, Adi's friends. How is everybody doing? Hi, my name is Tayser. They call me Mr. Cashier because I like to drink beer. Just kidding, I don't drink.

Adi Sideman: These types of live social networks have been happening in China for years. Over 100 million Chinese people are using such social video services every month.

The business model is virtual goods.

E-sports: the United States government just introduced a new visa category for e-sports people from all over the world who are now playing in tournaments just like regular athletes. This guy is making eight figures a year playing e-sports live on the Internet. This app that I showed you before, YouNow, which I'm the CEO of, is a top 50 grossing app in the iTunes store. This is real money that these folks are making. We can see some of them here.

FLIPPINGINGA

This is a 16-year-old kid making money by flipping live on the Internet. Mr. Cashier, whom we just saw, likely makes more revenue by broadcasting his job as a Brooklyn deli attendant than on his job. Snowmartie, a 16-year-old who draws on camera every day, is making more than just pocket money. YouTubers are making more money streaming live than on YouTube. A dentist in Cairo who has a daily talk show making five figures a month by broadcasting live and talking about culture, current events, et cetera.

The future of this is user-generated networks that are run just like TV stations except by the users, for the users. Right now, there are tens of thousands of broadcasters all over the world. We know exactly where they are. We can cut to the Champs-Élysées and choose one of the five cameras that are there, right now, and whisper in the guy's ears to turn his camera to the right so we can see the "I Am Charlie" memorial that's still going on right there.

We can be Tweeting right now that we want twins playing violin and we'll get a queue of people doing just that. If we don't, we can incentivize that through virtual currency, et cetera.

That's a little bit about live streaming and what's going on. Thank you very much.

26

PANEL: THE EXPLOSION
OF DIGITAL VIDEO

MODERATOR

Richard Greenfield, Managing Director, BTIG

PANELISTS

Greg Clayman, GM, Audience Networks, Vimeo
Emily Gray, VP, Revenue Products, Refinery29
Paul Marcum, Head of Global Digital Video, Bloomberg

Rebecca Paoletti: You have all perhaps interacted with some of the brands that are walking towards me. First of all, we have Bloomberg. I suppose some of you might watch that news. We have Greg [Clayman], who's responsible for my most addictive online experience, which is called High Maintenance.

Greg Clayman: Nice. Thank you.

Rebecca Paoletti: Thank you very much. From Vimeo, High Maintenance is a show that... People in this room are watching High Maintenance. You know what this is. Greg, you're going to have to explore that.

Moderator: Rich Greenfield, Managing Director, Media and Technology Analyst, BTIG
Panelists: Emily Gray, VP, Revenue Products, Refinery29; Greg Clayman, GM, Audience
Networks, Vimeo; Paul Marcum, Head of Global Digital Video, Bloomberg [L-R]

Greg Clayman: I'll tell them all about it. I will tell everybody that. Don't worry about that.

Rebecca Paoletti: Paul, who runs video at Bloomberg and is a long-time video expert, Greg, and Emily, who is leading video and all kinds of other revenue-generating opportunities for Refinery29. And Rich, he's a wonderful moderator and great analyst. Take it away, you guys. Thank you.

Rich Greenfield: Thanks. Thank you all for joining us. I think the big open question is that we've got this explosion of over-the-top video. There are so many more consumer choices than ever before. I think about how we've evolved. We started off in a world where video's always been locked and basically, whatever a local cable operator or satellite operator offered to you is what you got to enjoy as a consumer. As Adi just showed, there are so many choices for consumers now to consume video; every single day we turn around and we see a new way to consume content. Whether it's premium content or non-premium, user-generated content, it's all the same content to the consumer. As we see the bundle starting to crack in the US and we're seeing all these new players, it would be great to maybe get a sense from each of you... If you're sitting there as a consumer, what does the consumer actually want? What are you trying to provide to the consumer in what your own company's video efforts are? [Speaking to Emily Gray] You want to start?

Emily Gray: Yeah, I'll speak explicitly on behalf of millennial women, because that's really who we program for and who we think about. She's a consumer with truly competing needs and desires, which complicates it but also leaves a lot more opportunity. She wants things on demand. She wants them explicitly, the moment that spark of desire happens, so you have to make it very easy for her to find that thing, but she also wants to be led on a voyage. She wants to discover. It's a matter of programming to her based on time of day, based on the platform she's on, and understanding the secondary cues she shares about what you can give her. Some things are discovery and some things are more in response to a direct demand.

Rich Greenfield: Vimeo. So other than High Maintenance, what else are you giving or what do you think the consumer wants? Obviously every consumer in this room just wants more High Maintenance.

Greg Clayman: Which you'll be able to get on HBO. The explosion of video forms and formats and the explosion of the creation of video in different ways means there has never been a more exciting time to be in video. Consumers across the board seem to want various different things. One of the things that we focus on is, if you look at one end of the spectrum and you look at Netflix or Hulu, what you've got is, for lack of a better word, Hollywood content, right? It's *New Girl* and Hollywood films. On the other end of the spectrum you've got YouTube, which is everything under the sun. YouTube is generally ad-supported and it can be tough to make a buck there unless you're really one of the top creators. You look at Hulu Plus—it's generally subscription. In the middle what we're finding is a vast quantity of high-quality premium video that's hard to find at either end of the spectrum that consumers are actually willing to pay for.

Rich Greenfield: By pay for, let's just be explicit—meaning a transaction you pay for discreetly?

Greg Clayman: I mean transactional VOD content you purchase—a film costs $5 to rent or this thing costs $2 to buy or I can subscribe for $2 a month for this particular creator. One of the things that we sometimes hear about millennials is, "Oh, yeah. They don't want to pay for anything." We find that's not true. We find they don't want to pay for what they don't want to watch.

Rich Greenfield: Or what they aren't watching and have been paying for anyway.

Greg Clayman: Right. Absolutely. But content that they do want to watch and content they can't find anywhere else, especially content that is highly-tailored towards vertical interests... Action sports videos, for example, or even within that, surfing videos. We find people young and old do pay for stuff like that and we've been focused on that part of the market.

Rich Greenfield: Paul, you bridge the gap so you're the panelist that comes from both worlds. You're distributing through the traditional ecosystem but also going out onto your own. How does that fit into the lens of this concept of giving consumers what they want?

Paul Marcum: I'll speak on behalf of Bloomberg and then add some thoughts that are more than a mass market. In terms of Bloomberg, people want expertise and they want to know the full perspective, the full story like they do for any news story, but it has added depth when it's moving markets such as what we cover at Bloomberg. That perspective comes in many different levels of production quality and can be delivered just as easily on mobile as it can be delivered through your set top box and your 50-inch screen. To that extent, that accessibility is, of course, key to our users. Beyond that I think people want connections still and that hits the left end of the spectrum that Greg just described as well as the right end of the spectrum. I think that holds true whether it's tent pole content that you see from the Netflixes and increasingly Amazon and HBO, where they're creating these must-watch super premium opportunities out there, as well as sports, but also something just as much is true about the cashier. I think that those connections are what's going to build the foundation for these new experiences moving forward.

Rich Greenfield: When I think about millennials, let alone everyone beyond just millennials, watching TV ads feels like an increasingly archaic habit. In a world where nobody wants to waste time watching ads, just like they don't want to pay for things they don't watch, we've got models like Netflix where there are no ads, Amazon where there are no ads, Greg's model where I'm paying discreetly to avoid advertising. Is there a future for video advertising?

Emily Gray: I think the truth of the matter is that brands underwrite incredible content and they have since the TV set was turned on. The idea that we are going to continue to be able to create content that has the quality and quantity that we see today without advertisers in the model isn't realistic. I think the mechanism through which that advertiser message is delivered changes for sure.

Rich Greenfield: Meaning I won't have to watch 30 second spots with cars driving around mountains?

Emily Gray: Right. So a lot of the conversation around viewability right now—and all these things I think we're all a little bit in a frenzy about—goes away when you start solving for the real problem, which is millennials in particular are blind to those banners. They find a way to get around that message. So baking it in and making it less like a commercial and more like a service using data and better creative is probably the future and not so far off.

Greg Clayman: That's how we work with advertisers. For the most part, yes, our business is a premium business. That being said, and we don't have ads—we don't have pre-roll ads in front of our videos—but we do work with advertisers to create original content. We find advertisers from Intel to American Express to whomever who are interested in creating... In our case, it's often short films, and they're interested...

Rich Greenfield: How hard is that for the brands to do at scale?

Greg Clayman: It's hard. It is much easier to scale a pre-roll buy, for certain. Look, there's obviously still tremendous value in television advertising. I mean, 83% of the prime time schedule is still watched live, for better or worse. You can't launch a film, a major motion picture, without a TV buy where there's a big, beautiful trailer that fills the screen. There are an increasing number of advertisers who are in the content creation business. Everybody always points to Red Bull as an example of someone who's been doing it for a long time, but there are a lot of others. Marriott, for example, the hotel chain, has a content studio. And they make short films and they produce content as another major CVG company that's about to launch one. It is, to your point about doing it at scale, it does take time and effort to create content, whether you're a brand or whether you're an individual creator. But we're seeing a flood of brands move into that space. [Speaking to Emily Gray] I know you guys are seeing the same on your end, for sure.

Rich Greenfield, Managing Director, Media and Technology Analyst, BTIG; Emily Gray, VP, Revenue Products, Refinery29; Greg Clayman, GM, Audience Networks, Vimeo [L-R]

Paul Marcum: I don't think there's any overestimating about how much attention is being added to the world through the mobile experience. It was captured a bit in one of the prior presentations just in terms of how much viewing time we're all experiencing. I got a [Apple iPhone] 6 Plus. I don't know if anybody else has done that yet. I got it because of my increasingly weak vision, but it's now my first screen. I watch sports on the 6 Plus, I watch more content, more video on the 6 Plus than on any other screen that I have. At that massive scale, I don't see any other path that wouldn't include video advertising in the molecular form of the 15 second mid-roll, pre-roll, wherever the thing sits, that is just an entry point for virtually any advertiser and certainly something that can be addressed within virtually any context, whether it's a live streaming personal service or some sort of premium destination environment. The marketers are not going to abandon their 60 year love affair with video any time soon and that's always going to yield exciting opportunities for folks who want to get content out there.

Rich Greenfield: Evan picked an awesome week for this event because it comes just after the new fronts and just after the TV upfronts and so we got to listen, literally, last week to a whole host of television executives explain why they shouldn't buy digital.

Greg Clayman: You had a great video on that. Your little mashup.

Rich Greenfield: They have literally explained why buying Refinery29 content or Vimeo content or any of these people's content—generically, the online world—is a mistake. From viewability to it's non-premium to... I could go through a whole host of issues. It was like the 10 plagues of online video last week of why TV should continue to be dominant...

Greg Clayman: How about Cowboy Math? I love that.

Rich Greenfield: Cowboy Math was Turner. Literally, they're trying to convince an entire world of ad buyers of why not to shift dollars to digital and mobile. A: What are they so afraid of? And B: What should the brands know when they think about the comparison between TV and digital? What can you do that they can't or does it not even matter? That's kind of the big, fat softball.

Emily Gray: That's a big one. I'm gonna let him [Greg Clayman] take it first.

Greg Clayman: There are sellers of linear media that at least seem or sound more or less scared. I agree. When I watched that video that you had on your site, it seemed odd to have a seller come up and say, "Don't buy digital." That being said, one of the things that I think a lot of them were alluding to is the fact that the viewability issue, when it comes to online video and online ads in general, is a real one. I understand that when there's a television ad in your house and you're watching it doesn't mean that you're necessarily sitting in front of the TV.

Rich Greenfield: Even if you're sitting in front of the TV, I'd bet Emily's audience is staring at their phone even if the TV's on.

Emily Gray: Yes.

Greg Clayman: Yes, I saw some ridiculous statistic about millennials. I don't think there's a millennial who watches television without a phone in their hand.

Emily Gray: It's at 60%; the latest official stat that comes out of the companion viewership is 60. For our users, we just did an audience survey... 77% of them are actively looking at the content from the content, like from the same provider they're watching the show of.

Greg Clayman: Yeah, so that's funny because I saw a number that was closer to that; it was exactly right. Look, digital advertising is addressable. It's targetable, it's addressable, you can say, "I want to buy just this demographic. Psycho-graphic in this geographic market at this time, who has already seen something else." The re-targeting, right? "Oh, if they watched the video on this well then I'd like to catch them doing that." That is not technology that exists for broadcast television as yet or even really for cable. We're just beginning to see it in VOD. I think you'll see more of that technology creeping into those other channels. I think that you have to. I think there are some cable programmers right now who already can say, "Oh, so X amount of the people who watch this show where [they were] advertised [to] bought your product on Saturday night." It's the data scientists who are in this space who have had their focus on digital media, digital video. Now, it's incredibly trackable, it's incredibly programmable; there's also no shortage of inventory. So the whole concept of *I have to buy this show because there's only so many slots* doesn't exist. There's no scarcity. You've got all of this great targetable stuff at lower CPMs versus non-targetable at much higher CPMs. And we're going to, I think, see that balance.

Rich Greenfield: Is that really true, Paul? Do you think there really is no scarcity online? Or do you think key sites, whether it's Bloomberg or whether it's Facebook... Will we see certain sites actually...

Greg Clayman: It's Refinery and Bloomberg, obviously.

Rich Greenfield: The web is a very big, open place with a lot of scary corners. Will we see a differentiation of mobile destinations where there's winners and losers or will it be just there's so much supply that, as Greg's talking about, that we'll just see CPMs fall lower and lower because it's all generic?

Paul Marcum: I don't think anything is...

Rich Greenfield: I'm over exaggerating.

Paul Marcum: Yeah, and Greg doesn't say it's all generic, but unquestionably we're seeing...

Rich Greenfield: Except for Vimeo; Vimeo is distinct.

Emily Gray, VP, Revenue Products, Refinery29; Greg Clayman, GM, Audience Networks, Vimeo; Paul Marcum, Head of Global Digital Video, Bloomberg [L-R]

Paul Marcum: Yes, Vimeo, Refinery, and Bloomberg are your only three buys. Tell all your friends in the ad business. If something is exploding, that means the inventory's exploding along with it. What is going to create some scarcity out there in the marketplace? I would say that certainly you can look at the halo effect of certain contexts and that's what so many TV buyers continue to do. It's not just the great demos that they've got and it's also the fact that the show is great, and the buyers like the actors, and ultimately, it's a good context for their brand. Certainly the content halo and context halo is always going to be important. But I do think that, yes, despite this explosion, there will always be audiences that are particularly valuable to a particular marketer. Ours is certainly one of those for many, many marketers. There are, of course, others. Young millennial women is of course going to be incredibly popular and something where advertisers are going to find, at times, a scarcity of the right context with the right audience and perhaps at the right reach. Live sports has obviously continued to succeed with that at this point.

Greg Clayman: Context. It's interesting because one of the messages from broadcasters is that context matters and that's also true online to your point. When I was at News Corp, people would buy the *Wall Street Journal* because it was the *Wall Street Journal*, right?

Rich Greenfield: You wanted to be attached to that brand.

Greg Clayman: You wanted to be attached to that. You've got a play or if you've got a media property, having an ad in the Sunday *Times*, even though you can programmatically buy against that audience on multiple different platforms if you want, you can buy them when they're on crack.com, whatever, if you want to... But the people don't. I mean they do, right? They sort of give you both, supplement that with the programmatic buy. But you wanted *The New York Times* because you wanted around the other content that's in *The New York Times*, the neighborhood matters. And that is true in the digital space as well.

Rich Greenfield: Emily how does that play out as you think about distributed video versus video on your own site? The traditional media space is basically "You want to watch our content, you come to our channel or our place and our app," and it's all about them. You see a few examples, like John Oliver puts his content out or Jimmy Fallon puts his content out. You see huge success when people do that, but just given the way the money works in the traditional TV space, there really isn't much of a distributed ecosystem. How do you think that evolves online? How important is Refinery29 being the destination and the context of Refinery29 versus Refinery29 content being everywhere across the web?

Emily Gray: In our opinion, that train's left the station, so if you're chasing people to bring them back to your owned and operated property, you've probably already lost a really significant opportunity with that user. There are two use cases, of course. There are instances where it makes a lot more sense to watch something like a long-form video that's a look at trendsetters in Tehran. That's a documentary that we're making now. There's companion content that goes with that—articles with in-depth profiles on these women. It makes a lot more sense to watch that on Refinery29. Something like *Cut Above*... *Cut Above* is a stop motion Life Hack series we do that has universal, beautiful graphics on things like how to cut a pineapple or how to open a bottle of wine with a blowtorch, which you guys should all watch immediately.

Rich Greenfield: Are we going to do that afterwards? That sounds like an awesome post-panel demo. I want to see you do it.

Emily Gray: In case you can't find a wine opener, grab your closest blow torch. The idea being you can watch those in their entirety inside the Facebook environment and understand the R29 brand with the convenience of staying on Facebook where the conversation's happening and share that with your friends in a really native way. I think for us we're looking at a social mobile world where we understand that trying to get her back to our place is probably not the best way to become her most trusted source for premium content.

Rich Greenfield: But if I look at something like Facebook right now—unlike YouTube where there's TrueView ads—right now on Facebook, everyone's uploading content essentially for free.

Emily Gray: Yeah.

Rich Greenfield: There's no monetization.

Greg Clayman: The assumption is there will be, though.

Rich Greenfield: No, I understand. But when talking about distributed networks, right now Facebook's got an awesome model. They sell ads all around the content and make 100% of the revenue and don't share it. It's a great model for Facebook.

Greg Clayman: It's a good model for them, yeah, but I think that the understanding is that there will be and the rumblings that you hear are that there will be. But you're right. Today there's not and they can keep everything. But cut to a year from now or even six months from now, I don't think that'll be true.

Emily Gray: I think Facebook's candidly saying that they're rewarding great content creators and they want people to come there and use that as part of their business model. So to align with you, I doubt that they don't have something up their sleeve to make that work.

Rich Greenfield: Is YouTube, I guess as you all think about it... YouTube is this gorilla in the room in terms of video—massive scale, global. Probably every single person in this room has got a YouTube app on multiple devices they own including what's in their pocket in terms of a phone. Is YouTube friend or foe to your business? How do you look at the importance? Is YouTube scary because they have too much control or a big opportunity?

Paul Marcum: I would look at Facebook and YouTube and say that Facebook, with three billion streams a day, has become a gorilla virtually overnight. I think both of them are exciting and frankly, it's exciting to have two gorillas. I think an important point of distinction is that historically from a video distribution perspective, if you're doing a deal with the MVPDs, the Comcasts of the world, they had a different revenue stream. Instead, here, we're actually turning to businesses for distribution of our content that are advertising companies as well, and in many cases collect better data for the sale than we're able to collect from the platforms on which we sit. I think the fact that there are bigger options, or I should say a couple of big options right now, is good for content producers. I think a key thing for content producers, and what historically YouTube allowed and what Facebook doesn't so much allow, is that you have a better predictability in YouTube of how your user's going to find you. The challenge within Facebook is that based on the incredible success of the algorithms driving the feed, that's not always correct. If you're looking for any sort of marketing channel, predictability is a key variable that you need to track to succeed.

Rich Greenfield: But Greg, you made the comment before that most people aren't making much money on YouTube so how do you tie back to...

Greg Clayman: I've been really enjoying watching the artist draw the gorilla behind us; it' really, he's getting it. One of the things that Adi talked about in the last presentation... He showed a couple of YouTube creators who are making more money doing live streaming or you see people making more money selling t-shirts. In our case, there are YouTubers who will have a few hundred thousand views per video, which is a lot, right? But at a $2 CPM and then with 45% taken off the top and the sell-through, they're not making a lot of revenue at that level. YouTube is some of the larger MCNs are when they kind of add that all up, you can also sell for higher CPMs. When you guys are on YouTube, you're not selling at a $2 CPM. Other than the top—I don't know what the number is, few hundred thousand creators—it can be tough. As a place to build an audience, to build a fanbase, to learn a craft—maybe you make it up into that category and you are making real money in advertising. Or maybe you get a show on *E!*. There's a YouTube creator, Comic Book Girl 19. She's awesome. She does these 30-40 minute videos and she sells on Vimeo to her crowd. In general, if you've got hundreds of thousands of followers and you can convince 10,000 of them to spend 10 bucks on something, that is significant revenue for those types of creators.

Rich Greenfield: Or does it just force the brands to integrate themselves into the content so that goes everywhere and there's no need for a separate monetization mechanism? Emily's shaking her head so explain what have you done that fits into that world.

Emily Gray: 60% of the actual revenue that we create is directly linked to custom content, which means the brand is actually our partner in creation of that content.

Rich Greenfield: Give the audience an idea. What's one of the best examples of where it not only went viral but really resonated with what you're trying to do in terms of branded content or co-financed content?

Emily Gray: There's so many great examples, but for video we sell a program called Beauty Prep School every single time we make it. Beauty Prep School is literally a video that is under a minute that uses one product integrated to do a look. It's something as simple as how to get the perfect wet bun.

Greg Clayman: I find it incredibly useful myself—I mean, just speaking as a consumer.

Rich Greenfield: This is when your hair was long.

Greg Clayman: When I had hair.

Rich Greenfield: Yeah, I remember.

Emily Gray: The idea is it's really simple; it's something people really want to watch. You need a product. You would need a product whether or not P&G was sponsoring it, but because P&G's sponsoring it we spend more time with them thinking about how that product's cast in the content. We would never do a piece of video that they asked for if it didn't resonate with our user and we thought it wasn't really going to perform. It's sort of a win-win, which is a little bit of an easier instance. There's plenty of complicated ones that don't go as well, so you're constantly learning and learning with a brand. We have lots of brands who take risks with us.

Rich Greenfield: Experimenting, essentially.

Emily Gray: Yeah. We're doing a really incredible program with Method hand soap which will include original content and it's a different approach for us. We don't necessarily always talk to our user in this way; it's a different approach for them. We sort of had to learn together as we developed it, but it looks beautiful.

Rich Greenfield: We have four or five minutes left. Questions from the audience before I keep asking? Anyone with questions before I keep going?

Evan Nisselson: There's got to be questions.

Rich Greenfield: Evan, I'm trying.

Evan Nisselson: I know you're trying. The only other story... If you want to get noticed by people but you never talk and you hope they find you but they never know you exist... So here's your opportunity to ask a fantastic question. Oh, finally. Look at that.

Audience member: I sat here and kept telling myself, *I wish there was a pill that would kill all millennials.*

Greg Clayman: Thank you for your question. He wants a pill that kills millennials. Okay.

Rich Greenfield: That would have been useful last week at the upfronts.

Audience member (cont'd): Most of the stuff I heard is specifically about, and the talk before, about video streaming. When you're targeting the young, female millennials, as a user, I guess analytics or however you're getting your data to find out about their personal traits and so on, are you actually doing one-on-one tests with them? Do you ever talk to young millennials, as the young female millennials in your case?

Emily Gray: We spend 24 hours a day programming for her, so the data, you're right, is our primary filter. But if you walk into the office of Refinery29, it is 2 men and 200 millennial women.

Greg Clayman: You should really visit that office, by the way. I'm going later today.

Emily Gray: Everybody come on by. I'm exaggerating, but yes, we definitely lean our own staff ear to the ground. Then we just finished a six-month insights study that was custom-funded by us. We went into the homes of millennial women, so we do more broad editorial scrapes. But, actually, the best candid conversations we have with them is in our comments. Anyone who has seen any of the videos that we created about transgenders in America... The comments are very powerful, very opinionated, and they say everything about how our users feel about that video.

Audience member (cont'd): The reason I asked was, for example Vimeo has been around longer than YouTube, right?

Greg Clayman: It's about the same.

Audience member: Some couple months longer or something like that. I think I've been watching Vimeo footage, Vimeo content for years now and that it's the highest quality, beautiful, etc.

Greg Clayman: Thank you.

Audience member (cont'd): It turns out that I'm actually a millennial, but anything you guys would talk...

Greg Clayman: When did you discover you were a millennial?

Audience member (cont'd): I had to Google the dates actually, just now, to find out if I'm a millennial or not. Anything that I've been hearing today concerning millennials specifically is just like... It flies over me. Vimeo, for example, for me, is not targeted towards millennials. Are you guys thinking that your users create content that's just good content? We're a photo company so we like good content, you know?

Greg Clayman: It's funny that our filter is great content. You know it when you see it. We're not necessarily targeting content towards millennials.

Audience member (cont'd): My mom watches you guys.

Greg Clayman: That's awesome. Thank your mom for me.

Rich Greenfield: She's not a millennial.

Greg Clayman: I think one of the reasons that we talk about millennials so much when we're up here or in the upfronts or in the new-fronts is because of advertisers and because that is the main group that advertisers are targeting and that's such a large percent of the population. But content that we have on Vimeo, I think to your point is all... I mean, there are heartbreaking documentaries and there are instructional yoga videos and, yeah, there are YouTubers who are creating content and selling it, too. It's really across the board. It's not targeted necessarily to any one demographic.

Rich Greenfield: Any last questions from the audience? One more? Can we take one more?

Evan Nisselson: Yes.

Rich Greenfield: Now that you excited everybody.

Evan Nisselson: Yeah, yeah.

Rich Greenfield: In the back. Back left.

Audience member: I don't have a mic, but that's okay.

Evan Nisselson: No, use the mic for the video; there you go.

Rich Greenfield: You're going to be famous for this question.

Evan Nisselson: Depends on the question.

Audience member (cont'd): My question is: In the linear world we have Nielsen as the basic currency of advertising and measuring it, how close do you think the online world is to agreeing on a standard metric? Or are we never going to get there because you're going to have different metrics cater to different audiences?

Rich Greenfield: Emily's excited to take this one, I can tell. Online metrics for monitoring or monetizing eyeballs.

Emily Gray: I think, first, is agreeing upon a model that you would actually measure. Right now there are so many experimental models around the monetization of video that are being developed in real time. I don't know if you guys saw it today... There was a great Digiday article about all of the different ways that publishers, new media publishers... They used as an example. They're first creating hits and then they're retro-fitting what the ad model to that hit would be based on how people are watching it, where they're watching it, and what they're doing with that content. Until we agree upon a model, I think it's really hard to say there's going to be some uniform agreement on measurement.

Greg Clayman: Yeah, it's all over the map—I think is what you're saying.

Emily Gray: Yes, that's what I was saying.

Rich Greenfield: I think that about wraps it up because we're over our time slot and I'll turn it back over to Evan. And thank you to the panelists. This was great.

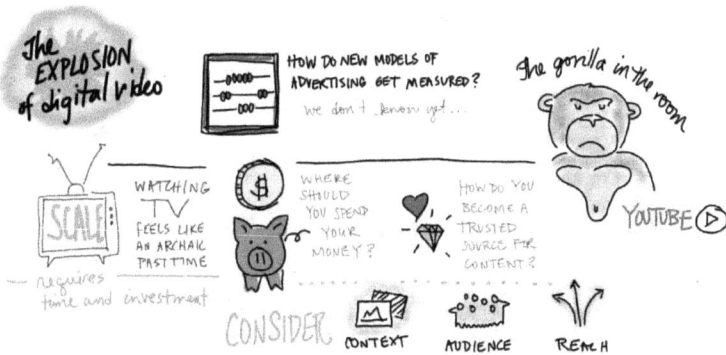

©Rebexlynn

27

THE REVENUE POTENTIAL IS TREMENDOUS FOR KEYWORDING ALL THE VIDEO IN THE WORLD

Matthew Zeiler, CEO & Founder, Clarifai

Rebecca Paoletti: Okay, so how many people have gone to their favorite website, news site, entertainment site, sports site, and wanted a particular clip of a particular last shot of the game or breaking news story and searched on that site and found it the first time? Zero. Okay, with that, I bring out Matthew Zeiler, who has a new and fascinating way to search for videos and actually find them.

I think we've been put onto this track of the conference, which is not the computer vision track, but more the business side, because this technology is ready for your business. I'd like to show you a demo to explain that. We understand images and video automatically using algorithms. It goes through a training phase where it understands the visual world and makes it easily accessible to you as a business or you as a consumer.

If we can pull up the demo, I'll show you exactly what that looks like. We have a live demo set up that has indexed thousands of photos and videos and allows us to search in real time. Maybe I'm looking for anything with a board in it.

But "board" is a very generic English word. It has so many different meanings. It could be a music board. It could be a skateboard. It could be a storyboard or a seaboard or even a billboard. All of this is automatically recognized. There is no human intervention in this system at all. No human looked at these videos and now they're easily accessible.

Maybe you went on a GoPro trip, took your GoPro camera, went surfing, went to the ocean, went scuba diving. All this is recognized to be related to ocean. Maybe you want to filter down. Maybe you want only the scuba diving shots.

So you're underwater, checking out cool new scenes. Our system not only tells you that it's somewhere in the video; it takes you right to that location. Instead of looking through a two-hour video that you know is about scuba diving, you can jump to the exact clip that shows it off perfectly.

Maybe you were surfing with your GoPro again, getting awesome shots right in the water. When you think of these devices like Narrative Clips, GoPros, Dropcams, they're generating 24/7 of video or images. It's generating so much data that you can't really leverage it, you can't make use of it, you can't find the interesting stuff. That's exactly what you want to pull up. That's what you want to show to your audience. That's what you want to share and make available to everybody. It could be that awesome shot with people in the sunset. That's what we do at Clarifai. We make this content accessible to you as a consumer or as a business. And it's ready today. You can put your own content in here, make it searchable, make it fully indexed, and bring this content to life. If we can switch back to the slides, I'll explain a few of the use cases.

This is a perfect segue after the morning because a lot of the conversations were around the new generation, people my age who are consuming and generating this content at a huge rate. There's so much content that you really need to be able to tackle it.

It comes from the creators originally; people like wedding videographers and photographers. You want to make an awesome wedding album to showcase that amazing event, that moment in time. It could be full-feature films that you record from multiple cameras, multiple angles, hours and hours of video, and all this is being put together manually. But we can automate that whole process and make these awesome films more readily available.

Media. This was a huge topic in the morning. There's just so much content out there. How do you find what your audience is actually interested in?

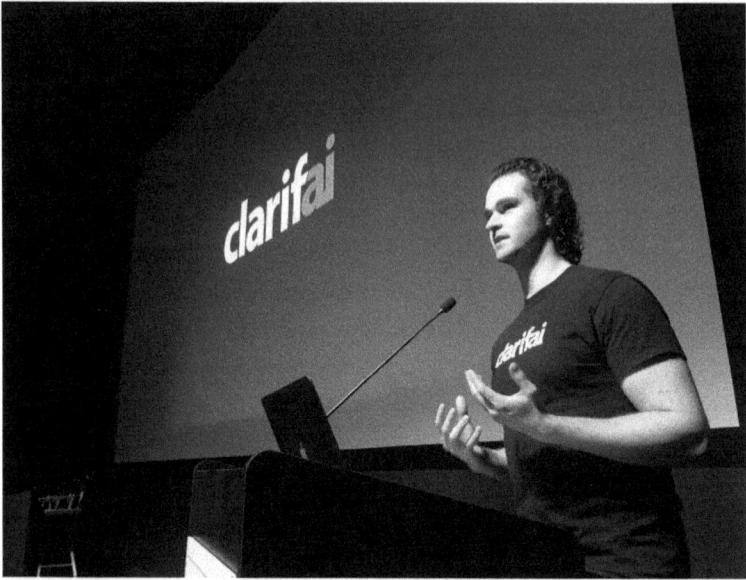

How do you find the awesome videos to show them? Well, we can do that. We understand what they want to look at. We understand the actual content. It doesn't rely on tags around it. It doesn't rely on people writing website descriptions to explain it. We can understand directly from the pixels, know that information, and get that right in front of your audience so that they're the most engaged as possible.

This directly applies to advertising. We literally just moved into a new office within the last three weeks and that was the picture I took. This advertisement would have been perfect if we could understand that I was taking pictures of a new office. Because we had to move a lot of stuff, storage would have been really useful. A new office picture is something different from what I'm normally taking pictures of, so this new event is exactly when you should be advertising this type of content. We can enable that for all of your photos, all of your videos, and all of your end-user content.

Finally, social media is a huge opportunity. Guys like YouNow are generating so much content that it's impossible to find it because... As an example, in Twitter, I literally did this last night and looked for #basketball right after the game. The top is what people hashtagged "basketball." The bottom is what we automatically recognize from the stream of Twitter content. Much more relevant and much more opportunistic for putting great advertising next to it and finding that content to show to your users. There are so many applications. We work across verticals and we built a platform so that you can connect to it as a business and leverage this technology in your applications.

28

PANEL: IS LIVE STREAMING A RECURRING FAD OR WILL THIS MEDIUM KILL TELEVISION?

MODERATOR

Rebecca Paoletti, CEO, CakeWorks

PANELISTS

Rushabh Doshi, VP & GM, Firetalk
Felipe Heusser, CEO, Rhinobird.tv
Lippe Oosterhof, CEO, Livestation

Rebecca Paoletti: We have three founders of three different live streaming platforms. As I've mentioned, I'm super fascinated with this space and very excited about them all joining us today. Different types of experiences but all about live video, live experiences with communities. They haven't all met each other before so here we go. And thank you guys for being here! Hopefully I have my questions somewhere here. It's never a good thing to be paneling with a computer that you haven't actually turned on yet. I should have stuck with my legal pad! Okay, so before we jump into this... And feel free to raise your hands at any point during this session, but know that I am going to call for questions at the end. Very happy to have you all [in the audience] jump in while they're talking about particular topics because we want to keep this super interactive. If you guys could just each introduce yourselves very quickly with one line about who you are and one line about what your product is actually doing today.

Felipe Heusser: Sure. Well, good morning everyone. My name is Felipe Heusser. I'm the founder and CEO of Rhinobird.TV and also a fellow at the Berkman Center for Internet and Society at Harvard. Rhinobird is a live video platform. It's an open live video platform that delivers video in real time. And by "real time," I mean real time with a lag of zero and not 20 seconds late as Periscope and Meerkat do. We also deliver live video without the need for apps to be downloaded or plug-ins, so it's basically coming straight from the browser using WebRTC. We also focus on delivering live video with multiple angles. Basically, we focus not just on the video streams but in events or places with different angles in the same website.

Rebecca Paoletti: You're over your one sentence.

Felipe Heusser: Yeah.

Lippe Oosterhof: There are a lot of commas in that sentence.

Rebecca Paoletti: You can use semi-colons.

Lippe Oosterhof: My name is Lippe Oosterhof. If you forget that, it's a difficult Dutch name... I'm based in London. I run a company called Livestation, which was one of the pioneers in live streaming. We started in 2009 aggregating all of the world's linear news channels and put them on one consumer platform. We've recently pivoted into more of a Periscope-type model. We can talk about that later.

Rush Doshi: Good morning, everyone. My name is Rush Doshi and I'm working on a new initiative called Firetalk, which is an always-on, interactive video platform. We're building this for a new generation of broadcasters to help them engage with and monetize their audiences.

Rebecca Paoletti: Okay, before I go to the next question, does anyone in the audience work at Meerkat or Periscope? I was going to call on you if you did. The other question is: Has anybody here ever used Meerkat or Periscope? Everyone has live broadcasted themselves or one of their close friends recently? Somewhat familiar? All right, good. Generally speaking, the audience should all be using your platforms.

Lippe Oosterhof: Who's been interacting with the people that were watching you? About half. That's pretty good.

Rebecca Paoletti: Okay. You guys can go in whatever order, but my original question is, based on having spent most of my entire digital career in video, what brought you to live streaming? Why is live video streaming more compelling, driving you to all found these products? What did it for you?

Felipe Heusser: In my case, I come from an advocacy and Freedom of Information scholar standing point, not from video or media. I'm from Chile, and back in 2011, there was a huge protest taking place in Santiago, and the story that was being told by the mainstream media was really a story of violence and riots, and not really the peaceful event that was taking place. People were taking to the streets claiming their rights to education. What I wanted to do at that point in 2011, in a pre-drone time, was to share what was going on in the streets from an authentic and credible perspective. At that time we used Tweetcasting to broadcast. Because drones were not available in 2011, I bought a bunch of helium balloons and tied a string to a phone, sending the balloon up in the air and giving the string to different people throughout the protest. The first live broadcasting balloon we sent up in the air got 10,000 viewers organically, no money into Facebook or whatever, to bring attention. That did it for me in terms of saying, "There's this thing about live in a network that was terrible." Thousands of people... 3G in South America was not the best context ever for live streaming. Nevertheless, it was powerful, with thousands of people watching it, because it was credible, because it was genuine. Basically, Rhinobird for us is the latest iteration from that idea of saying, "How can we build an infrastructure to create this kind of people-driven TV network that is open?"

Rebecca Paoletti: That's amazing. Okay, so helium balloon, first drone, you heard it here. Go ahead, Lippe.

Lippe Oosterhof: For me it was the Arab Spring. I got a phone call in the beginning of the Arab Spring from a company that I had never heard of in London. They said, "Our company is blowing up. We need someone to run it. Are you interested?" I thought that was one of those fake LinkedIn emails that you get. Turned out that this was real, and these guys were running all of the news channels—BBC, CNN, Al Jazeera—and they had unique rights to stream these channels in 2009, which was pretty pioneering. Of course, as the Arab Spring blew up, everybody was watching these channels on this particular platform. I have news and a video background. I've done a bunch of startups in that space. I was immediately intrigued. I think live is just super exciting. We had to overcome a whole lot of hurdles with distributing linear feeds on the Internet, but we did. We managed to build, pretty substantially for a startup. 20 million uniques a month was pretty decent. Then we were asked by lots of channels around the world: "Please help us with our live streaming technology." So we did. That's how we originally got into this. We solved a problem of the diaspora, if you will. Lots of Arabs around the world, but also other nationalities, were interested in watching news from home and they couldn't because traditional satellite and cable doesn't allow you to do that. The Internet is a great way to distribute live streaming video. About a couple of years ago we recognized that more and more of our partners were doing their own live streams, sometimes using our technology, sometimes using competitor technology, making our proposition a little less relevant. Our news channels were saying, "Hang on. Nobody watches our channel anymore. We have the data, and 24/7 news channels are kind of dying. It's really tough to..."

Rebecca Paoletti: Don't say that to this room.

Lippe Oosterhof: It's true.

Rebecca Paoletti: Especially not to Bloomberg.

Lippe Oosterhof: I can only speak for the government-backed channels as we work very closely with them. It's a pretty open secret that it's very tough to make a 24/7 news channel work financially. About a year ago they came to us and said, "Look, the aggregated reach of our journalists and correspondents on social media is bigger than our own organization. What if we give our journalists the ability to live stream on the Internet from their phones?" We started working on that, and we launched something recently which is along the lines of Periscope and Meerkat, but blended with old fashioned linear news channels. We're kind of taking the old model and the new model and blending them together.

Rebecca Paoletti: That's great. Rush, do you want to talk about entertainment since we're talking a lot about news? I have a feeling that's where this gets started. Go ahead. Where the passion of live video streaming came for you.

Rush Doshi: Sure. My background is mostly in social apps. Over the past 15 years now, I've seen it all. We started with just basic text and GeoCities-types of sites over to the Myspaces and evolving to Snapchat and Twitter, and now live video. I'm always trying to figure out ways to build communities and keep them engaged. Live streaming is especially fascinating. The company that I work for, Paltalk, has been in live video chat for about 15 years now. What's interesting about them is people come together who don't know one another, turn their cams on, and connect with others around the world. But that's decentralized. Firetalk gives these broadcasters the ability to build their own communities. I think Periscope and Meerkat are really interesting, but one of the challenges is they stream maybe once or twice every week. What happens when they're off-air? How do they keep these people coming back? How do they build a sustainable community? That's something that we're trying to figure out.

Rebecca Paoletti: You were all here for the earlier presentations, and we had some very (arguably) super premium content providers on stage talking about big brands, and big interactions, and a lot of money flowing in this space. Do you feel that anyone can be a video broadcaster, anyone should be a video broadcaster, anyone can create content? How do you feel about all these people who are just getting in front of their cameras and broadcasting out to the world?

Felipe Heusser: I think anyone can be a broadcaster. Absolutely. If you think about very important images or footage being taken with phones lately... For example, a very important one was in Ferguson when Michael Brown got shot. Of course, we don't know who took that footage. Nevertheless, it was hugely important. There's so many examples that we can think of, of very, very important footage that was taken by just anyone. I think the challenge is not just about who can be a broadcaster; I think the other huge challenge for live video, which also connects to what you were saying, is the idea of who can help curate and organize all this content that's coming in. That's something we're focusing on with Rhinobird. We allow people to curate live video as it's coming in. We focus, again, not just in streaming individually, but collectively a broader picture. Let's say, for example, in this event, if you would be using Rhinobird, using #visionsummit, you could see all the different streams inside the room. And people could switch from one site to the other. We also created what we call the "VJ tool," which is basically enabling people to curate and switch from one site to the other and share that in a different UI. Basically, you allow people to not just create content, but to be editors in a context where there's so much video going on that you need help. Of course algorithms can do part of the trick, but also people who understand what kind of content is out there can also help organize it in a way that is clean and it's worth consuming.

Rebecca Paoletti: Lippe, you've been at this the longest. How do you feel about this curation around creating more premium content out of "accidental" broadcasting?

Lippe Oosterhof: I think curation and discovery are probably the hardest problems to solve, and nobody's cracked it yet. I think the announcement of Google and Twitter yesterday, where Google is now publishing tweets, I think is going to help Periscope and Meerkat to discover, because that's what people do. They search. I was at Google Newsgeist in Helsinki last weekend, where we met with people from Google who are trying to address this, because this is a hard one to crack. Going back to your earlier question about whether anybody can be a broadcaster: I think anybody can broadcast, anybody can shoot a YouTube video, anybody can take an Instagram picture. But is it interesting? Hmm. That's the key question, right? Who can master this new medium? We've seen this on YouTube. We have years of cat videos and boring videos and now we have YouTube stars. These are kids who really understand how you engage an audience and how you build an audience. I think we're only scratching the surface of this new, interactive, live-streaming format. I think it's going to take months, if not years, before we see the first stars emerge. Probably months, actually.

Rebecca Paoletti: Do you think that's the technology effect, that's a generational effect? Are kids less afraid of being out there in the media space than they were before? We're better as parents helping them do that? Why is there a change?

Lippe Oosterhof: You saw it with YouNow, right? It's dominated by teenagers because they're the first to try new stuff out and they're not afraid to make mistakes. The millennials, they're used to being in the spotlight and they're very comfortable with this idea of sharing their personal lives. We did a pilot with the BBC just before the UK elections where we had big-name journalists use our technology to follow political candidates in the field as they were campaigning. It was almost laughably bad because they were using this new technology in an old way. They were broadcasting. We had a guy in *Charlie Hebdo* in Paris saying, "I'm Dana Lewis, and I'm live on the scene." Like he was talking to his CNN audience, he was working for CNN. Then the first chat messages came in, and these journalists were really confused about how you interact with an audience. They never had to do that before. On the flip side, the younger kids... Tim Pool is a good example. He invented the genre of live casting on livestream.com. Maybe there's a generational thing, but I think at the end of the day, people will get used to this. We see in the UK and in the States some older journalists who use this really, really, really well. I don't think it has anything to do with technology. It's just a mindset.

Rebecca Paoletti: Just getting used to it. All of you are getting used to it. Rush, what do you want to say about interactive? I know Firetalk is all about the community that exists around the stream.

Rush Doshi: Yeah. About your question about whether anyone can do it, again, just to add, I think anyone can live stream, but it takes a great deal of dedication and commitment to become really good at it. On YouTube, they don't just become a YouTube star overnight. Most of these people who are making a lot of money have been doing it for six to seven years. It doesn't just happen in two to three months. I think that it's just about commitment. What was your other question?

Rebecca Paoletti: Just about the interactivity around it.

Rush Doshi: Live stream has been around forever, right? It really has. Livestream, Ustream, they're all amazing platforms that publishers are using. It's all about the content and not as much about the interaction. I think that changed when Twitch really took over and people started learning about the true formula for success. Content drives people to a destination, but community keeps them there. I think that can happen across a variety of different categories where, if people start knowing one another, and there's a certain level of intimacy that's formed, you'll see these people coming back and this becoming a long-term relationship between the broadcaster who can serve as a community organizer or could serve as a producer with their audience or their supporters.

Felipe Heusser: Just to add to what you were saying—I totally agree, and I also agree with Lippe when he was saying that we're really just scratching the surface of this space. One of the great things that I'm excited about, and that our team is excited about in the live video space, is that we all see the trend of video taking over the Internet completely. The head of Google Research was saying a couple years ago that by 2020, 90% of the bandwidth was going to be video. Also, real time is another huge trend.

Rebecca Paoletti: It's 84% of Internet traffic.

Felipe Heusser: 84%, okay. But still massive.

Rebecca Paoletti: We learned of this yesterday. More than 50% of mobile, and that's just going up and up and up.

Felipe Heusser: Right. Then, on top of the huge video trend, you also see the real-time trend taking over the Internet on text and audio. If you see both trends combined, you will probably say, "There's a huge space for real time video." That said, in spite of the recent excitement around Periscope and Meerkat, it's still super tiny compared to the potential of the space. The day when Periscope launched was the day of huge attention on the space. Twitter was of course pushing for it, and lots of coverage about this sort of battle between those two companies. They had about 50,000 mentions on Twitter, and that same day—there was a *Wall Street Journal* article that mentioned this—that same day a guy resigned from the boy band One Direction, and that had ten or fifteen times more Twitter mentions than Periscope and Meerkat combined.

Rebecca Paoletti: Boy bands always win. Always.

Felipe Heusser: It gives you an idea about the space, and that we are really just getting started. There's a huge trend for video and I think it's just great and exciting to be here trying to figure out what is the best equation to take advantage of it.

Rebecca Paoletti: I obviously agree with all that you said. I think from an engagement perspective, and there's a lot more to talk about from an engagement perspective for later. Clearly where people want to be is in front of video and interacting with video and broadcasting video and sharing that whole live, emotional, rich experience. How do you make money? In a live stream environment, we're not talking about pre-roll. You're probably not talking about some of the other stuff that was prior to this and that you needed that space, but just very curious. You actually have to make money because you're the oldest standing. You're obviously new, but there are obviously new models coming on. Rush, you're obviously looking at new things. I'm very interested to hear, and probably everyone here is like, "This is all really, really great, lots and lots of hours of interaction whether it's millennials or beyond, but how are you going to make money?"

Lippe Oosterhof: I think the key word there is you mentioned "engaging," right? This is a live video that only lives by its audience. If there's nobody watching, you will stop the live stream. In other words, there is an audience, you can see how engaged they are through their chat activity, and they can't skip anything. They have to be there right then and there. The engagement of live video, the next generation of live video is what we're talking about—the Meerkat-, Periscope-type stuff. I think that provides massive monetization opportunities. We're based in London, but I'm flying to LA this weekend because we're setting up an office in LA to work with the MCN's of this world because we think there's a tremendous opportunity to do interesting stuff there. Our journalists back in London and Europe have already told us, "Look, this is great, but how do we make money from this?" We've been thinking about this for awhile. On Livestation we already charge for content, so there's a pay-per-view. If somebody does an interview with David Cameron, they can stand outside and say, "Look, the first 200 to pay me 10 pounds can come in with me and ask questions." Premium content is something that we've been doing for years because that's how we got CNN and BBC distributed, and this is just the next version of that. But then you have advertising. You said pre-rolls. I don't believe in pre-rolls on live.

Rebecca Paoletti: Anywhere?

Lippe Oosterhof: Anywhere. If you want to watch a movie or some quality television content, why not sit through an ad? That's how this stuff gets paid. But when there is, say, breaking news on Al Jazeera, then the last thing you want to do is suffer through a 30-second ad or any advertising at all. However, a live feed is very engaging in the Periscope, Meerkat sense. What we're working on on Livestation is mid-rolls. Think about it. This is how linear TV got monetized in the old days. You would announce, "Okay, we're about to commence this performance of this artist after the break." You can do this on Periscope or Meerkat.

Rebecca Paoletti: If you're producing it yourself?

Lippe Oosterhof: Well, you are.

Rebecca Paoletti: That's always the challenge when you're broadcasting live.

Once upon a time when I worked at Yahoo, we did a very big, live video streaming of Michael Jackson's funeral. Not an awesome topic, but three hours of Michael Jackson's funeral streaming from the front page of Yahoo.com and it did not break, I'm happy to say. Did not break, but there was really no good way of monetizing. The display ads around the page and other things are happening, but now I think the notion that you'd interact with a live event with some sort of advertising is tough. What do you do around that?

Lippe Oosterhof: On user-generated live stuff, you can ask the broadcaster to push a button and they insert an ad. On linear feeds or events or anything that is broadcast in a traditional sense online, it's really tough. You can't really do it. The models there are very simple. You either charge the user or you experiment with overlays. Anything else, you can't do mid-rolls because you'll interrupt the experience.

Rebecca Paoletti: Rush, what do you think?

Rush Doshi: I think, and Adi mentioned over at YouNow and the success they've had with virtual credits... And that's not new. That's something through which a lot of these companies in Asia, including YY, have made hundreds of millions of dollars a year. I've seen it with what we're working on as well, where virtual goods are something that people are willing to buy. It's a form of tipping, but also a form of self-expression. It's a way for you to express what you're feeling towards the broadcaster, whether via a rose, a chocolate, a trophy, or whatever. Those things can also be a form of native advertising as well. We did something with Miller where they were doing something around the Super Bowl, and they wanted these stickers to be sent around, almost like a messaging app that you'd see online. It's the same thing that can happen while somebody's live streaming where they're showing their approval or expressing themselves through a sticker or a gift. Saying "Go team" or "I love that scene" or "You go." Whatever it is. Those things can be branded and it's really a new way of brands inserting themselves into this experience. Again, they're coming from the audience more than they are in this traditional sense in terms of ways to monetize. Of course there's subscription, which is if the content is worth it, or there's something where only, like you mentioned, if only 200 people have access to this live stream, then those 200 people that are your biggest fans might be willing to pay for it. You may have five million followers or subscribers on YouTube, but you want your live streaming channel to be complimentary and be something that helps keep your top 1% or top 5% of fans engaged.

Felipe Heusser: In our case, yeah. We have outlined the different models that are out there of course; ads, search, it all depends on the size of the audience. It's also very important to see what the audience is willing to take, to understand the audience and basically wait for the demand before supplying a specific kind of offer. What we're aiming to do right now, and we believe there's a big model there, has to do with content distribution and creation. We're working with some media organizations this way to help them curate, organize, and be sourced-in by their audiences via mobile phones that actually get there first when news breaks. Users broadcasting with Rhinobird can flag content coming their way in real time, and media folks can use our VJ tool to actually switch in real-time videos as they come in. Then you can toll for that. Maybe you can even share that toll with the audience that is contributing with their broadcasts. That's definitely an avenue that we're exploring.

Rebecca Paoletti: That's awesome. Okay, just to wrap up, we're going to do a quick-fire, one-word round robin. I will throw one to you, you can throw one word back. Rush, you get "LIVE."

Rush Doshi: The future.

Rebecca Paoletti: Lippe, "MEERKAT."

Lippe Oosterhof: Good luck.

Rebecca Paoletti: You get "PRE-ROLL."

Felipe Heusser: Old.

Rebecca Paoletti: Rush, "FACETIME."

Rush Doshi: Periscope.

Rebecca Paoletti: "NATIVE."

Lippe Oosterhof: Profitable.

Rebecca Paoletti: "YOUTUBE."

Felipe Heusser: Respect.

Rebecca Paoletti: You all get to answer this one. The future of video is...?

Felipe Heusser: Live.

Rebecca Paoletti: The future of video is...?

Lippe Oosterhof: Super exciting.

Rebecca Paoletti: The future of video is...?

Rush Doshi: Interactive.

Rebecca Paoletti: The future of video is HERE. Now I will take questions if you guys have any out there. Good.

Audience member: How do you deal with improper content which is still live?

Rebecca Paoletti: Great question.

Felipe Heusser: That's actually what we were talking about yesterday, right there next to the coffee place.

Rebecca Paoletti: Was that a plant question?

Felipe Heusser: Search can definitely play a very important role in terms of organizing content. Definitely, for us, curation is something that machines can do a lot. YouTube is doing a lot of that as well, but we're also relying on people as well as organizing that content in a way that, again, makes sense for people to consume. Curation, hashtags, and VJs are some of the tools we have for that.

Rebecca Paoletti: This is big news for me because the YouTube kids channel is being reported for inappropriate content. Obviously, YouTube already is in some trouble for this. Lippe, what do you think?

Lippe Oosterhof: Crowdsourcing, right? This is kind of how everybody tries to solve it at the moment, and it's what YouTube did for years. You let the audience flag it. If it gets flagged five times, remove it and then you review. That's kind of how we're looking at this.

Rebecca Paoletti: Another chat? One more question?

Evan Nisselson: I can't believe. You're not all that shy.

Rebecca Paoletti: I know. This is about live broadcasting, we're AMAs. Ask me anything. What's going on?

Evan Nisselson: I bet following you'll come up to me and say, "I'd love to meet.... I've got this question." Ask it now.

Felipe Heusser: Is anybody broadcasting this? Maybe there'd be a virtual audience question.

Evan Nisselson: We were broadcasting it earlier over here in front. Anybody know? [Audience member indicates desire to ask a question.] Take it away, finish it up.

Rebecca Paoletti: There we go.

Evan Nisselson: Okay, I'm coming. [Evan runs over with microphone.]

Felipe Heusser: This is how you keep fit.

Evan Nisselson: That's right.

Rebecca Paoletti: He's also filming all around the room.

Evan Nisselson: Can you stand up?

Rebecca Paoletti: Thank you.

Audience member: Just curious if you're building any tools into your streaming services like DVRs or replay or anything like that?

Rush Doshi: We're giving that option to the broadcaster after every episode. They can either make it ephemeral and disappear, or they can select and make it something they can add to their archives. We're also making it very easy for them to put it onto their YouTube channel with one click.

Lippe Oosterhof: We record everything and keep it forever. We've actually done something pretty innovative, I think, because we record it in the Cloud like everybody else, but we also give the user the option to record a local version on their device in 720p, and that version then gets uploaded to the Cloud once they're back on a good connection. That means that if a journalist or somebody else is shooting something live, if somebody doesn't catch that in time and they see the recording an hour later, they can see that in crisp HD quality.

Felipe Heusser: In our case, we're also keeping our archive by default. People can then erase it. Another important factor for us is actually gaining the trust of our community. People have control over their data, and if they choose to eliminate it, it's totally gone from our servers.

Rebecca Paoletti: We're 57 seconds over so we will stop. Thank you so much, all of you, for coming.

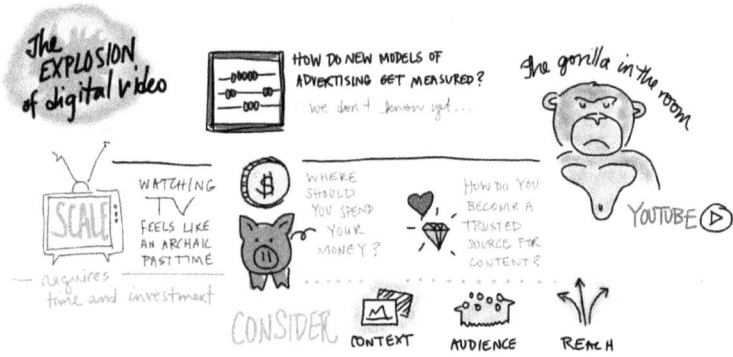

The EXPLOSION of digital video

HOW DO NEW MODELS OF ADVERTISING GET MEASURED? We don't know yet...

The gorilla in the room

YOUTUBE ▶

SCALE

WATCHING TV FEELS LIKE AN ARCHAIC PASTTIME

requires time and investment

$

WHERE SHOULD YOU SPEND YOUR MONEY?

HOW DO YOU BECOME A TRUSTED SOURCE FOR CONTENT?

CONSIDER CONTEXT AUDIENCE REACH

@LDV VISION SUMMIT :: MAY 19-20, 2015 :: @rebexlynn @VizWorld

©Rebexlynn

29

BARK & CO. IS THE 'DISNEY FOR DOGS'

Matt Meeker, CEO & Co-Founder, Bark & Co.

Evan Nisselson: Our next keynote, Matt Meeker, co-founder of BarkBox, is a good friend. This company started when I was in a coworking space called Dogpatch and they were starting the idea. It's an unbelievable success story. He's a serial entrepreneur, but more importantly, a good friend. Come on up, Matt. We're going to do something before, just for fun, because it's a visual tech event, right? Come over here, Matt, for a second.

Matt Meeker: Okay.

Evan Nisselson: Come here.

Matt Meeker: Oh, no. What are we doing?

Evan Nisselson: We are doing a selfie, with a selfie. Hang on.

Matt Meeker: To the right?

Matt Meeker, CEO
& Co-Founder,
Bark & Co.

Evan Nisselson: Ready? Okay. Take it away.

Matt Meeker: Oh, that's it?

Evan Nisselson: That's it.

Matt Meeker: I'm going to talk about how we invent new products at Bark & Co. It's not specifically to photo and video and the topic of the conference, but we'll touch on that a little bit.

But, first. This is a picture of me from a few years ago. I've known Evan since I was this tall. Right? No, really we were looking for a photo of Evan because he's in that room, somewhere, back in the day, helping us pack boxes for BarkBox at the very beginning. We needed Clarifai to find a photo, but we didn't have time. But we've known each other quite a while and we wouldn't be where we are without his help in the early days.

Just to give you a quick overview on me: I've been working in early-stage startup environments for about 17 years now. I started back in the '90s with a digital ad agency here in New York called i-traffic. We were acquired in December of '99. I did a stealth startup. If you hear people say that, it's a kind word for "we don't know what we're doing and we failed."

Then I started Meetup.com in 2002 and ran that for seven years. I met Evan at Dogpatch Labs, where we were hosting a variety of companies who were starting new businesses, new technologies, and having a lot of fun helping them get on their feet and be successful. Probably the best known, and very relevant here, we were host to was Instagram, before they were Instagram and during. They were Burbn and then they turned into Instagram in our house. Then, for the last few years, with Bark & Co.

Here are a few of the products that, when you think of us, you associate. I think most people who know us would hear of BarkBox and know what that is. We're a company that builds a lot of stuff all the time. In addition to BarkBox, we have all these other properties that are out in the world, in addition to several that aren't listed here. The boxes is a retail subscription, retail commerce business. But something interesting, for example, and to give you an idea into how we think of invention and innovation, we have this thing in the lower left corner called "BarkPack."

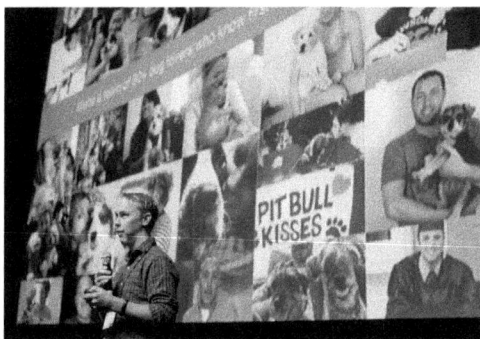

BarkPack started as us reaching out to influencers who are dogs on Instagram and saying: "We'd love to work with you somehow." Then those people point customers to our other businesses. It has sort of mushroomed in many, many different directions, where we follow what's working and what the audience, or that core group, wants. It is now, in some ways, acting as a talent agency for some of the most popular dogs of Instagram, YouTube, and other platforms out there. You start with a very small kernel of an idea and it grew from there.

That brings up the dog here, because this is often how, in our experience, that invention or innovation goes. You set off on a journey. You're going somewhere. You have a vague idea of where you're going and then you get there and you discover it's something completely different than what you had in mind. Sometimes it's better. Sometimes you're in outer space on the moon. Sometimes it's disappointing: it's not made of cheese. But it often goes that way, and one of the joys is you have to be willing to accept that.

I was going to talk through three things that, when we think about how we do innovation, are very important to the products that we invent and how we go about it. The first is that it's certainly important for me, and important for our company, that we have a tremendous amount of passion about the things that we work on.

1. Passion Whatever you're working on needs to be personal something you care deeply about.

It needs to be deeply personal. I believe that great product people or great inventors have a tremendous amount of empathy. It helps a lot because pretty soon everyone around you is going to care a lot about that. If you don't, it will become very annoying to you very fast.

This is what I'm passionate about. This is my dog, Hugo. He makes me very happy and I live the company through him. I want to make him happy and create products and services and new things for him all the time. That's what keeps me going, keeps the passion going. That's my guy.

We have a team now of over 90 dog lovers. This is some of them, with their dogs. We generally only hire someone who really, really, really loves their dog and has that passion and understands that customer or core user. That's a very important part of great invention, great innovation, all the time.

2. Invention. You need to use your creative muscles everyday & create lots of products.

The second is: we do a lot of stuff. Most of it doesn't work. Most of it the world doesn't see. But it's very important we use those creative muscles every single day. Oftentimes, similar to the dog going to the moon, it doesn't turn out the way you thought, but it uncovers something very new. Just the process of being creative and coming up with new ideas and coming up with new products every day gets you into that mode of habit and thinking in creative ways that will lead to new and important discoveries. In addition to having a lot of passion and empathy for the customer, you've got to do it all the time. It's not something you can turn off and turn on with a lot of ease.

I'll give you an example of some of those. We have created everything from the core products that we talked about, BarkBox... We have a content property called BarkPost that serves more than 10 million unique visitors every month. We have a Tinder for dogs that's called BarkBuddy. We opened a retail store for a weekend where we connected rescue dogs to people in SoHo. We have dog-shaped paper clips and probably another hundred or so things that are at some various point of either being out there in the world or in development. Again, we're exercising that creative muscle all the time.

This is a recent example. Everything you do doesn't have to be a huge thing in that invention. This is actually a chrome extension called BarkBuddy, so every time you open a new tab in Chrome it shows you a rescue dog near you that's available for adoption. It's small but it speaks to our passion. And it's just a fun thing.

Finally, to come up with anything original, you've got to be prepared to be wrong. Because you will be. A lot. The key is just be wrong very quickly. Be ready to accept that, move on, get on to the next thing and learn from it. But you have to be prepared to be wrong a whole lot.

3. To come up with anything original, you have to be prepared to be wrong (just be wrong quickly)

When we were getting started, back when Evan was helping us, the initial idea for BarkBox was just that. It was an idea. It lived in my head and I would walk around Dogpatch Labs and I'd go to Evan and tell him the idea and he said, "Wow. That sounds great." I got tired of hearing, "Wow. That sounds great." Here's something that's absolutely true. Your family and friends will lie to you. If you say, "What do you think of this idea?", they will say: "It sounds great." It's amazing. They will lie to you to make you feel good. Evan probably lied to me, and that wasn't nice, but I had an answer. I pulled out a Square and I plugged into my phone and I said, "That's great. Then you can buy one." That's when the truth comes out. When you put it to someone right then to swipe their card: "Tell me if you really like it or if you're just saying that."

At that point, all of his objections came out. *Well, I don't even have a dog. I thought it just sounded like a great idea.* But he was one of our first customers. He bought two boxes for friends who had dogs. You really have to figure out that you're on the right track, or not, in a real way very quickly.

Those are our three concepts that we follow. We explore that personal passion. We invent a lot of stuff and use the creative muscles very often. We're always prepared to be wrong and we very frequently are wrong.

Do I have more? Nope, that's it. That's me. That's how you get in touch with me, if you'd like. Oh, I'm over time. Ten seconds over.

Thank you.

30

EMOTIONS ARE CORE TO HUMANITY AND EVERY ASPECT OF OUR LIVES

Nick Langeveld, President & CEO, Affectiva

I'M THE CEO OF A COMPANY CALLED AFFECTIVA. OUR WHOLE REASON FOR being is to understand emotion. Understanding emotions wherever they occur and really focusing in on the digital notion of an emotion occurring: the interaction that someone has with a mobile device, a digital experience, a gaming experience, content of any kind.

We want to be the analytics platform that has the ability to drive that insight—to understand how strong is that emotional reaction, how engaging is that experience?—and allowing our clients to use that to improve the experience and make it better. What happened to the deck? I'm having a negative emotional experience. There you go.

I managed to get out to the movies last week, which is pretty impressive I think for me with three young kids at home. But I managed to see this movie, *Ex Machina*, and I found it fascinating. I don't know how many of you guys saw it, but it's about the notion of AI and this kind of director's, writer's view of where it could go. I found it fascinating because it really spoke to a piece of our technology, which is understanding emotion. This robot here had a relationship with a human being, and that relationship was really driven by the emotional connection that the human being thought they were having with this robot. The human in the movie was comfortable, was able to really connect with this robot under the impression that it had this level of emotional connection. With that sense of emotional connection, the human felt like, *Well, there's an ability for me to understand what is going to happen next*, which is really what emotions allow you to do. They allow you to have a nuanced, instinctive insight into predicting what's the next step. I won't ruin the movie for those of you who haven't seen it, but the guy was wrong at the end of the movie.

Emotions matter. They really do influence everything from how we choose to live our lives, what cars we buy, who we marry—these are all emotion-driven decisions. That's really what drives us to improve our technology and to really allow for it to become a tool that can help our clients understand and improve their customers' lives. Quick stat: the fact that we're living and continuing to live in an ever-expanding world of connected things, an Internet of Things... By 2020, 26 billion devices will be connected. In our view, they will be connected and also emotionally enabled, having the ability to understand how the user interacts with the device and improving that experience in the process. Our technology is focused around understanding the face, which I'll show you an example of in just a moment, and using optical sensors to understand the facial gestures that occur and how they map into emotional states. We do this in a universe where we drive insights and analytics to our clients, and we also allow for real-time interaction.

If you look at this slide here, it depicts quite well how we view the world, where the consumer is in the middle and all of the different digital experiences are around the consumer. The emotional layer that exists between that consumer and those devices and those experiences is where we want to deliver that insight and that real-time interaction. How we do it: using the face and looking at the facial landmarks. We look at the forehead, the eye region, the nose and the mouth, and all of the different points on the face, and we map those into emotional states in real time. I'll just give a quick demo here using our mobile-enabled device here. Can we switch over to the demo?

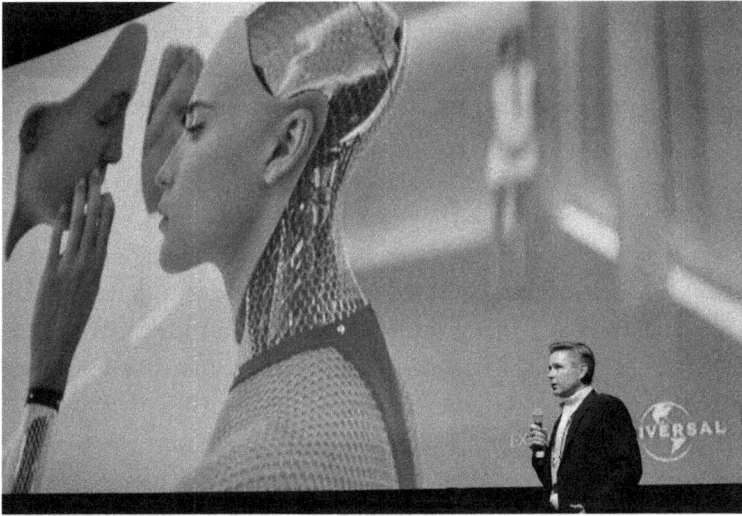

Nick Langeveld, President & CEO, Affectiva

What we have here is my face. This is a real-time depiction of the metrics that it's collecting, so when I smile, the smile metric goes up. Eyebrow raise, furrow, valence, which is a measure of net positivity or negative reaction, comes out and engagement is lean in, frown. This is all real time, and all in the background—I'll switch back to the slides now—in real time, allowing for our clients to be able to take that data and to improve experiences and drive insights into what's happening.

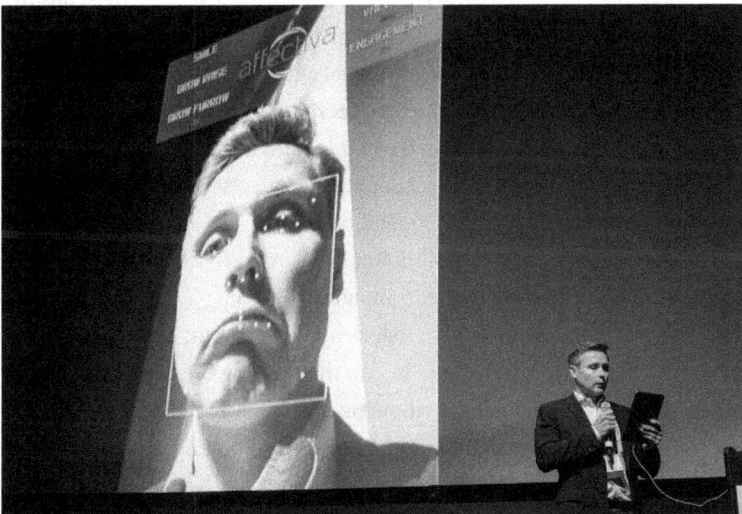

Deep learning is a tool we use to become as accurate as possible. Within the past year, we've been really increasing the accuracy of our classifiers by an order of magnitude, 10 points as you see here, just by using the massive amount of data that we've been able to collect over the past couple years of being in business.

The data that we have collected is growing rapidly. Here you see a quick snapshot of where we're collecting data, 75 different countries, working with 1,400 brands, 2.8 million faces have been measured, 11 billion individual emotion data points, 14,000 media units, and working with one-third of the Fortune Global 100. Why do emotions matter in business? We see the purchase intent accuracy of understanding whether or not someone is going to buy increase to a level of 76% when our data is used. When understanding the experience from a positive or negative standpoint, we're able to have a prediction on outcome in sales. Again, the interaction that people have with their devices is only improved when the device actually understands its user. Online communication is much more effective when there is the ability to understand levels of engagement that go along with it.

How are we being used today? A lot of our clients are still looking at the streaming content that they deliver and they want to understand the level of engagement that goes along with that. Upper left, we have an example of our CBS work, measuring audience engagement; upper right, our work that we've done with the campaign around dancing babies for Evian water and helping them understand what drove the virality behind that campaign. Purchase intent with the Mars company on the lower left, and the Kellogg's work we've done, which is focused on selecting which ad had the most positive engagement, the right level of engagement that drove their decision around airing that ad and where and against what audience to air it.

We are continuing to deploy the technology into new areas. Video communications is obviously a tremendous fit for us. Online education. The whole notion of virtual environments where classrooms are no longer physical but virtually established, and an online session occurring where the teacher doesn't have the ability to turn around to the class and look at the students' level of engagement. Imagine just using a dashboard on the side of the screen for the person delivering the content, getting a read out quickly of how engaged the class is. Are they getting the content? Are they bored? Are they falling asleep? Allowing for that to then be a tool to speed up or slow down the content. Looking at political polling, allowing for our clients in the political space to derive insights into how well quick campaigns are engaging the target voters or the swing voters. In gaming, as I said, the ability for a game to be that much more immersive as a result of having our emotion analytics being part of the tool set.

Here's a quick example of how Hershey's is using our technology. It's actually an in-store example. It's a smile sampler that gives out candy for the right smile. Walking up to the kiosk, showing your pearly whites gets you stuff. We're doing work in the interactive ad campaign space as well with Disney. They're actually coming out with a movie about emotion called *Inside Out*, and our technology is powering their interactive experience where they allow for the user to come onto their site and it tells the user which emotion is going to be the character that they're going to be carrying through the app by detecting the emotion seen on their face, which is kind of fun and cool. We'll see where it goes.

Again, our view of where this goes and our view of where this can go is really just about anywhere there's a digital experience. As we all know, optical sensors in cameras are just growing in ubiquity, so seeing kiosks that provide recommendations based on your reactions, smartphones that react to your mood, cars that sense emotion and engage people in it, games that respond to the player, social robots, and then telemedicine as well are all areas that we're currently exploring and currently really getting excited by and really figuring out how our technology fits there. I think I'll leave it there.

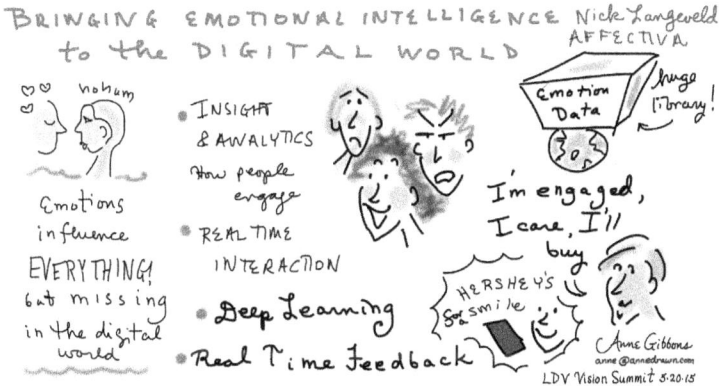

©Anne Gibbons

31

PANEL: CHALLENGES AND OPPORTUNITIES FACING ORIGINAL CONTENT CREATORS

MODERATOR

Jocelyn Johnson, CEO & Founder, VideoInk

PANELISTS

Inbar Barak, SVP, Digital Strategy & Programming, Martha Stewart Living
Nathan Brown, GM, SVP Video, Huffington Post
Tom Wilde, CEO & Founder, Ramp

Jocelyn Johnson: Thanks. So, quite an interesting topic, talking about the opportunities and challenges that are facing content creators right now. With all of the distribution opportunities—and we can actually, if you're tracking the news, add Spotify to that list of distribution opportunities now, too—I'd love to give you each a chance to talk about your vantage point on the space, very quickly. I think maybe the audience isn't familiar with RAMP. I think what Martha Stewart is doing on the video side, our audience could probably use a little primer on that.

Panelists: *Nathan Brown, GM, SVP video, Huffington Post; Inbar Barak, SVP, digital strategy & programming, Martha Stewart Living; Tom Wilde, CEO & founder, RAMP; Moderator: Jocelyn Johnson, CEO & founder, VideoInk [L-R]*

Jocelyn Johnson (cont'd): Huffington Post is a little more self-explanatory, but it would be interesting to talk about what you guys are up to as well. Let's just run down the line very quickly and set the stage with that.

Tom Wilde: Sure. We tend to operate behind the scenes with a lot of major media brands like NFL and FOX Sports and Meredith magazines, helping them get more value from video content. A lot of the challenges they face is really how to make their video more discoverable, how to make it more engaging, and how to generate some kind of ROI from it. Typically advertising, but it could be eCommerce or Lead Generation or things like that. So our technology platform enriches all the video content they've produced with a much bigger set of metadata, better approaches to publishing and social, and helps them ultimately generate more ROI from the video content that they produce and match it to contextual offers or related content as well.

Jocelyn Johnson: Can you very quickly give an actual-use case scenario of one of the clients that you're working with and how they're putting that into action?

Tom Wilde: Yes. *Better Homes and Gardens* would be a good example, one of the Meredith magazine's properties. If you watch any video on *Better*

Homes and Gardens, that's entirely powered by RAMP, and while you're watching the video there's a card that gets displayed to the right of the video and it becomes synchronized with the contents of the video. So if someone mentions garlic in the video, it will show related recipes including garlic. If someone says "food processor," you'd get an ad for a blender immediately to the right of the video. It's all synchronized to the timeline in that video. We do the typical commodity stuff of hosting streaming and coding but make it much richer in terms of that interactivity. In addition, all those videos get published with a complete set of transcripts and tags, which helps with the video SEO and the video sharing as well. That would be a little case study, in terms of what we do.

Jocelyn Johnson: Awesome. Great.

Inbar Barak: For us at MSLO [Martha Stewart Living Omnimedia], the company's background is rooted very much in TV. Some of the things that we've been tackling and really interested in in the last couple of years—I would say four to five years—is really shifting from the perspective of creating/producing video for TV only and really taking advantage of the opportunity, both on owned and operated, but on distribution channels, which is the topic of our conversation, and really shifting the mindset of one thinking of the video and the monetization models to include not just owned and operated, not just channels that you fully control, but also shifting to create content, video content, that lives well and creates engagement on those various platforms.

Jocelyn Johnson: From a programming perspective, are you guys really looking at specific formats? I know that a lot of the content is tips and tricks and how-to and some of them DIY, but how are you approaching some of the scripted or any of the other formatting?

Inbar Barak: It's a really good question. I mean, the background of the type of video that we created for TV was very service-oriented, and part of what we're learning and tackling with the different distribution platforms that we work with very closely is how some of our content becomes more entertainment-focused and less how-to and tips.

Jocelyn Johnson: Okay, got it. Let's talk Huffington Post.

Nathan Brown: So, at Huffington Post, it's all about creating great content, obviously, and telling great stories. We're really focusing now on programming that content across, as Inbar said, across multiple distribution points and being very unique and specific about the kinds of content we make for each of those platforms. So programming is very important, distribution is very important, and then, obviously, monetization is an important topic, too—probably not the fourth in that order, but certainly something that's an important part of the process.

Jocelyn Johnson: One of the interesting themes that I saw come out of the new fronts was this idea that all of the social platforms are really helping with marketing, helping drive awareness of your content. Let's start with Huffington Post and we'll definitely move on to the Martha Stewart side, but how are you guys viewing the content everywhere? Are you distributing your content everywhere or are you splicing that up and using each one of the platforms in a promotional manner?

Nathan Brown: I think it started out as promotion and now we're really trying to utilize getting the right piece of content in front of the right person at the right time on the right platform. So, just like every publisher, we were really at Huffington Post, and most of the others—I can't speak for MSLO, but I'm sure it's the same, based on what you said—it was all about driving traffic, using these social platforms to drive traffic back to our O&Os. I think the switch has been made, that it's really about using distribution and reaching that person on the platform that they're consuming content, rather than just trying to drive them back. There was a big fear with publishers—that it would be cannibalizing their rich property that they have—but really in my experience here and at other companies, it's really more of an accretive experience in terms of revenue and in terms of audience growth.

Jocelyn Johnson: From that side of things, on the revenue side, a lot of these distribution platforms don't have their monetization baked in just yet. Even a massive platform like Facebook doesn't have any type of rev share model baked in right now. I would anticipate that's probably one of the challenges that we're supposed to be addressing on this side of things. So when you're looking at monetizing across all of these different platforms, how are you measuring the ROI and the value takeaway on some of that, if there's not even economics really baked into it just yet?

Tom Wilde: Is this one my take?

Jocelyn Johnson: Either one. Any of you.

Nathan Brown: I can take it first. Yes, that's a problem. There's no monetization on something like Facebook. It is coming. You can guarantee that Facebook's going to figure out how to scale a massive audience and then have it flip the switch and start to monetize that audience. They are in those discussions now. You see some post-roll things that they're doing, I think, with NFL content and very soon I think they'll start to beta different monetization strategies with publishers like ourselves.

Jocelyn Johnson: How does that impact what you guys are doing when you're evaluating the different platforms that you want to distribute content on, when there's really no inherent revenue that will come back to you just yet?

Nathan Brown: Finding an audience with content is always a good thing, and Facebook... We're the number one social publisher on the Facebook platform, so we have a massive audience there. For us, it is about audience; it's about building awareness, it's about tuning in, it's about driving traffic back to our platform as well. So for us it's looked at as that, as sort of an audience development promotional platform. Once monetization gets turned on, that'll be really helpful for us, obviously, but we want to be there and have the audience in scale when it does.

Jocelyn Johnson: Inbar, how are you guys approaching that strategy? Even Snapchat or Vine or Instagram... All of those as well don't have their revenue model baked in yet.

Inbar Barak: Right. The distribution channels vary greatly in terms of what you can do on them. Some we have direct deals with, some—Facebook is a good example—the economic models are just not there, or they don't match what the advertisers want as of yet, right? If you could start focusing on click-per-view versus other stuff, then you could be making money off of Facebook and you could be distributing and monetizing successfully. So I think partially it's really working very closely with our partners at Facebook, our partners at Snapchat, and on Vine to figure out how those can happen. And that's exactly what we're doing.

Jocelyn Johnson: Tom, on your side of things, you were talking about discovery and enabling discovery of video, it sounds like primarily on someone's

O&O, on their owned and operated sites. So what are you seeing in terms of how your clients are leveraging other distribution platforms or solving their own discovery on a site that's not owned and operated by them?

Tom Wilde: My background is 20 years in the search business, so you've seen this fascinating shift from users looking for content. If you go way back, a viewer looking for content might tune to ABC at 8pm on Wednesday. And then there became so much content online that Google and others showed up to solve that problem, but it was still very much, "I'm looking for something. I'm giving you a keyword expressing my interest." Now it's completely inverted, where the content is actually looking for users, so that content from brands like these guys being pushed out to things like Snapchat and Facebook, out to where the audience is... It used to be optimizing for brand, you know, "I need to make sure I'm promoting my brand," to now, "I need to optimize for the audience, and where they are." The problem is it's a bit chaotic and the advertisers are all, a little bit, saying, "Help us figure out what to do now because we understood the old model [but] we don't quite understand the new model." The users do. They're happy to find the content where they happen to be, and if it's relevant and somewhat curated, either with humans or algorithmically, all the better. You're going to see more engagement there. When we hear from the folks who we work with, they need help in the scale side of this—because the reality is, especially if you're in news or sports or infotainment, you can't curate your whole collection with humans. It's just impossible. You can't do it fast enough and that's another key component. So you have to combine some kind of algorithmic bootstrapping, which can either help an editorial process or even replace an editorial process to get that content in front of the user where they're looking for it.

Jocelyn Johnson: So if a publisher like Huffington Post or a Reuters is looking to distribute their content and have their content discovered on a different platform, do you think there's a particular strategy that they should employ to make sure that they're being seen on all of those different platforms? Or should it all drive back to their O&O in some capacity, and we're using it as a kind of marketing vehicle?

Tom Wilde: No, I think that ship has sailed. You'll still have an O&O business where you want to optimize for what happens when people come to your site, but the nirvana is, if there was a Venn diagram, you'd have "knowledge of content" on one side and "knowledge of audience" on the other side, and that intersection is the sweet spot where value gets created. So when you're a content creator, understanding how to write good titles, how to have complete metadata, whether it's transcriptions or tags or thumbnails, all of that becomes raw material for the merchandising of that content—either on your property or send it out with a piece of content of these other properties so that they can make use of it. What we hear is that when content producers are talking to these distribution points, the distribution points are coming back to them and saying, "We need more and richer metadata and annotations and thumbnails so that we can plug that into our content discovery and recommendations technologies." Because the reality is there's going to be many, many of these. Each has its own silo of recommendations technologies, whether it's a Netflix or an iTunes or what happens on your own site. So making sure that that semantic information and that audience information travels with the pieces of content and reports back to you what's happening with that piece of content is really critical.

Jocelyn Johnson: Interesting. On the data site, how are you guys tracking where your content is being seen and how effectively it's doing on some of these new platforms that are emerging? It seems like that also is a little bit of an uncharted area right now.

Inbar Barak: Yeah, it's really challenging. I think the whole idea of customer behavior and customer data is something that's really challenging when you're working on distribution platforms. It doesn't matter what the platform is; it's pretty much a black box. The audience data is proprietary, so you can do a lot of work on your end to tag, to meta, to match up with somebody else's taxonomies to make sure that your video is as discoverable as possible, but you're not actually getting the user behavior information where you can optimize what you're making and have an ongoing conversation with the audience that way. That's definitely a big challenge.

Jocelyn Johnson: Because I find this fascinating about a television industry. Even where the data is correct, or incorrect, there's still a pretty instant feedback loop. We know that it got this rating and it's performing at this type of rating for so long, or not, and we decide to renew or cancel, or however that process might flow. But that seems to not be so easy in the digital space. How are you guys driving content renewals and production decisions when you... YouTube maybe has a direct feedback loop. You can look at view counts. But how are you benchmarking that and how are you guys making decisions on what to renew and when and how long to wait?

Nathan Brown: It's a combination of analytics and instinct. I think we have a pretty good idea of a hit when we make it, just because it resonates so well with our editorial staff. But sometimes we're completely wrong, too. But I do think the data is very shallow. Clicks are clicks and watch time is watch time, and so you can get some relatively good data and audience metrics behind that, and we just basically use that and the gut-check for how we renew and how we program.

Jocelyn Johnson: So does that end up getting back to the content creator? Does that end up being potentially more favorable for the content creator, because you're committing to production regardless of performance? Whereas somewhat in television, it can be a tricky mix. Sometimes we're like, "We're committed. We're going to make this and it's going to go forward, and we're going to do a 12 episode order or a 6 episode order and that's going to get made." And other times it's not the case. They'll pull the plug on something. So that's tough from the creative side, I think, in Hollywood. Would you say that it's more favorable, then, for the creators looking to do digital? Because it's a commitment, we're just going to gut-check it and do it?

Nathan Brown: Well, I think so. I mean, the budgets are smaller than television, so there's a lot less risk from where we sit. I do think that if you're a creator, there's a lot more buyers out there, so for that reason alone I think it's a good, healthy marketplace for people who are making content. For that reason, I think they're in a really good seat.

Inbar Barak: I think that the challenge is... You can get direct response of whether what you're creating is working with the audience that you have. Where it's really challenging is developing the audience, understanding where they came from, who actually watched it, really figuring out if you can increase production to a specific segment. That's the information that's really missing. That's really the black box when you're talking about distribution channels. All of that information you have on O&O... And I completely agree: the ship has sailed. Nobody is looking at themselves, the publishers that I know, as just creating content for O&O. You really know now that the audiences don't jump from platform to platform, necessarily. They stay on platform, and you want to serve them there. But then, figuring out how you get to grow that segment, that love of a specific show that you did, is really very problematic.

Jocelyn Johnson: Do we have any questions from anyone in the audience so far on anything that we've been covering? Yes.

Audience member: For those of us who are trying to make new forms of media content, is there anything from an analytics perspective while the product is soft that you wish you could ask for, rather than trying to go back to existing products and then saying, "Can you tell us this, can you tell us that?" What kind of metrics are you hungry for with respect to video?

Inbar Barak: One of the things that would be super interesting is to know a little bit of the information behind the click—where users are coming from, what is the overlap with other channels, where you can do real audience development initiatives instead of just look at the video that you're creating as pure one-liner marketing. Because the distribution platforms—again, not all of them—but if we're talking about the social platforms, the economic models are still emerging. If we can get more of that information, it makes it that much more worthwhile.

Tom Wilde: I think that has to be, for at least larger media organizations, I think there's a little too much focus, even today in the short-term, on figuring out what the rev share should be. I think that's actually a problem much later down the road. I think the first negotiation should really be: What data are you giving me back about my content? I mean, it's great to get tons of views on Facebook, etc. I think the value exchange negotiation with any of these third-party platforms should be: I need data. Just what you said—"behind the click"—that's how we're both going to build value together. If you give me that data, I can create more content that is relevant to your audience and eventually we're both going to make money on it. I think that doesn't get the front-billing that I think it should when all these different distribution discussions are on the front page.

Inbar Barak: Thank you for that question.

Jocelyn Johnson: Any others? Great. One other thing that we talked about on our pre-call, the discussion, was how do you really evolve your storytelling on each one of these platforms? Can you tell the same story? It seems like there are different parameters for each one. 15 seconds, 5 seconds, live, lives perpetually, doesn't live perpetually, as long as you want, as short as you want, right? Different audiences, per caps migrating, different age demographics for each one of these platforms. How are you approaching storytelling and content creation for each of those platforms? I'll start with Inbar.

Inbar Barak: For us, the length of the content, the length of the video, is less of an issue. Again, figuring out what type of content works on what platform and why is more important. But the thing that we've been tackling, and I think the maybe unforeseen issue that really is the thing that we're grappling with, is the quality of the content. A brand like MSLO is so known for the beauty of the visuals, so known for the beauty of the photography and the video. There's a lot of care and a big crew and perfect lighting and high-def video. And that doesn't necessarily work.

On some of these platforms, authenticity is translated into lower production to not make it feel as done as you would expect it on TV or even on YouTube or even on distributional platforms like AOL. When you're really talking about the social platforms where there's such tremendous opportunity to get to an audience, the quality needs to be redefined, and what the brand stands for, and what the quality of the brand stands for needs to be redefined. So it's not taking away from the quality of the brand, but how that manifests through video is something that we're really testing out.

Jocelyn Johnson: When you're looking on the technology side of that, are you going native into the actual platform and saying: "Okay. If people are creating content using their cell phones, then we need to create content for Snapchat using our cell phones. We can't create video elsewhere. All we need is an iPad or a different technology."? What are you guys doing on that side to maintain the authenticity?

Inbar Barak: iPhone is a perfect example. Really quick, very authentic, of-the-moment videos. I mean, there's not one answer. If we teach to roast a chicken, it will be a very specific type of video, and if Martha has a quick comment to the audience, it will be a very different type of video. But there's room to explore different types of technologies and different types of qualities, and the immediacy that that gives is very exciting.

Jocelyn Johnson: Do you build that into a daily programming strategy, then? Is there someone who's just, "I'm going to be posting on Snapchat; I have agency to do that"? Or do you guys pre-program and pre-plan? How orchestrated is that?

Inbar Barak: We're pretty pre-planned, but as opposed to what it used to be, where there's one group, production group, that's very much about TV-type video, that was the only agency to create video in the company. We now have multiple arms of abilities to actually produce video and publish them. So, in every group, the marketing group produces video, the social group produces video, the TV group produces video. So it's widely distributed and everyone does what, or what we think, the video should be doing on the different platforms.

Jocelyn Johnson: And Nathan, you've been in a few different media companies as well. How do you guys approach that from a programming and pre-planning perspective?

Nathan Brown: I think my counterpart at AOL, Dermot McCormack, who runs video there, he says that platforms are the new day parts, which I think is very interesting, and I think that that's something really compelling for how we program. We look at how our audience is consuming content on what platforms at what times and we try to program based on that. If it's in the morning, 7am to 9am, and someone's commuting to work, and they're probably looking at their phone—hopefully not driving, but on the subway or something... We program very specifically and try to launch videos that they will be consuming at that time of the day. We do that throughout the day, whether it's longer form during the day when they're most likely in front of their desktop, whether it's something that's geared more towards a tablet, when someone's home on the weekend on a Saturday. We try to be very specific about the kinds of content we're making and when we're programming to meet that.

Jocelyn Johnson: Interesting. I think we had a question in the back?

Audience member: Sure. I wanted to ask you about native advertising and the editorial line. A lot of publishers have a hard time deciding what line to cross and where to go with native advertising, and your thoughts on that.

Nathan Brown: This is probably one of the biggest struggles for most publishers, for good reason. The editorial staff wants to make the kind of stories that they want to make. They don't want to be seen as showing or making commercials for a brand. But I think that there is a place, there is a world that exists where everyone can win in that category—where the content is really good, it's very compelling, it's very editorial in nature. It takes a lot of faith and a lot of trust from the brand, and a really strong salesperson to help the brand understand why the content matters and why it fits their strategy. When we make content that solves both of those things, then everyone wins. We win as a publisher, the brand wins, and the audience wins. If any of those things break down, then everyone loses. It's a very complicated dance, but it's certainly very necessary. I've seen it work, and it works really well when it does.

Inbar Barak: I completely agree with that. I think there's room for it, but it has to be done in a way where the brand that's working still has editorial control; the advertiser that is being utilized to create the content, or you create the content with them, is really a choice and a true fit. And then the end user also wins. But that line is going to be something that's negotiated all the time, which is why this is a solution that happens a handful of times a year and not something that's just run-of-the-mill, like pre-roll.

Tom Wilde: I think that's building on those points. It's a question of relevancy, right? Something becomes more like content when it's more relevant to you and when it's less relevant, a lot of the time it feels just like advertising. Being able to make sure that that piece of native advertising is well-matched to that user, and even done so in real time, increasingly will be important because then satisfaction with that content experience goes up if it is really relevant at that moment for you, whether it's targeted to a specific gender or demographic or time of day, whatever it might be.

Jocelyn Johnson: It feels like we are at an interesting time as well because a lot of the publishers are somewhat becoming glorified creative agencies, right? Even on the editorial/tech side, video and editorial both are now blurring the lines of our creative agency and is that what we are doing? Or are we a media brand and content production and publisher? Right? It's kind of an interesting time for that. Any other questions out here so far? No? I did want to ask you guys, dovetailing right back into this concept of creating content that feels authentic, and also what Inbar brought up a little bit earlier on the actual production side of things—the devices side. It seems like quality on cameras and smartphones is getting higher. There is maybe not necessarily the need to have full production teams and boom cams and this whole built-out production. How are you guys approaching that side of things? Are you using any interesting technology to innovate on the actual device side?

Inbar Barak: Not more than what we spoke about before.

Jocelyn Johnson: Okay.

Nathan Brown: We are trying to use everything. We have a pretty massive digital studio here in New York where we have spent several million dollars with amazing 4K cameras and the whole bit all the way down to exploring some of the newest technologies that are coming out that probably were talked about in earlier panels, like Rhinobird and some of the interesting things that they are doing from a capturing perspective, and everything in between. We want to have the full stack available to us so that we can try a lot of things out and see what works. And then keep doing what works.

Jocelyn Johnson: Cool. We have a few minutes left. I want to go down the line and ask you guys what you think is the biggest challenge right now. It might be something you already mentioned casually, but biggest challenge for storytellers and content producers given all of the distribution opportunities, and biggest opportunity. Challenge and opportunity.

Nathan Brown: For creators?

Jocelyn Johnson: Mm-hmm.

Nathan Brown: It's kind of interesting because I think it's almost the same thing. The challenge is the fact that there are so many people out there creating. There is no barrier to entry anymore as a creator, so there are a lot of people out there trying to tell a lot of stories. The opportunity on the flip side is that there are so many people buying content right now. Whether it's someone like Huffington Post or Netflix or YouTube or whomever it might be, I think the challenge is that the barrier of entries are low but the opportunity is that there are so many people out there looking for great content.

Jocelyn Johnson: Inbar?

Inbar Barak: I think the opportunity that is created with all of these social platforms being video enabled, all of these people looking for video, is unbelievable. We can create more video to match more people's desires than ever before. I think the real challenge is figuring out how to do that sustainably until the economic models meet and catch up.

Tom Wilde: I think the biggest challenge, I touched on it earlier, is understanding who your audience is now. It used to be something that you could capture in a box and analyze and now the audience is so dispersed and distributed. I think it's a big challenge to make editorial decisions about your content if you don't know who is responding to it and how they are responding to it. I do think the opportunity is content consumption continues to skyrocket. It's not that people are consuming less content; they definitely are. If you have something of a brand, being able to use that brand as a proxy for curation and quality is a way to cut through a lot of the noise. As we even heard on the live panel, potentially have tens of millions of little, live broadcasters that are going to come online here in the next few years.

Jocelyn Johnson: Any other questions from the audience? Yes, right in the middle. I can repeat it as well if we can't get her. Or maybe we can get this one first, Becca. And then we can give our mic to him.

Audience member: When we think about content, we still use terms like "audience," but people right, are on digital media... We can potentially, using AI, machine learning, and computer records, customize and create individual films for one person. Individual ads. We can completely customize everything in order to be a perfect feed, but somehow I don't see that happening in my lifetime. Perhaps some examples to look like a customized newspaper but we haven't reached enough in the video sphere. Do you think you see signs of this happening anytime soon? Of having complete customization where you film an ad created just for me completely by computers?

Jocelyn Johnson: We clear on the question?

Nathan Brown: I think I understand it. There are some technologies out there, like Watchit is one that comes to mind, that are really good at using data points and creating customized videos in real-time based on the data they are receiving. I think to your question, where when is it going to be fully customizable in video scenario, I could foresee a world where that does exist, but I agree it's going to be perhaps not in our lifetime.

Tom Wilde: I like the raft example earlier where it's not "the asset gets created for you" but "you get to guide the asset" more, the particular video that is put in front of you, with optional storytelling and interactivity. That's maybe a step in that direction. I think the knowledge of the audience data is getting to the point where instead of just thinking about the audience as this blob, which if you go back to broadcast television, you have these huge audience blobs, now to a particular viewer and even having almost an actual CRM relationship with that viewer ultimately into the future. I think that is the next step before you get what you described, which is very cool, but probably a little ways off.

Inbar Barak: I think we also have a couple security issues on the way to solve for that. Right now anybody who is looking at audiences is looking at them as segments and not as individuals. We need to surpass that hurdle also.

Jocelyn Johnson: There is also an interesting approach that Mode is doing in terms of the way they are distributing everywhere and customizing video delivery. You might want to take a look at that because the way they are approaching discovery and distribution is interesting. We will see in the long tail if it works, but it's kind of an interesting approach. In the middle here?

Audience member: Jocelyn, you brought up the question of content creation. Nathan, you showed us that there is no barrier to entry. Tom, you said that we are going to have millions of people coming on, doing these sort of things. Proliferation of content is maybe not a problem. Can anyone make a living doing it? Or what is the percentage of people who are creating this content where there is no barrier to entry? What is the percentage of people who are actually going to be able to make a living doing it?

Nathan Brown: I think some of it doesn't change. How many starving actors are there in Hollywood? Not everyone can do it because not everyone is good at it. At the end of the day, the content still has to be good. It just means now that there are many more places where you can potentially get your content seen. I think the barriers are lower for sure, but at the end of the day, content is about good storytelling or about being relevant to an audience. I am not sure if that barrier has changed because of the technology.

Tom Wilde: Yeah, I would agree. I think it all comes back to the quality of the filmmaker and the quality of the story. There are people like Freddie Wong and others on YouTube who are rubbing two nickels together, and to make these films that have been incredible and you have seen them rise to historic levels... I think it all just depends on the storyteller themselves and the quality of the story.

Inbar Barak: I completely agree, and then there is also the whole other layer. I know we are talking about content creation, but even if you are an amazing content producer, and we see this happening on YouTube where they are amazing content producers, they don't necessarily make enough money to compensate them for what they are doing because they don't have an entire ad sales team behind them and a platform to actually do programmatic buying for all the remnant views. It's those two sides—it's creating amazing content, getting the inventory there, and also being able to capitalize on it.

Jocelyn Johnson: I would say my view on it is that there are always going to be the early adopters who will succeed and they will probably be the 1%. If you think in political terms, the 1% on YouTube is maybe 150 creators. I think there is a 1% on Vine, and the faster people jump and can lateral their audiences and become a master of that platform, I think those are the people that will build businesses first and foremost. Brands will go to them, audiences will find them faster, and after that 1% has been established it's just going to get harder and harder. Any other questions?

Evan Nisselson: I think we are ending up.

Jocelyn Johnson: Oh, sorry.

Evan Nisselson: No, that's okay. Bring it to a fantastic close. Go.

Jocelyn Johnson: That was an amazing panel, guys. I really appreciated all of your contributions. And it's time for lunch.

Evan Nisselson: No, I think it was fantastic, but I am sorry—we are a couple minutes over. Fantastic. Thank you very much.

©Anne Gibbons

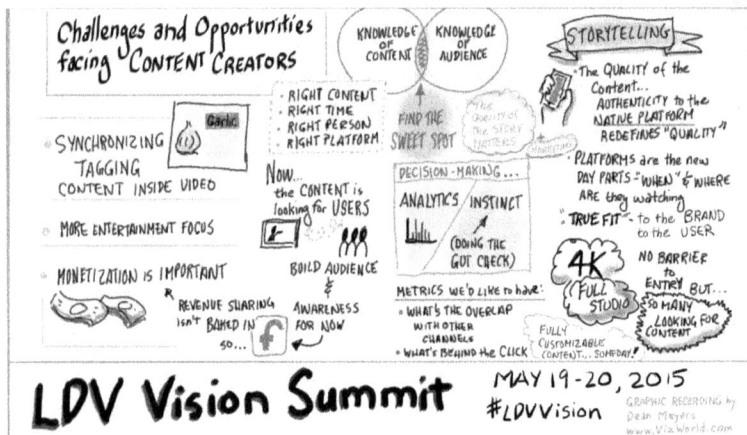

©VizWorld

32

CAN DYNAMIC IMAGES AND VIDEO EXPONENTIALLY INCREASE ROI IN EMAIL MARKETING?

Vivek Sharma, CEO & Founder, Movable Ink

EVAN, YOU LOOK EXACTLY THE SAME AS WHEN WE SAT NEXT TO EACH other at General Assembly except you had a full head of hair [laughter]. It's funny, we moved into General Assembly around the same time, about four years ago, and I remember sitting diagonally across from Evan. The common thing we noticed about each other is we were a few years older than others in that group at GA. I was working on a marketing software company and Evan is a marketing genius. If you kind of do the *Mad Men* analogy, he's the bald Don Draper to my Indian Roger Sterling.

At Movable Ink, we're a contextual marketing company that happens to do email really well and we're headquartered here in New York. We think a lot about the power of imagery and the power of photos to compel people, to get them to engage, to get them to interact with you. Where we think marketing is headed, is it's really about creating an experience over selling a product. Creating an experience that delivers on the promise of a brand. I want to talk a little bit more about how we see imagery and photography changing and being used in more creative ways within marketing.

Let me start with a survey right here. This is a campaign that Allen Edmonds, a clothing retailer, ran recently and there were two different types of creative that they wanted to test out. Creative A that you see on your left side is a nice fabric background. It's an interesting copy. And creative B has a few more product shots in a different type of imagery.

The Quiz

Option A

Option B

How many of you think creative A performed better? Only five hands are showing. How many think creative B performed better? The vast majority of you. This is the kind of thing that happens in email marketing departments and people go by gut feel and decide, is this going to be effective or not? Well, most of you were wrong. As it turned out, in this particular case—and this could be completely different in the next campaign—people are voting with their engagement, with what they choose to view and read and click on. Something that was impossible for was that ability to test creatives on the fly. This is a use of dynamic creative where you're able to A, B, and multivariate test. While an email campaign is running, your audience is telling you what's working really effectively and it's switching that over to the winning forms of creative on the fly. These are types of things that happen in other areas in web marketing, but email marketing has been a little stagnant and it has been formerly impossible.

A huge lift in click-through and this is one example: images change in real-time. The analogy that I use: if you were to walk outside here and head over to Times Square, you're surrounded by giant billboards and for the most part, everyone has seen the exact same message. Millions of people walk by and you're seeing the same creative that's being used. Contrast that to the experience of walking down Park Avenue into any boutique. The second you walk in, somebody notices you, they see you're a little harried in the middle of day. You might be in there buying something for your boyfriend or girlfriend. They're kind of sizing you up and figuring out how much you're likely to spend, what kinds of products you're interested in, peppering you with a few questions back and forth. So that's a very different experience. That's an opportunity where a real person is sitting across and sensing and responding to your context and understanding that and tailoring the message to you. Unfortunately, that's very one-to-one and it's been impossible to do that at a massive scale, especially when creative bottlenecks exist.

We know imagery works in marketing. Over and over, the statistics prove that even including an image in search results gets people to engage. It builds a trust. It gives people a sense of what they're buying. Especially if you're a consumer brand, if you sell any sort of physical product, if there's a fit and finish and feel to it, today, the best way to communicate that is through imagery. Images are important, but as we've mentioned, for the most part in email marketing today, they're very static. So there's a challenge in bringing that dynamism that you might see in the real world into a marketing program like this.

Your context is changing. You might be very similar to me. When I get up in the morning, one of the first things I'm checking is my email. I have my iPhone. I'm running into work. I'm at my desktop during lunch. Some of you may be scrambling to get into the flash sale in time and you get a big, giant screen where you're making your purchases. Then in the evening, you're back at home and you're reclining on your sofa on your tablet. This is a new world. The next step to that world: I'm wearing an Apple watch and now there's another way to reach me. Consumers are choosing how and when they choose to engage with you. This is really difficult. If you thought things were tough 15 years ago, keeping up with the vast number of devices and different channels that your customers use is incredibly difficult. This adds a creative burden to any team that is thinking about marketing content and copy and photography and that sort of thing. But it's imperative to think about some of these.

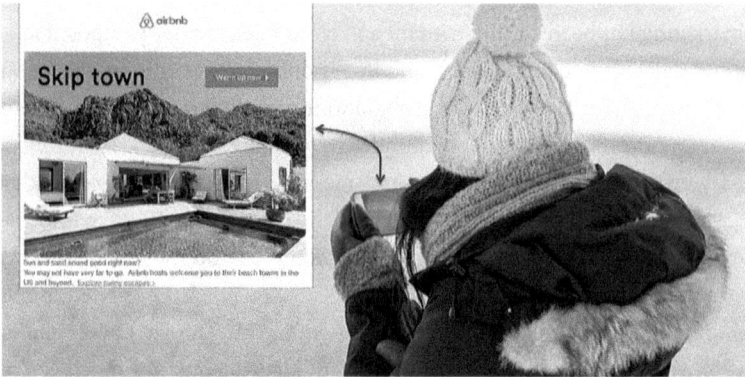

Skip town

Sun and sand sound good right now?
You may not have very far to go. Airbnb hosts welcome you to their beach towns in the
US and beyond. Explore sunny escapes ›

I want to share just three ideas today that you could use to tailor an experience and create a very contextual experience based on some cues your customers are telling you the moment they choose to engage. Let's think about the weather. We actually have a giant chalkboard drawing of this cartoon right now. Spring just started and every season we're changing that up on our blackboard. But the weather changes so quickly and especially retailers have to think about this. Unfortunately, there's a huge amount of information to crunch and to decide and to figure out how to tailor offers depending on the weather in your area. Here's one example. We work with Airbnb and it's possible you've seen some of these emails and are completely unaware that the imagery and the creative and the copy are being tailored for you based upon the weather outside.

In this case, it was very important to them, for their customers who happen to be in very cold-weather areas... So if you lived in New York or Massachusetts or anywhere in the northeast this winter, you are right in the bucket, right in the segment that these Airbnb marketers were looking for. It would detect that the moment you open that email campaign, it happens to be snowing outside or it happens to be 15 degrees outside and it'll give you a very sunny destination that you could think about and houses you could rent. You get a totally different message and creative if you happen to be down in Florida or in California. Similarly, this could be done in retail.

Allen Edmonds again. Most of their customers are in the northeast and into the midwest where it's very cold. But again, the types of offers they put in front of you would vary significantly based on the weather outside. In this case, a very simple weather-targeting rule. If it's above 41 degrees, we want to show you rain gear and umbrellas. If it happens to be colder than that, let's show you something that keeps you nice and bundled up and you might actually get bundled up and get outside of the house. So winter boots,

winterizing your wardrobe, and thinking about those types of things. This was something that was formerly very difficult and in some cases impossible because even if Allen Edmonds had someone on their email list, perhaps they never purchased from them before. So you don't even know where that person lives necessarily. So there's this vast amount of data, these real-time signals that are impossible for you to even collect, impossible to even have and to be able to tailor against in real time, thinking about the changing creative and the changing nature of your customer.

Let's jump to a device. When I sit at my desk at work, I've got five computers right around me. I have my desktop, I have my tablet, I have my iPhone, I have my Apple watch. There's probably someone's old laptop sitting in the corner. It is very hard to guess at where someone is likely to be and how they choose to engage with you. You have to be very responsive to your customers wherever they are.

This is one of the things we did for Comedy Central where the Stephen Colbert show, rest in peace, was incorporating a video into email, but the iPhone experience was very different. It was actually a call to action at the top to download the Comedy Central app and that would only show up if you happen to be on an iPhone.

American Eagle did something very similar. Why waste that valuable real estate and show irrelevant creative if someone is just not going to engage? Again, on an iPhone—and at the time, they only had an app for the iPhone—you'd see a banner and if you clicked on that, you'd go to the app store and have a chance to go download the American Eagle app and engage directly. On a desktop or an Android phone, the creative is the same. It might be mobile optimized, but it's completely changed up. This was incredibly effective for them. They actually saw a 230% lift in app downloads by simply having that very relevant call to action and having a new way to engage with their customers—mainly, via their app.

Finally, location. Where you are matters. The second you're opening the email, it's powerful to be able to see offline brick and mortar stores where you can transact and you can literally get a digital marketing campaign, walk outside the door, see the nearest Steve Madden—especially for products where fit and finish and trying it on really matters—and drive people into your stores and be very tailored in that approach. Steve Madden is one example and the other example, Avaya, which is a UK-based company that lets you use your Nectar reward points and lets you see, the second you open this, where the local restaurants and businesses happen to be. All super tailored and interestingly, that map accounted for 31% of the click so people are really enjoying and looking for geo-targeted content.

Finally, to wrap it up, contextual marketing really has to be about you providing utility for your customers, creating a totally tailored experience, and thinking about the outcomes you want to achieve. That might be something on your website, it might be a more content marketing approach where you're not hammering someone over the head to buy from you every time, but giving them valuable content.

That's us. We're Movable Ink. We're a four-and-a-half-year-old company right here in New York with offices in London and Buenos Aires. We're here to change marketing with the use of brilliant and timely photos and images. Thanks, everyone.

DYNAMIC IMAGES, VIDEO & EMAIL MARKETING Vivek Sharma, Movable Ink

Images in marketing

PEOPLE VOTE ✓ with ENGAGEMENT

Images change in real time!

Images STATIC in email.

changing NATURE of your customer!

Consumers CHOOSE when to engage with you

How to tailor message based on weather?

Why waste valuable creative?

RELEVANT CALL TO ACTION GEOTARGETED!

Where you are matters!

· Utility

· great customer experience

BRILLIANT & TIMELY PHOTOS & IMAGES

LDV Vision Summit

MAY 19-20, 2015
#LDVVision Anne Gibbons
anne@annedrawn.com

©Anne Gibbons

33

HOW PHOTOGRAPHERS & BRANDS CAN BUILD AN EXPONENTIAL VOICE & COMMUNITY ON INSTAGRAM

David Krugman, Social Editor, BBDO

JUST BY A SHOW OF HANDS, HOW MANY PEOPLE IN HERE HAVE INSTAGRAM on their phone? Yeah, so a lot of people. I'm not going to really explain what Instagram is, but I'm going to explain a little bit about how brands are starting to engage in this space and to tell their stories on this visual storytelling platform. I'm at BBDO New York and one of the things that I'm tasked with is, how do we use this platform in a way that is native to the experience? We don't want to interrupt people's experience of this app. People are following their friends. How do we put our brand into that community in an authentic way?

I'll start with how I got started. Well, I'll start with key points about Instagram, actually. There are a couple key things that I always say to brands, the first of which is understanding Instagram as a platform. It's not just a broadcast tool in the way that a print ad or radio is. Instagram is a subscription-based, highly visual editorial platform. It's a place for stories and it's a space where you need to convince people that you're going to make it worth their time if you stick around. It's not hard to convince your friends to follow you, but as a brand it's a harder sell. It's like, *Why would I follow a Nike or something like this when I know that I'll mostly be served advertisement-type stuff?* It's a demographic that has been advertised to their whole lives. They grew up on Facebook and we are looking for ways to speak to them in a way that isn't directly an overt advertisement.

Another thing to note is that there's an ecosystem of highly followed, what are referred to as "influencers," in this space. You heard from Natalie this morning. If you have a channel with 250,000 subscribers, that's a pretty significant messaging space. We're starting to see these people coming in as almost fragmented media channels and we're hiring them instead of going the traditional route to magazines and different TV stations. Now there's this whole new media landscape that we're starting to tap into.

These people have often built up their own followings by providing exactly what we're trying to provide, which is stunning content and a good story. There's a lot of efficiency in bundling your content and your media and saying, "Hey, can you go create content and then can you amplify it for us?" Also, adding that layer of this advocacy of these influential photographers gives you a lot more authenticity on the platform. It makes it a little more palatable to the community who's watching.

The bottom line here is that advertising is a narrative thing, right? We're telling the story of brands. But you can reinforce that story by taking people who already have that following and interweaving it into your story. The brand says one thing and then hires these other photographers to give their take on it and drive their traffic towards the brand feed. These arrows represent two different stories that are united by a common message. The way I got into this world goes like this. I was at the Metropolitan Museum one day and I was looking around and I was just bewildered, because every single person I was looking at had their phone in front of their face. It was like, a statue, a phone, and then a person. I was just like, "This seems like of all the places in the world to keep your phone in your pocket, this would be one."

But it wasn't really until I took a closer look that I noticed they weren't ignoring the art. They were engaging with the art and sharing it out to their networks in a way that I had never really seen happen with this degree of magnitude before. I've seen people take pictures and post them to Facebook but almost every single person I looked at was posting to Facebook or to Instagram and the Instagram stuff really made me realize that they weren't being distracted by technology, they were actually using technology to enhance and share their experience of visiting that space.

I got in touch with their social media manager, Taylor Newby, a really smart guy over there at The Met, and we came up with this plan to take influential photographers, give them complete, unrestricted access on private tours of the spaces, and they'll create content and in exchange we'll just have them basically give a shout out to The Met's page. This was really fun to do because it was a situation where there was incentive in it for everybody. The Met got tremendous exposure, the photographers got access that they really, really wanted, and then I got to be there and see all the great stuff that was coming out of it. And I also got to be in The Met as well.

@newyorkcity is a friend of mine named Liz Eswein and she has such a massive following. She was one of the first people we brought in. She took this picture that you see here and you see it has, I think, 17,000 likes. And then right next to that it says "Thanks to @metmuseum" and that links right to The Met's page. And then on The Met's page, they have some of the content that we generated. So we're really sharing traffic and cross-pollinating here. The story is, *Look at these amazing spaces. When you come to New York, don't miss out on this.* If I had more time I would dive a little deeper into the comments that were being left and stuff like that, but a lot of people were tagging their friends saying, "Let's go see this when we go to New York." And that was something that we were really going for.

It's easy in a space like The Met Museum, because there's so much visual content. It's an institution dedicated to visual content. But how do we approach this for brands that might not be so easy?

These are a few of the shots. This is the team on one of the first times we went through The Met.

So, for a brand like AT&T, which does not have as much... They don't have 3,000 years of art to display and for you to share on Instagram. How can we add value to this community? How can we incentivize people to follow a brand where it might make them think of their phone bill? Let's embrace the creative community in a way that is natural to the AT&T brand and is natural to the space that we're playing in, which is Instagram.

Our narrative for this story was, *You can learn the craft of mobile photography and sharing and using our network from influential photographers by following AT&T.* So we hired people like Ravi Vora. I would highly recommend you follow him. He's doing some of the best branded work on Instagram. We hired him to create a series of six lessons telling people the basics of photography and the basics of smartphone photography and how the community can do what he does. But we had him tease those and then we'd host them on the AT&T channel. He would drive all this traffic to the AT&T channel where they could see exclusive content. We're incentivizing his massive following, saying: "You can see more of what he does only with us." That to me was an interesting solution to this problem and at the same time we're getting really stunning content from him for the AT&T page. We're getting content, reach, and adding value to people who might potentially follow the brand.

In the interest of time, I'm going to run through a few other brands that have done this kind of technique of hiring influential photographers to tell stories that accompany a brand's story.

Monster.com—this is also a BBDO project— they did a campaign with us called "Find Better." We went around the country... We sent these influencers to different towns and we said, "When you hit the ground, just go out and try to find people who love what they do." And the whole point is that if you're not happy in your job, there are resources and there are tools to find a better one. You can see more of this project with the "FindBetter" hashtag.

NYonAir—I don't know if any of you are familiar with this account but this is a helicopter company that does visual imaging. They signed contracts with influential photographers. They said, "We'll fly you for free if you give us content and you give us reach." This actually got them to over 400,000 followers on Instagram now, and that happened all in the past year which is, to me, a pretty amazing leap.

Zagat is another one. They did this thing "Dinner With Zagat" and they're inviting influential photographers to come eat at some of New York's finest restaurants and that same exchange of influence and content.

The takeaway here is to thrive as a brand in the Instagram environment, work closely with connected creatives who have proven their success. Hire them to create content that fits the platform and to amplify your brand narrative. We live in an age of unprecedented visual literacy, so work with those who are fluent. And just to wrap it up, I've put together a list of some of the people in the community who are doing the most branded works. I'd say the best way to learn about this is to follow some certain people and I'll just put those up as I walk off the stage here. These guys are really working with brands more than anybody I know and they're doing really great storytelling on the platform.

Thank you so much. Thanks for your time.

David Krugman, Social Editor, BBDO

$$34$$

PANEL: NEW VISUAL LITERACY IMPACT BY MODERN TECHNOLOGY

MODERATOR
Charles Traub, Chair, MFA Photography &
Related Media, School of Visual Arts

PANELISTS
Marvin Heiferman, Curator & Writer, Art in America
Cheryl Heller, Chair, MFA Design for Social Innovation, SVA
Elizabeth Kilroy, Chair, International Center of Photography

Charles Traub: Hello, everybody. We're going to change the dialogue and format a little bit. We're going to be showing a lot of images and each one of these wonderful people is going to talk to you about projects, educational initiatives, and the very issue of visual literacy that affects us all.

This issue has changed radically. Frankly, I'm not sure we know what visual literacy is. This is how Google illustrates it in many different diagrams. It's so confusing. It's so illiterate. It's so graphically unpleasing. I don't know where we're at.

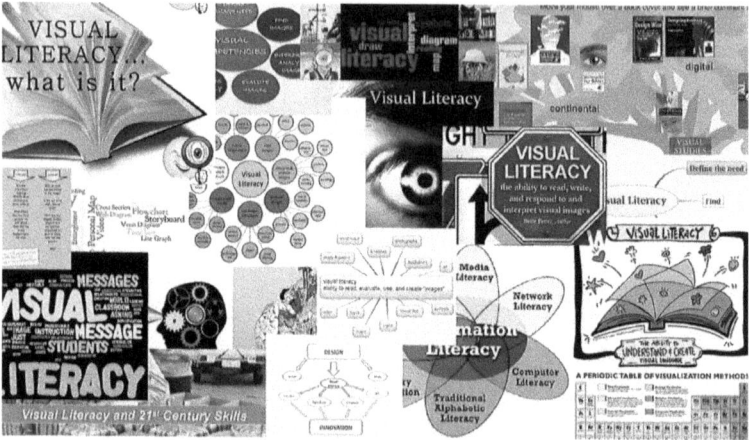

The idea of visual literacy has been around for a long time. You all may be surprised to see this panel—this panel which understands the word "branding" as something to do with cowboys, and that data sets are something we learned in statistics. All these network things are pretty anathema to the days when we were younger and then we had to learn it. That said, we are all educators. We are people who are concerned with training new people to do bright things, to try to change the world, and to change the world using the technology in a humanist way.

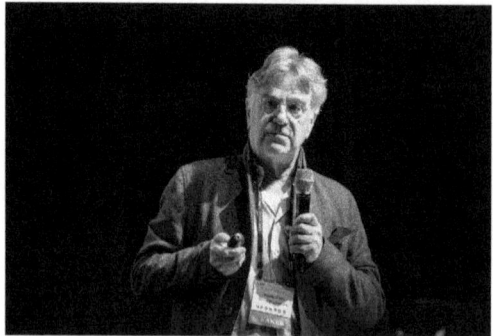

That's our big concern. All of you out there, you great masters of the realm of the Internet, the realm of the circuit: you've achieved a lot. You're of an age that grew up with it. You're an age that takes it for granted. You're also very privileged. You're educated. You are visually literate. You know all this technology, but you have to remember that you're an elite. The world is divided in a very peculiar way. We're divided in a way in which we have two things to deal with: the great economic divide and the fact that we're basically creating an economy that sells for the wealthy. What do we

do about the other half, the other three-quarters?

Literacy is a big issue. Thinking about how to get young people into the world of imagery, to manage it, to understand it, to know how they're being manipulated, and to have something to say with it. It is a narrative. It's a story. It's always been a story. That's what we're going to try to parse for you a little bit.

This is early visual literacy: a photograph from the 19th century that is supposedly Julia Dent Grant, the wife of President Ulysses S. Grant. She's a very literate person and she spoke three languages. I just like the idea that photographers always went back to the book.

Then somewhere in the 1950s, we got this kind of divide going on. This is the beginning of a new kind of literacy. The kids on the right and this great divide between them.

Then this is something I found in Brooklyn: the history of communications, the street art being the final statement in that sequence of events.

Finally, this is what the new literacy is about. By the way, that is General Grant's tomb. Anybody know who's buried there? (That's an old Groucho Marx joke.) We're of a generation that grew up with him. We read Marx, the other Marx, Freud, Darwin, Joyce, Bart, even McLuhan. Now, we're in another generation.

First of all, I want to introduce my colleague Marvin Heiferman. Marvin has been teaching with me at the School of Visual Arts and the ICP for many, many years. He's a thinker, creator, gallerist, curator, and writer whose recent book is about why photography matters and why it matters in every discipline. I'm going to begin with Marvin.

Marvin Heiferman: Great. Thank you, Charlie. The project that Charlie mentioned is something that I was asked to do for the Smithsonian. They had 14-to-15 million photographs there and they wanted to monetize them and to figure how to get them out to the public in a better way. It proved to be a big challenge working across 19 museums on the mall in Washington. What we wound up doing instead was an online project called "Photography Changes Everything" that looked at how photography across a hundred different disciplines had transformed every discipline that it came in contact with. I was kind of fascinated yesterday to hear people speaking about how imaging is progressing and developing in terms of facial recognition, object recognition, and scene recognition.

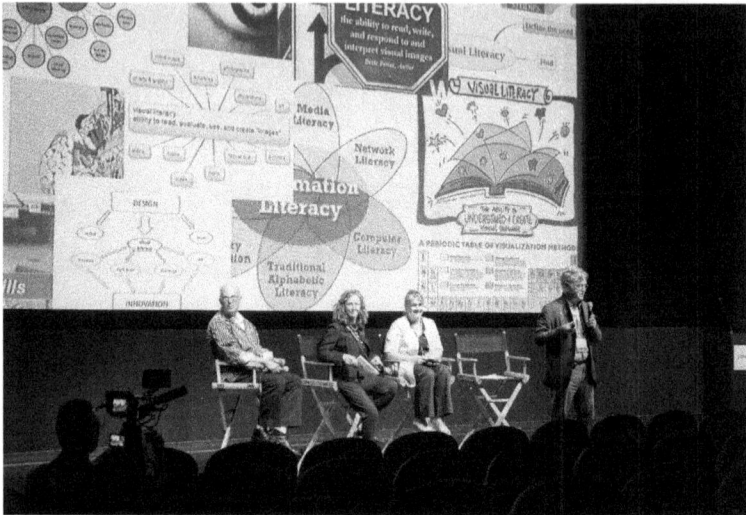

Panelists: Marvin Heiferman, curator, writer, contributing editor, Art in America; Elizabeth Kilroy, chair, International Center of Photography, New Media Narratives; Cheryl Heller, chair, School of Visual Arts, MFA Design for Social Innovation; Moderator: Charles Traub, chair, MFA Photography & Related Media, School of Visual Arts [L-R]

One of the things I realized while working on the Smithsonian project and having the opportunity to interview people like Philippe Kahn, who invented the cell phone camera, or Jos Stam, who was the developer of Maya software, was that there is not a big enough discussion across disciplines about how images work. Yesterday, the talk was about images as data. Today, the talk is very much about images as tools for marketing and communicating. But it's the meaning of images and the dialogue around images that I'm particularly interested in, so I wanted to talk about that briefly today.

This is a photograph of an installation that has been done a number of times since 2011 by a man named Erik Kessels, who was a book publisher, curator, and ad man from Holland. He's been doing these installations where he prints out all of the million photographs uploaded to social media in the course of any single day. It's an extraordinary physical manifestation of photography at a moment when photography is changing, when it's dematerializing as the audiences for it and the dialogue around it change a lot.

An installation that has been done a number of times since 2011 by Erik Kessels

One of the things that's most important to me in terms of visual literacy is not so much mining images for data and using it to sell shoes and umbrellas and drive customers to corporate accounts, but for ways for us to communicate about our lives and understand how we use photography, how we understand photography. It's something that, frankly, we don't do. Again, the talk yesterday was about photography as data. People say photography is a universal language. It isn't. Photography is about images and how we read them, what we see in them, and how we make sense out of them. It depends on the knowledge that we bring to them.

I got very interested in visual literacy a while ago and started doing a little bit of research in it and found that visual literacy as a field is fairly young. It is not so old. The term "visual literacy" was used for the first time in 1968. It wasn't until the 1990s that educators got to the point of where they realized that there are multiple literacies and we need to understand how people make and use images. Then, that kind of fell away. Nobody took any action on it. In 2004, Adobe came up with a white paper about visual literacy that started talking about how important it is for people in general, for businesses, for institutions, to understand what it means to make images, to comprehend images, to share images.

Then, again, nothing much happened on that front... To the point today, where visual literacy is something everybody pays lip service to and doesn't really do much about. I am very interested in the issue. I'm really curious about how and where we can talk about and teach visual literacy. It's become a code word in some universities for "art history." They talk about visual culture, meaning art history and art culture. We need to look around at all the images that we make and all the images that we use. Estimates for the last year were anywhere from 1.5 million to 2.5 billion images being made every day. What do they mean? What do we think when we see them? How do we make sense out of them?

I started a project after I left the Smithsonian that's called "Why We Look." It is a Twitter- and Facebook-based project where each and every night before I go to sleep, I scan the media for stories about photographic images and visual culture and post links to them the following day. I'm fascinated by stories on subjects like those we heard yesterday about facial recognitions, surveillance, and scene recognition—all of that stuff. I'm also interested in images that make people talk, that start conversations around photographic imagery. I just wanted to show a few of them today.

This is an image that's gone viral that happened in the last couple of weeks, a follow up to another viral image in which viewers had to decide if they were looking at a blue dress or a black dress. The question here is whether the cat is going up or down the stairs. This is an image that generated a tremendous amount of conversation, and you can approach this image itself in terms of the psychology of the way we see and in terms of perception. I was fascinated to see millions of people engaging and talking about this image.

Is the cat going up or down the stairs?

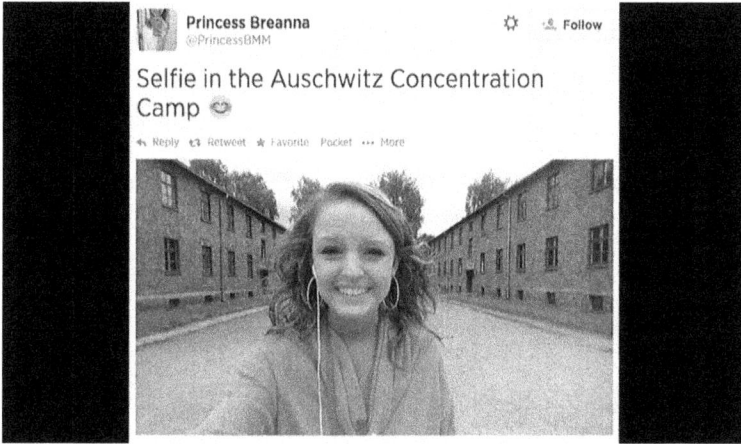

© @PrincessBMM

There are other images that generate very different kinds of conversation. This is an image from a year ago, made by a woman visiting Auschwitz, who had talked about the Holocaust often with her father, who had died a couple of months before she went to the site. She went to Auschwitz and was happy to be there, even if the experience itself was upsetting, because it made her think about her father. She took a selfie. She put it up online and she got creamed for being a narcissist and the most despicable person on earth. It was another very, very interesting example of how an image can go out there through technology and have many, many readings and many meanings and spark intense reactions for different people.

©Manda Moore

This was an image that went up a couple of weeks ago after the National Guard was called in to Baltimore. The woman who took the picture put it up on Twitter. It was picked up on Reddit. Within a day or two, 2.5 million people saw the picture and everybody was commenting on it and saying: "Oh, this is cute. This picture is cute." The woman who took the picture was horrified by that response, and her retort to it was: "This isn't cute. This is sad. This was sad for Baltimore. This is sad that the National Guard had to be called to an area. It's sad that we have to explain situations like that to our children." I'm fascinated by the kind of initial public response to this and the possible meanings of a picture like this one and what kind of dialogue can we have around them?

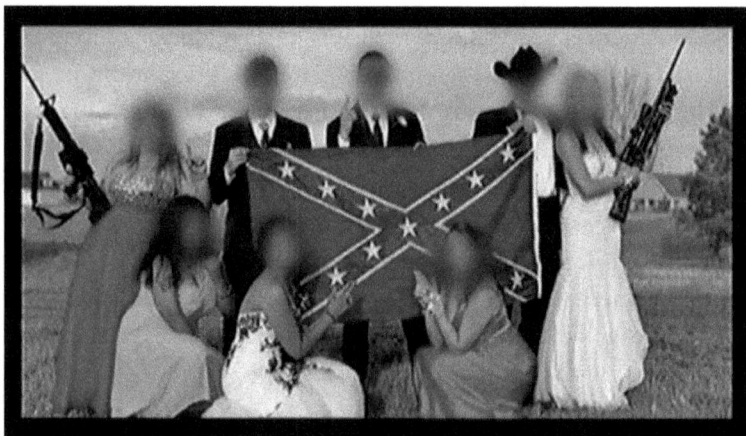

Prom picture of Chaparral High School students in Parker, Colorado

This is yet another instance of many. If you follow these kinds of stories, they appear all the time. This was widely distributed online about two weeks ago. Some seniors in high school in—I think it's North Carolina, I forgot if it's North or South Carolina—put up a prom picture. It's prom night. They went out there with a Confederate flag and guns and then put the picture up on Facebook. By the next day, when the media started covering this story and the kids, their mothers and their school were all saying they didn't mean it. The kids didn't know what they were doing. The kids didn't realize what it meant to put an image out there and in this way. I'm fascinated about images that go out of control because of technology, because of media.

©*Pete Souza*

This last image is one that fascinates me, too. This was shot by Pete Souza, who's Obama's personal photographer. This is a photograph from Obama's trip to Jamaica last year. I put it up there because it's so symbolic and such an amazing, iconic image. But it also points to how the media and technology changes the image we see as newspaper fire photographers. The media resorts to picking up images that are provided to them by content creators hired specifically for that purpose, which should make all of us think twice about the images we see. It kind of skews the media. I'm fascinated by that. I just wanted to show these images to open up the discussion a little bit to talk about how images operate in culture, and what visual culture means, and how complicated it is to both work with, show, and live with images.

Charles Traub: Marvin, thank you. We're going to continue with questions after everybody has presented. I just was thinking about the TV pictures and all the kinds of things that we've engaged through our generation of image making. This idea of programming was something you read in *TV Guide*. Now, programming is the lingua franca of everything that we do in communication. I'm sure at the very base of visual literacy is the need to make sure that every school kid learns the essence of programming and that it's part and parcel of a new education that needs a big technological boost from those who can do it.

Our next speaker, Elizabeth Kilroy, is a graduate and teacher at NYU. She graduated from ITP in the early days of that new and exciting program. She teaches in the Department of Photography & Imaging at NYU Tisch School of the Arts and is the chair of the New Media Narratives Program at the new ICP [International Center of Photography]. This is an exciting program that asks how can visual storytellers use new technologies to create interactive, narrative experiences that engage their audience and deals with the very nature that image makers, creative people, need [in order] to learn how to deal with the narrative. Every image has a story in it. That's part of what we need to parse in order to become more visually literate. So, Elizabeth?

Elizabeth Kilroy: Hi, everybody. I'm going to start with this rather scary look- ing image of strategies for the kind of site maps used for building nonlinear interactive stories, apps, or games on the web.

I started designing websites back in the 1990s when Netscape had a gray background. I've always been fascinated by hypertext and hypermedia, not just as a linking strategy but in how we think about the space of the web. To connect this idea of space and to move this into the realm of photography, let's look at some images.

Elizabeth Kilroy, Chair, International Center of Photography

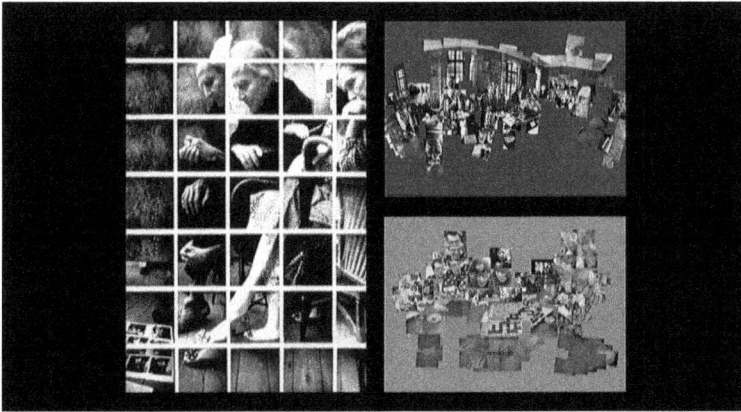

These are images by the artist David Hockney. They're from his "Joiners" series. The image of Bill Brandt and his wife Noya is a collage of Polaroids where Hockney is playing with the idea of time and space in photography. They are actually looking at the images of themselves being edited as Hockney is making this collage. They look like Instagram images. He moved away from Polaroids to creating these Pentax prints, which didn't have the borders, didn't have the edges, they didn't look like Instagram. These works to me are a form of nonlinear narrative storytelling using flat photography, printed photography. His images tend to be very meta: photographers watching other photographers.

This is his image of himself photographing Annie Leibovitz in the Mojave Desert in 1983. She's actually photographing him as well.

©David Hockney

Another photographer who sort of plays with this idea of nonlinearity and playing with space and surface in images is Jan Dibbets. He's actually a bit of a curmudgeon; he really doesn't like the work of very many photographers. He's constantly criticizing almost all photographers, except perhaps himself. He thinks that there are far too many "nice" images in the world.

The last time I checked, we had about 1.5 million sunset images on Instagram. Do we really need any more? That's just in a few years. How many sunset images are we going to have in another 5 years, another 10 years, another 20 years? We have to think about what's happening with all these images. How do we make sense of them? We've got too many of them. There's such a proliferation of images. Who's making these images? Who's looking at these images? How do we make sense of them?

These are some stats from just a couple of days ago when I checked to see who were the top 15 influencers on Instagram. There isn't one single individual photographer on the list. There's a bunch of Kardashians, there's a bunch of football players, there's a lot of singers. There's Instagram itself. There's one brand, Nike, and there's National Geographic, which is a media organization with a lot of really good photographers, actually.

346

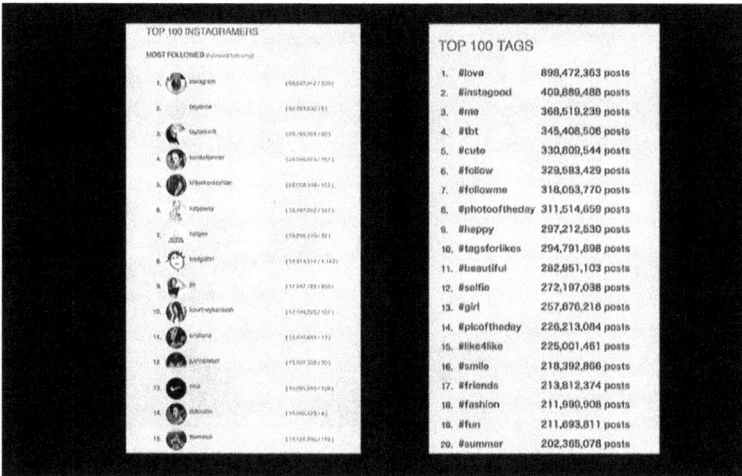

Then, on the other side, you've got a list ot of the top hashtags just from a few days ago. We all know how many images are posted to Instagram every single day. I was rethinking about why these influencers, people like Beyoncé, who is the number two influencer on Instagram... Why is she so successful? What does she do so well? Part of what she does is that she bypasses traditional media. She even bypasses her own publicist because she creates her own image and she controls her own image. She blocks the media from creating the image for her. And all of these top Instagram users essentially understand that. Instagram is a set of data points. You can use these different data points to get your message out, to control your own message.

I'm not going to talk about selfies. There's lots of conversations out there about selfies. I thought this was a really cute image of the queen, because the queen for the last 70 years has been really brilliant at controlling her image in public. She's not a big fan of selfies but she's in at least one, which is this one we can see over here.

I don't know how many of you heard about the Chelsea Handler incident on Instagram. Chelsea Handler got booted off Instagram about six months ago for collaging herself into an image with Vladimir Putin, where they were both topless. Instagram has this policy that you can't be topless. Chelsea has been fighting this battle with Instagram ever since. Referencing the John and Yoko image on the cover of the 1968 issue of *Rolling Stone*, she posed with Nick Offerman on the cover of *Esquire* magazine. She posted the cover shot, then she talked about it on Instagram. I can't really see it from here but I think she had about 120,000 likes.

Here she is. She's back on Instagram now with a new image. This one they haven't taken down yet. I think she called it something like "A Muslim allowed a topless Jew to sit on his camel."

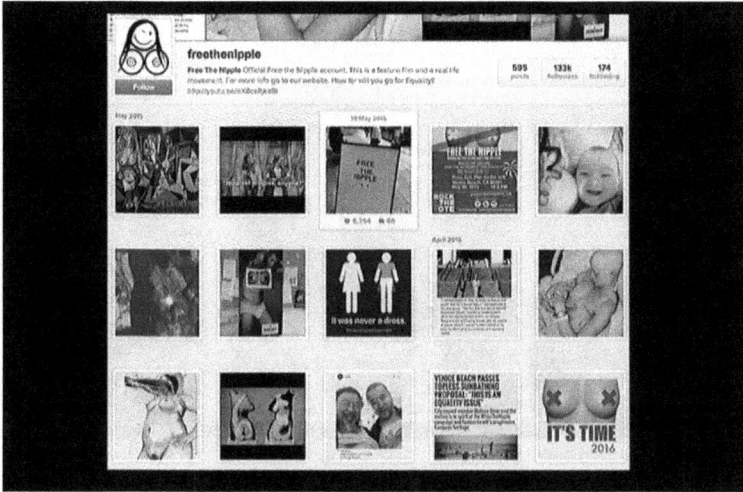

We could talk about hashtags all day. I just wanted to pick one hashtag to talk about today, sort of continuing the theme. I don't know how many of you heard about the "Free the Nipple" hashtag. Has anybody heard of that? A few of you? Okay. This has actually really spiked in the last month. It's become massive. It was a hashtag that was started last year by a filmmaker when she released her film. Just a few weeks ago, there was a young Icelandic girl who took a topless picture of herself, posted it on Instagram, and she got slammed by her friends. She started to use the #FreeTheNipple hashtag. She's a young feminist. It sort of took off and became viral; it just went crazy. Everyone's been posting these topless images of themselves and Instagram can't really do anything about it. It's kind of gotten out of control.

Then Fox News jumped in on the act because Fox is trying to be provocative. This is a screenshot from a TV show last week. This Picasso painting was up for auction. If you notice, they blurred out the breast area to post it on the TV.

Ai Weiwei got in on the #FreeTheNipple hashtag act. He's very active on both Instagram and Twitter. He has his own little Chinese character hashtag, which is not really a word. My son speaks Mandarin so I asked him to translate. He said it sort of means like "hmm" and it implies active listening, but it doesn't really mean anything. Ai Weiwei communicates with hundreds, thousands of fans all the time, using this personal hashtag. He interacts with other people's hashtags and accounts on Instagram. He started another hashtag—#flowersforpeace—which encourages the sharing of lots of rather banal images of flowers, but for a good cause.

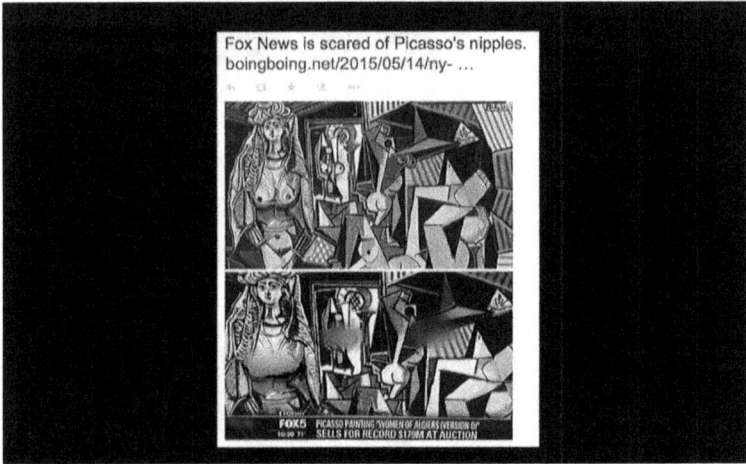

So Beyoncé is the number two influencer on Instagram. Here she is pictured with her husband Jay Z in front of the Mona Lisa. She is deliberately trying to recreate the enigmatic Mona Lisa smile and her slightly tilted facial expression.

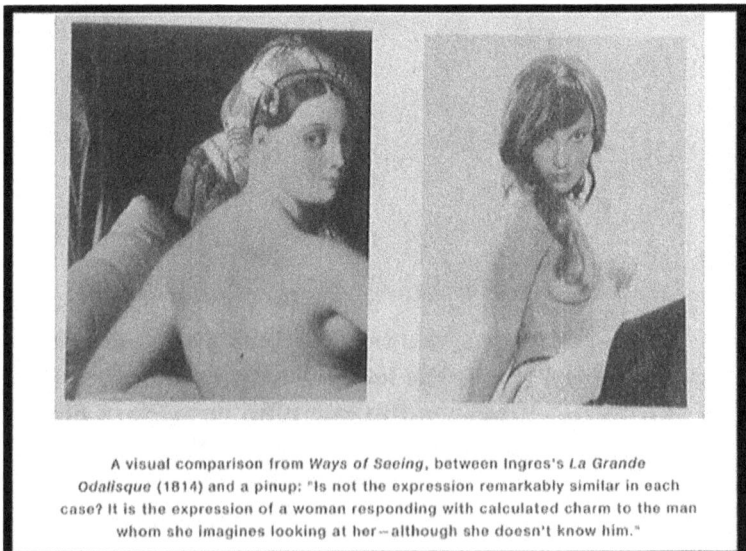

A visual comparison from *Ways of Seeing*, between Ingres's *La Grande Odalisque* (1814) and a pinup: "Is not the expression remarkably similar in each case? It is the expression of a woman responding with calculated charm to the man whom she imagines looking at her—although she doesn't know him."

This is a screen capture from John Berger's book *The Ways of Seeing*, which I think is from 1972. It's quite old, anyhow. In it, he compares the Odalisque image with pinups and the female gaze and the male gaze and so on.

We all know Kim Kardashian is really huge on Instagram and Twitter as well. She's also a lot more clever than people give her credit for. Here she is referencing a Frida Kahlo painting, wearing a dress designed by Balmain, when she announced her engagement to Kanye West. She often recreates artistic images: the Three Beauties, the Gauguin, the Two Sisters.

This image is from a series of images about the artist Wade Guyton. Have you guys heard of this story? He's been making digital images and paintings for the last 15 or 20 years. His work has started to get very valuable and he didn't really like that. So he set up this hashtag called "Burning Bridges," which references a song and also an installation he created. In this case, he's burning bridges with the art world. They were going to sell one of his paintings at Christie's. I think it ended up selling for about $2.5 million. To sort of upset the art world a few days before Christie's had this auction, he printed multiple copies of the image in his studio. He posted them all in Instagram and he gave a few of them away.

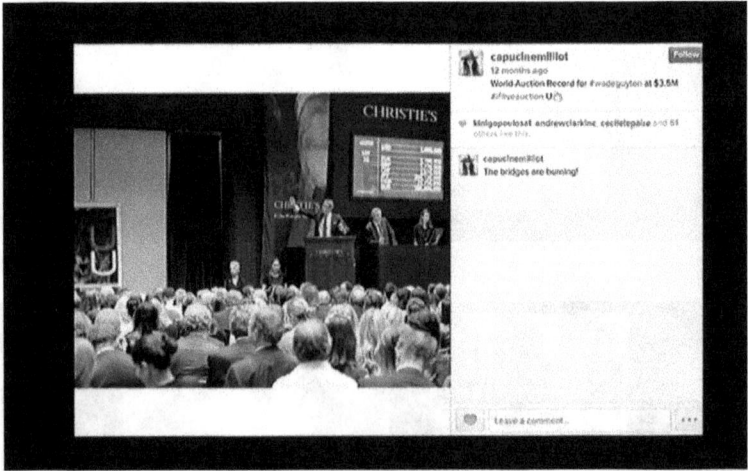

Am I out of time? Sorry, I'll have to go a little bit faster here. Lots of great photographers are using Instagram really thoughtfully, like Matt Black. I have to go really quickly to get to the end here.

This is a former student of mine, who posted on his Facebook page. Within three or four hours, he had 10,000 likes.

This is an image that I particularly like on Instagram. It's a Jehovah's Witness stopping somebody on the street and I really like the way that even the dogs are kind of walking away.

There are a lot of artists, or a lot of photographers, who were looking to Instagram and online photography to share very creative work and to respond to the cultural effect of Instagram through their own work. Elad Lassry, for one, he doesn't even have an Instagram account. Thomas Demand is another one. This work references the kind of images that could be seen on Instagram.

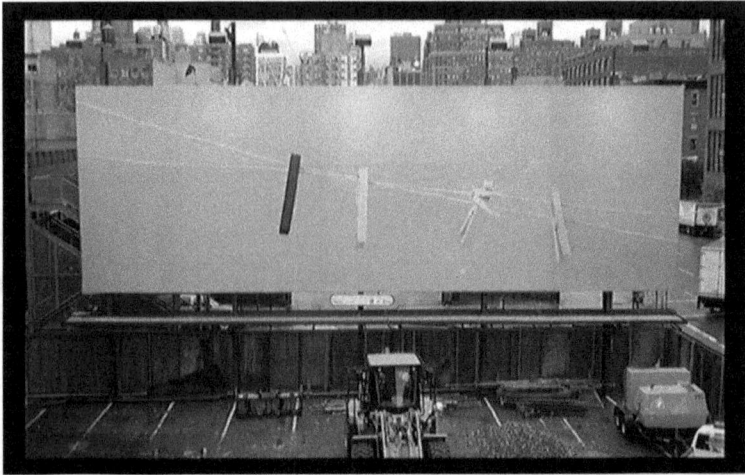

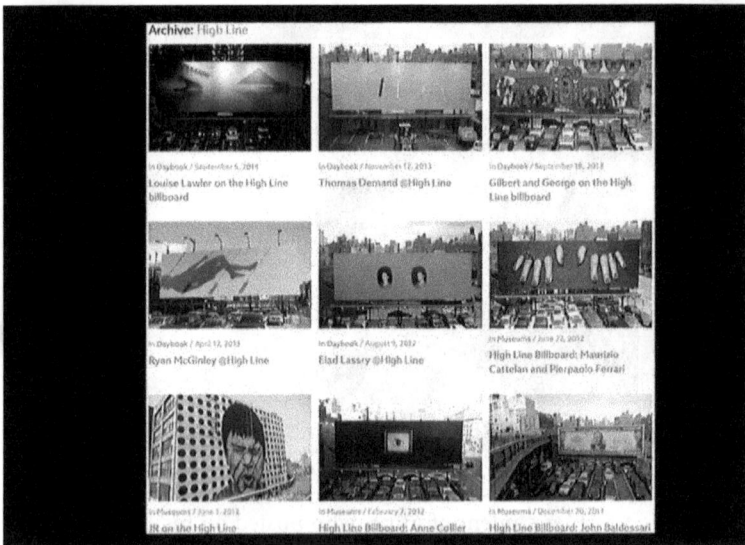

This is actually a really thoughtful image on the High Line in New York. That second image down is Elad Lassry's. He believes that images should be sort of banal and ugly—that ugly is okay and you can challenge the orthodoxy. Everything doesn't have to be beautiful. Everything doesn't have to be "nice."

Who's doing interactive on Instagram? The Toronto Film Festival is creating this Choose Your Adventure interactive experience on Instagram.

Then, this is an interactive video that pulls images from Instagram.

And these are my students, who are teaching the rest of the class how to use Snapchat. That's another conversation. All right, we'll leave it there.

Charles Traub: Thank you. One of the issues is that all of these people whom Elizabeth has shown are using these platforms for remarkable interventions into their experience and to our experience. They are influencers. They're leaders. They're rebels as well. These lessons have to be taught to a younger generation of aspiring image makers to create a deeper and more meaningful conversation through all this imagery. Last, I'd like to introduce my colleague at SVA, Cheryl Heller, who is head of a remarkable program—she created it and has driven it—called Design for Social Innovation. She comes from the design and advertising world originally but is concerned with the social meaning of all these concerns for the betterment of our environment and our people.

Cheryl Heller: Thank you, Charles. Thanks for giving me a chance to think about this. It was fun. How many of you know what social innovation is? Do I not see a hand? Okay. There are some hands in the back.

Social innovation is the development of any new idea or model that strengthens society and the environment. It is by definition always done within an intended outcome. You set out to solve a problem and you create a program to do that. What it means is that the visuals used in social innovation always have an agenda. They're never neutral. In fact, they are a whole lot like the images you would use in advertising and marketing. No one in social innovation will ever admit this and the outcome is always measured by something beyond how many products are sold or how many clicks you get or all of that stuff. It's intended to have an impact on the whole system and be responsible for it.

The way that images have been used has really evolved with social innovation itself. What used to be called social innovation was, for the most part, charity. You would give money to a cause and some big NGO would intervene and take care of the issue. Buy our products and we'll make a donation to a cause or take care of some disease or some program. Now it's hugely based on activism due to technology.

Another shift is from, as I said, this all being done by institutions. People couldn't get involved. There are now many examples of programs that are principally citizen-led without the oversight of institutions. Technology has allowed everyone to have a voice and become an activist in it. Another transition is from storytelling to things that are much more data driven. The kinds of visual literacy that are needed to understand them has shifted.

Cheryl Heller, Chair, MFA Design for Social Innovation, SVA

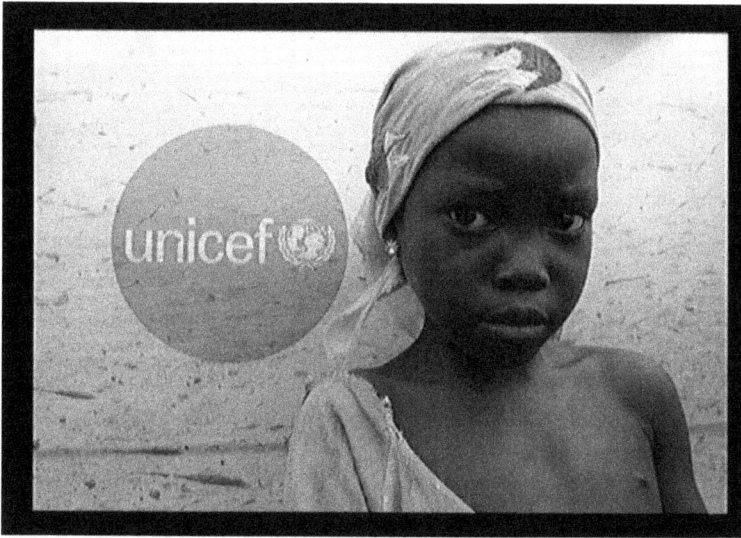

This is an example of sort of the classic big NGO image: an iconic, recognizable communication. This is how you know you are being asked to give money to something. The world of NGOs is highly sophisticated in the way they use images, understanding what triggers charity, and that it is a human fact that people will support more often than the cause itself. Contributors don't give to causes. They give to people. This image represents an entire genre of image making and symbolizes human charity of helping people in need.

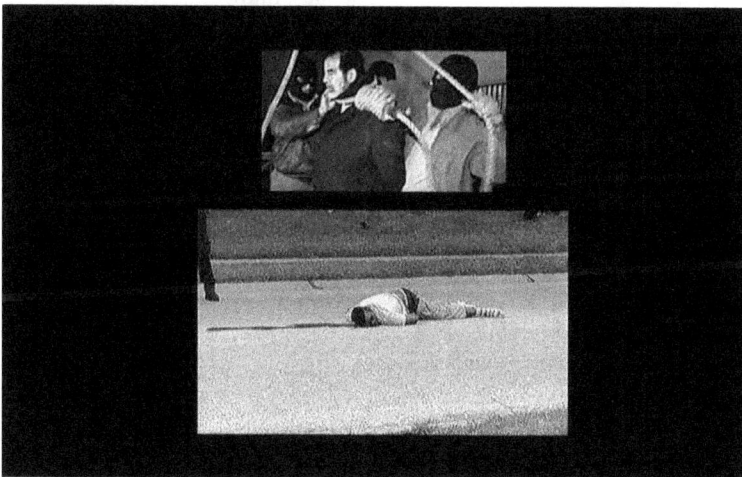

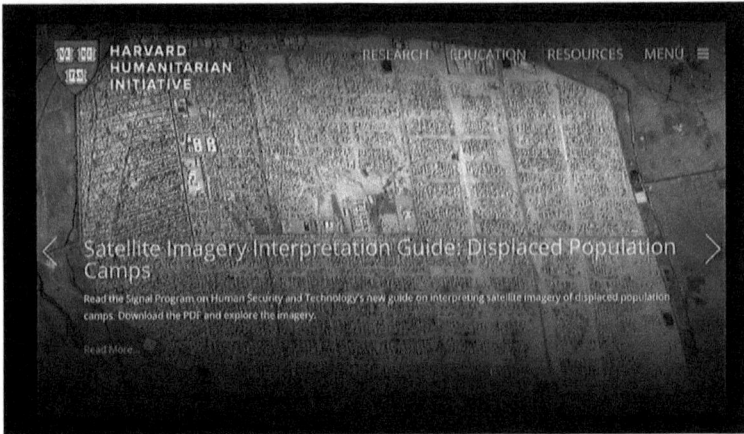

This image was created by a group at Harvard that deals with using surveillance for humanitarian purposes.

This really is a shift due in large part to technology. These are the stories that we see now, that are wildly different. These are not the pretty and hopeful side. They contain a narrative rather than iconic symbols. They are unposed, unfiltered, not carefully crafted or staged. It's the responsibility of activism now to publish images like these and to disseminate them around the world.

This is really interesting. Surveillance has become a big part of social innovation. Surveillance is a form of invasion, as we all know. It's also a source of relief. This image was created by a group at Harvard that deals with using surveillance for humanitarian purposes. What they've been able to do is track the spread of polio; track and alert humanitarian organizations to the location they look for refugees; they're able to offer relief in ways and to find people in need of relief in ways that have never been possible before.

There's a whole category of technology-based imagery used in social innovation now. If anybody knows Ushahidi, the relatively new mapping technology that translates text messages from mobile phones in rural areas to a visual map on the Internet... In this way, stories can be geographically located and communicated directly to massive audiences. Maps are being created by the stories that people tell on the ground. It's being used to address violence during the Kenyan elections, to find victims of earthquakes and disasters, and it's now being used for a wide variety of purposes all over the world.

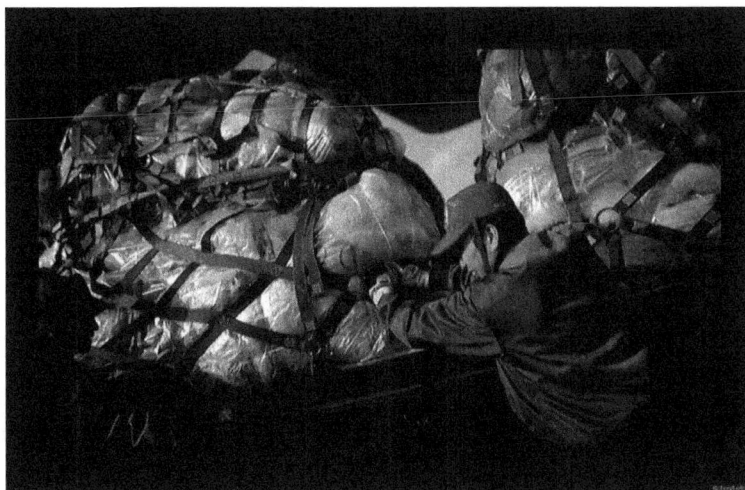

This is interesting. It's a very different kind of picture. An NGO or a relief agency might publish this. It uses evidence of what they're doing in the world. In some cases, it is evidence that programs they engage in are working. It can also be a way of raising funds. In other cases, it can backfire, providing evidence of failed programs or insensitivity, such as organizations that have delivered blankets to people who are living in an overheated tropical environment or building huts that people didn't want to live in. This documentation, providing evidence and insights into the lives of the people in need of aid, is, for better or worse, part of the visual landscape of imagery now that we see in social innovation.

Surveillance is also obviously a source of content in its own right. This is a video installation by Trevor Paglen. He collected more than 4,000 code names from the archive of leaked documents that Snowden published. In this installation, as the viewer stands in front of the piece, those code names crawl up the walls around them.

Then last, an example of the most traditional form of visual activism. This is Salma Ashaybah and a series of photographs of warheads. It's a much more traditional aesthetic but again, part of the landscape of visuals in the broad category of being used to make people aware of issues and hopefully become activists themselves.

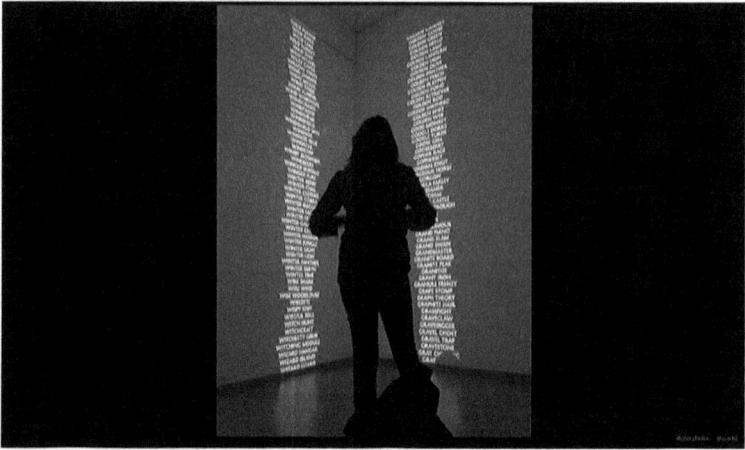

Video installation by Trevor Paglen

©*Salma Ashaybah*

Charles Traub: Cheryl, thank you very much. Thank you all. I just want to sum up a little bit in saying that we live in an age that is so proliferated by imagery. All of you are generating more and more imagery, helping us find it, make it, create it, parse it, store it, archive it. All the things that all of you are concerned with. The ultimate question is: What do we really do with it? How do you get meaning from it? How do we allow individuals—not computer marketers, not branders, not even educators, but ordinary people—to find meaning in their own stories and narratives in those images? To be able to get to them and use them productively, other than just for the obvious entertainment needs and for the obvious commercial needs, but rather for the subjects that we are touching upon only briefly but we hope stimulatingly. Thank you very much.

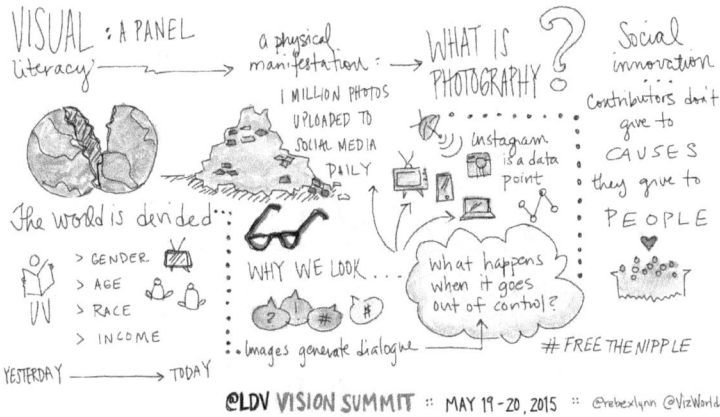

©Rebexlynn

©Vizworld

35

WIKIHOW IS THE WORLD'S MOST POPULAR HOW-TO WEBSITE

Thom Scher, Sr. Director Marketing & Visual Content, wikiHow

GOOD AFTERNOON, EVERYBODY. THANK YOU SO MUCH FOR HAVING ME here. I want to talk about wikiHow and this idea of visual content on our website, which didn't used to have it. This all started about four years ago when we sat down one day and looked at our mission, which is basically to teach everyone, everywhere how to do everything. We realized that's almost impossible to accomplish without images.

The reality is, we looked at our site and there were very few images at all. There were some bad Flickr images that had been loaded in over time. There were a few user-uploaded images on their articles. The site as a whole was very, very dry.

Thom Scher, Sr. Director Marketing & Visual Content, wikiHow

Four years ago, we started an initiative to start producing this visual content ourselves. The idea here is that when you take a step like this, which is basically an abstract from our article on how to make an origami rose, it means nothing. Even if you followed steps 1-11 of this article, which I tried to do before I came in today... And without the images it was a disaster. My rose looked like crap.

At the end of the day, this step means nothing. But when you add an image to it, it becomes instantly understandable.

You don't even need the full text. You understand exactly where to put the tweezers. You understand exactly how that rose is supposed to look. This is a really obvious use case, right? Slap some images on your craft articles and BAM—it's easier to understand them.

We went a level deeper and we realized that on softer topics, adding curated, step-by-step visual content can actually make a huge difference to user engagement. It makes the content digestible.

This is an illustration from, I believe, "How to survive a long distance relationship." This particular illustration style is something that we've realized we can use on a broad range of our relationship articles and really move the needle with it.

In the process of trying to figure out what came next, as of four years ago, we started illustrating relationship and mental health topics. We thought to ourselves, *Why not just add stock photos?*

Well, when you add stock photos, you get things like this. That's really bad. When you're looking at cooking articles, for example, nobody wants Mr. Chef-hat and the menu board. They want the actual process shots. They want to understand how to cut the tuna. How to make the dressing. What that sauce looks like every step of the way.

Then, at the end, what that product looks like. That became this guiding mentality for us as we moved forward with this process. Four years down the road, we learned a couple of things.

It started with the first photo shoot we ever did. I was on set with this photo shoot and we got into the studio and I sat down with Elise, who is a Google marketer who moonlights as the wiki-

Curating visuals
means curating text

How fitness model. We immediately tried to read her the article and have her do it. That didn't work—at all, actually. We started reading her the steps and she didn't understand the steps. She started asking us questions in abstract: *What do you need me to do? Get into the position yourself. Show me.* We found ourselves rewriting articles as we started shooting. With time, this turned into a process where all of our photographers and illustrators began weighing in on the visual content. The article ends up utilizing the images just like the images are part of that text.

The text and the images have to work together. I think frequently publishers look at things like this and they think images are just supplemental. We quickly realized that they can't just be supplemental. They have to actually work together. Four years down the road, we started to think: *How do you actually measure success when you're adding images like this?*

These are from "How to make macaroni and cheese." Version one over here was from the really early days of the wiki visual project. We thought it was so cool. It's not. It's a really bad illustration of blocks of yellow being put in water. Here's the thing, though: it still led to an increase in traffic on the page. We were being rewarded for adding images, regardless of whether or not the images were crappy. We were still being rewarded.

We then go to version two, and replace those with photo. When we replace them with photo, we saw engagement on the page change really dramatically. People stayed on it longer. Their click-through to other articles went up pretty dramatically. We were able to actually track that when the right visual content matches with the right written content in the right categorical bucket, you could move the needle for a food that was actual photos of food.

Then, when we replaced that with video, what we found was not just that we got the traffic boost and the engagement boost, but we also got an authority boost. People told us that the articles were more helpful. They viewed the video as providing an actual, verified element to the article.

I want to take a step aside for a second. Not very many people click through on the videos. That was something that really surprised me. All these publishers go through all this work to create all this video content, and for us the data says not that many people are actually engaging with the video. You still get the helpfulness bump the second you slap the video in the page, but people aren't actually using it.

I started diving deeper to try to find out where they were using it. One of the things that surprised me was that they're on topics for kids. How to hopscotch. How to subtract fractions. These kinds of topics kids actually click through on the video. I would challenge publishers, if there's anybody in the room that is a publisher, to think about the reality here, which is that kids are expecting the video and actually engaging with the video. As publishers, we have an obligation to provide that to them down the road.

Over time, those three metrics—traffic, engagement, and then helpfulness and authority—became the metric that we evaluated the project on. There are obviously costs here, though. Big ones. Obviously there's a production cost in producing huge quantities of visual content, but beyond the production cost we actually took a revenue hit every time I did my job.

I would do my job and put visuals on a page, but everybody's eye is drawn to this really cute illustration of all the various diverse characters loving one another and not the ad at the bottom, which has effectively become invisible. We had to accept what was a cost as well as a revenue hit, and believe, which four years later we've seen, that long-term that was worthwhile.

We believe that the benefit to user-engagement and actual storytelling experience is worthwhile. It's worth those costs. It was so worth the cost, we did it at scale. Over four years, we ended up doing 1.4 million custom visuals. Those are live in over 150,000 of our articles. That's a lot. It's taken a lot of manpower to make it happen. We're constantly trying to move the needle even further on that visual content. That's where those costs actually came in.

I want to close by talking a little bit about this concept of building a visual brand. For us, people think wikiHow and they think, *Okay, so my toilet is overflowing. I'm going to quickly search something on my phone and try to get it to stop overflowing.* When we started doing the visual content project, we began to be able to take our content and market it as evergreen content. We were able to take it and put it on social media, where people were engaging with it and browsing the content. They ended up coming back to the brand and engaging with the brand in a much more abstract way, and not just specific to a use case.

That's been a huge win for us. Separately, it all boils down to the same underlying premise, which is: the visuals have to mean something. It's very easy to believe that they're just splashes of color on a page as a publisher. They mean a lot more to individuals. That's why we've seen visuals rise to what they are on social and rise to what they are in terms of user engagement. For us, that means making the visuals instructive. It can mean a lot of things to a lot of different publishers.

Thank you.

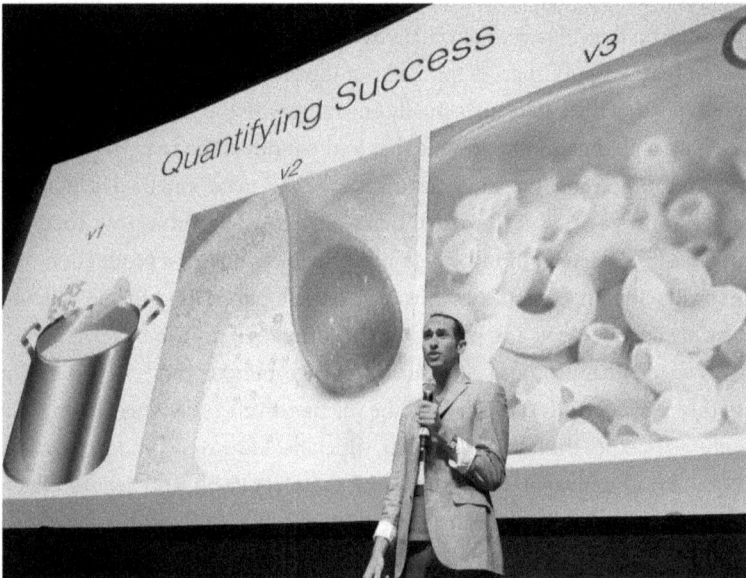

Thom Scher, Sr. Director Marketing & Visual Content, wikiHow

36

VISUALIZING USER-GENERATED CONTENT TO BETTER UNDERSTAND HUMANITY

Dr. Lev Manovich, Professor, The Graduate Center, City University of New York

EVAN SENT ALL US PRESENTERS AN E-MAIL WHERE HE SAID, "TRY NOT to use too much text. Use as many images as possible." And I think I went and took it too literally because I actually don't have a single line of text on my slides; I even took the slide out of my name and title. But I try to be a good boy because he is just such an amazing leader.

We all hear this expression "thinking outside of the box." For me, the box is this one [holds up smartphone], which enables amazing experiences and changes the world, but it's also very limiting as a mechanism for visual experiences.

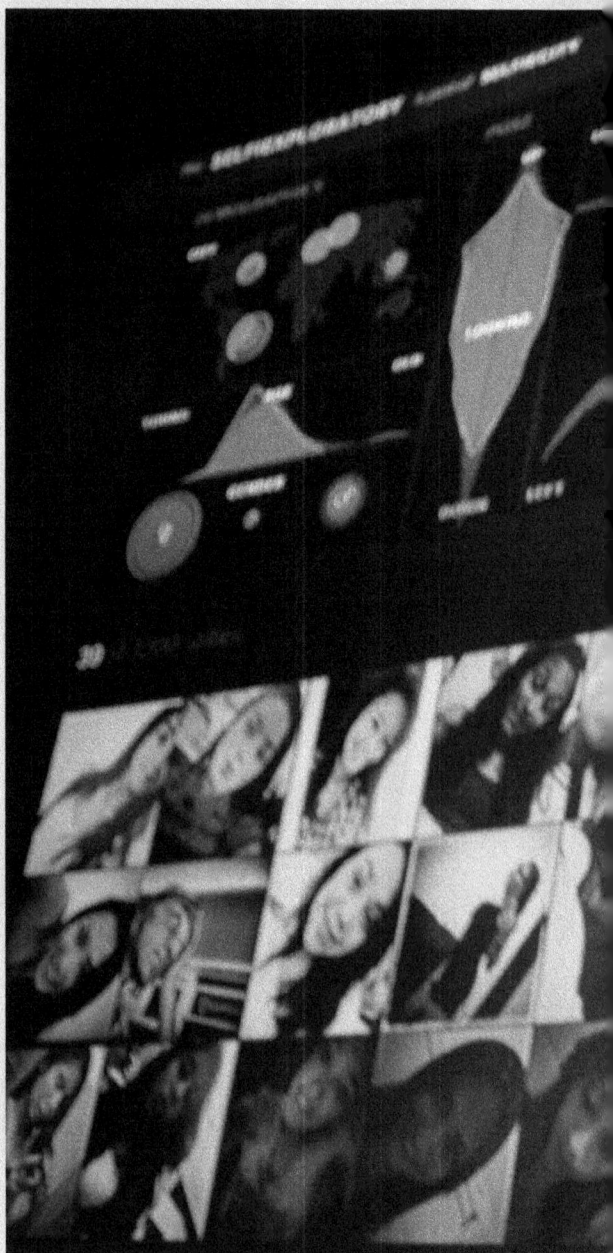

Dr. Lev Manovich,
Professor, The
Graduate Center,
City University
of New York

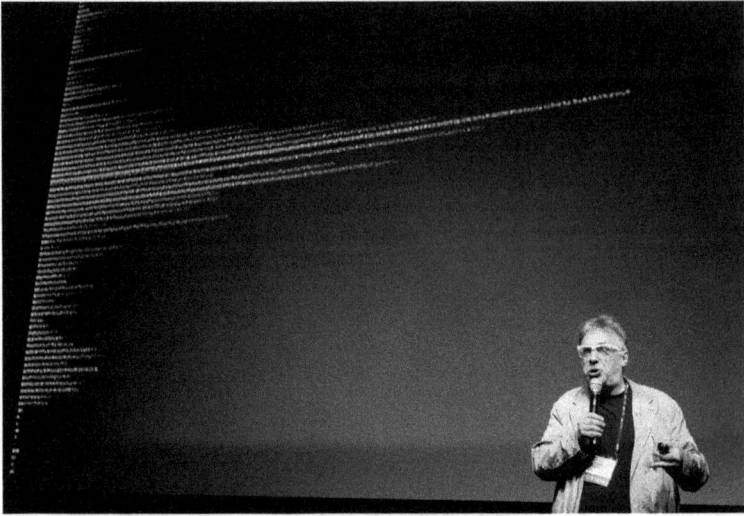

Dr. Lev Manovich, Professor, The Graduate Center, City University of New York

I was a professor of visual arts at the University of California, San Diego, and we opened a new research center called California Institute for Tele-communication and Information (Calit2) which brought together some leading researchers building new platforms for information visualization, such as this one, which consists of dozens of flat screens. When I saw it in 2005, exactly 10 years ago, I had this "aha" moment and I imagined: *What if I could take all digitized Renaissance paintings from, let's say, 1500 to 1600 in Italy, and put them on this wall, and use computer vision to sort them by color, content, composition, and actually have a better understanding of what the Renaissance was?* This was even before social media exploded.

For the last seven or eight years, me and my collaborators worked on such projects in our research lab. Now it operates between the University of California-San Diego and here at the Graduate Center at CUNY, where I am teaching since 2013. Meanwhile, I also made the transition from being professor of art for 20 years. Six years ago I started learning Excel, then R, and now I'm professor of computer science teaching PhD-level computer science students data analysis and art. Just one step away from my students. We developed further this idea of looking at large image collections and using visualization techniques, and also using computer vision to understand various cultural datasets. I'll show you a selection of five projects. Three of them will be using contemporary user-generated content, such as Instagram, but before that, I will show you a couple of projects where we look at collections of digitized historical imagery.

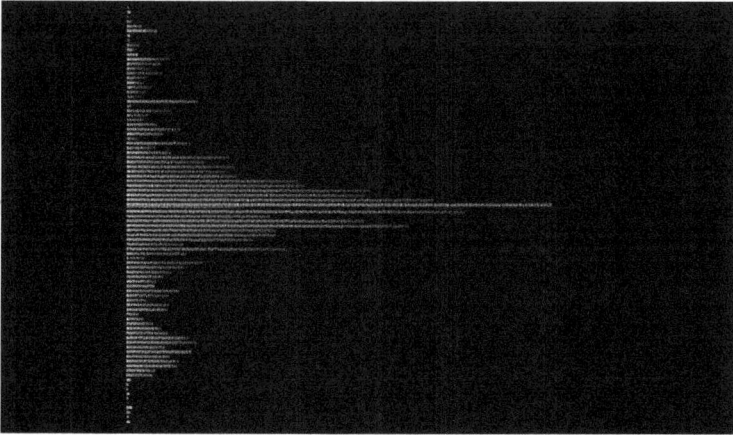

Two years ago, we were invited by the Museum of [Modern] Art in New York to look at their whole photo collection. Of course, they have everybody who's anybody from the 1840s until today, but nobody has ever visualized the collection as a whole. So here we applied the most simple 200-year-old technique of bar graph.

It's a bar graph which shows all the images of the photo collections, about 20,000, but it's a bar graph which is made from all individual photographs in the collection. When you zoom in, you can start to look at the whole collection in a single image. You realize that it's a very particular representation of the history of photography.

It's all about the great people, but not only about these great people. It's about the people which MoMA deems important. For example, from 1920s to 1930s, you have lots of modernist abstract photography and you don't have lots of photojournalism. When you look at this photo collection as a whole, you can also see that MoMA is very strong in 1920s and 1930s. That's why you have this big spike. There was a little bit of a spike in 1970, but other decades are not represented very strongly. Now I will show you another project we have done in the lab: analysis and visualization of one million manga pages. I said: "I'm interested in visual style of manga." Manga is one of the most popular cultural forms today and it includes Japanese, Chinese, French, etc. comics but it started in Japan.. The pages are sorted by one dimension of visual style. On the one end of this dimension are the pages where you don't see lots of detail, not lots of texture, image are all black and white images.

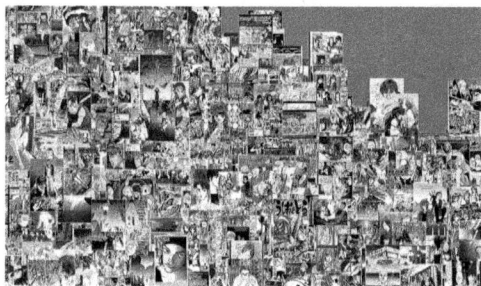

On the opposite end, we have opposite style—lots of texture, 3D, lots of labor. Using the very basic techniques of computer vision, we measured some visual features such as entropy, and standard deviation and so on. Then, using a little visualization tool which I wrote (I'm very proud of it, because it turns out there was no actual tool to visualize lots of images together), we simply sorted all those images. That was done in 2009. It took about two days to render, and now it would take a few hours.

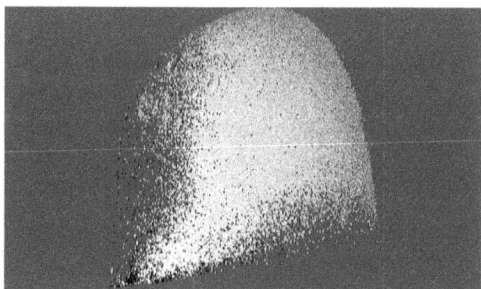

Here's one million manga pages, which correspond to the most popular manga titles around the world, organized by two visual features. As you already saw, the feature which organizes images on the vertical axis is entropy. We get images of very high entropy, lots of detail on the top of the cloud, while images with very low entropy and opposite at the bottom. It's like discovering a new continent. To use a metaphor, it's like people coming to America 500 years ago and discovering this new continent and mapping it out. What is the map of this continent called manga? What is the most popular manga being created? What manga people did not create? If I'm a young artist and I want to make something original, I should operate in the lower right corner, or maybe the upper left corner, because for whatever historical, cultural, evolution reasons, there has been much manga created in this style. It's kind of diagnostic; it's like an x-ray of a cultural field.

Now I'll show you briefly three projects we've done in the last few years where we applied these concepts to look at user-generated content—specifically using Instagram. For our first project done in 2013 (it was called "Phototrails"), we simply started in a very broad way. We downloaded 2.3 million Instagram images from 13 global cities, and then using very simple visual techniques and this idea of mapping big image samples together, we started to compare visual signatures of different cities.

Dr. Lev Manovich, Professor, The Graduate Center, City University of New York

Here, for example, you have 50,000 images of San Francisco compared to 50,000 images from Tokyo organized by image brightness and average hue. And we can also organize images by time.

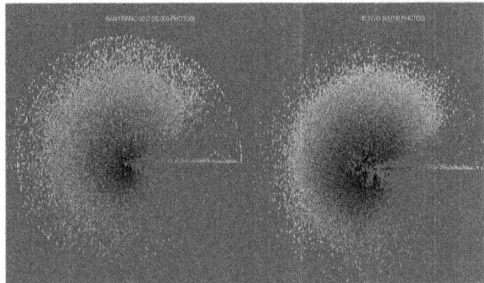

Here, for example, you have one week in Tel Aviv. The images are organized by hue and by time and day of week, so you get these little slices which correspond to each day and you see these visual routines.

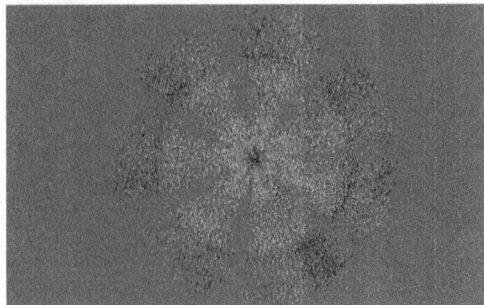

Even when people are taking very different images for very different purposes, almost every day looks the same. But you have a couple of days at the bottom where something exceptional is taking place. In fact, these are the two most important national celebrations in Israel, and you see something happened. But you also see reflections of some exceptional events are just like a weak signal, let's say, in the sea of, I wouldn't call it noise, but in the sea of everyday visual routines.

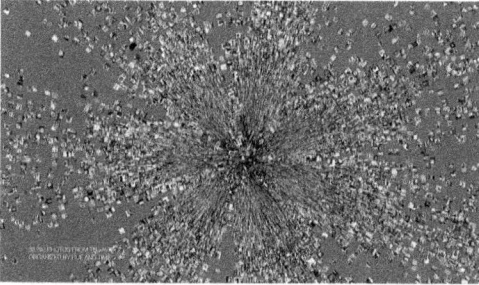

This is a close-up of the same image so you can see that it's all made from individual photographs.

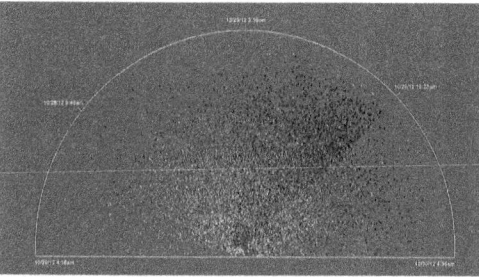

This is a little bit close to home. This is 24 hours on Instagram of publicly shared images in Brooklyn, which we also downloaded during an event which took place a couple of years ago, Hurricane Sandy. So the time goes clockwise—the images were organized by time and simply by average hue. And you know what happened at 10:23pm on October 29 where Con Edison Station went out of service, and a big part of NYC suddenly went dark. You see how much these kinds of dramatic events are immediately reflected both in the quantity and the different kinds of images being shared.

In the fall of 2013, we said: "Now let's make a new project." As opposed to comparing images from different cities which is a bit like comparing apples to oranges, we said: "Let's compare images of a particular kind: selfies." We started to work this a few months before the selfie became very famous. e decided to download thousands of selfies from around the world and use various visual techniques and a bit of computer vision to compare them.

You will see a video in a second, but the key part of the project is this interactive, web-based interface, which we'll call "Selfexploratory," where you can navigate a database of 3200 selfies from five global cities using various metadata.

Interfaces is in fact the key purpose of our research. We don't want to reduce the variety or the richness of the visual world to bar charts and pie charts, right? We want to be able to combine media browsing with graphs because every image is not just a data point. An image is a whole visual universe. I think this problem of how we can actually do it, for example in the mobile device, hasn't really been addressed yet.

So this is our latest project, On Broadway. It's a computer-driven installation which is installed for 13 months in New York Public Library on the 46-inch touch screen. Here we're focusing on

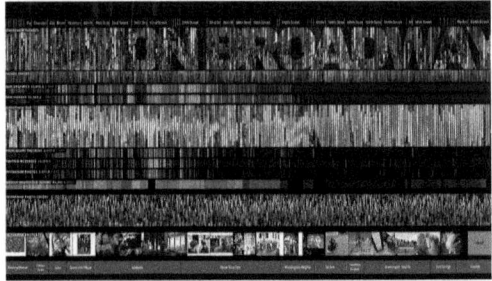

visualizing all of New York by using a street, which is Broadway, and combining about 40 million data points and images. The task is how to combine multiple layers of data and images into one interface. So we have everything from Google Street façades, façade images colors, information for 22 million taxi rides along Broadway last year, median income, Google Street View again, Instagram, and neighborhood names, etc. The idea is to create a kind of interface where you can navigate this, building by building, or you can zoom out using familiar touch gestures and go through the whole thing.

The idea is to downplay statistics and focus on images. Also we wanted to explore the idea of: *Can we create interface for a city which doesn't use a familiar map?* So we don't have a single map on the interface.

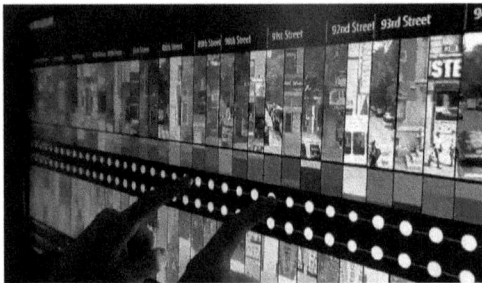

This is just a screen capture with video and you can interact yourself if you go to New York Public Library main building on 42nd street and Fifth Avenue. It shows navigation, so at any point you can also click the button and get selection of Instagram images at a particular point. You can also zoom out and see the whole city or you can zoom in and navigate the city one building at a time. It is a zoomable interface, which is our attempt to think about how we can put together urban data, potentially hundreds of thousands of layers of data, and present it to users in a visual interface.

37

KEY METRICS: A STRATEGY FOR BUILDING A SUSTAINABLE DIGITAL VIDEO BUSINESS

Rebecca Paoletti, CEO, CakeWorks

THERE ARE THREE REASONS THAT I'M NOT SHOWING A VIDEO RIGHT NOW, even though this session is all about video, and I'm talking about video, and there are a lot of reasons why video would of course be great. Number 1: If I put a video up here, you wouldn't listen. I show videos all the time in presentations. No one listens. Number 2: I was going to livecast my message, but then I wouldn't be able to actually be here with you, so that video option was out. Number 3: I am actually terrified of video fail. It's true. I can't tell you how many times I've been standing in front of a huge room with so many people and it's time to play the video and it hasn't worked. Epic fail. I hate that! So, for the purposes of this chat, we're all just going to pretend that a live video is happening…

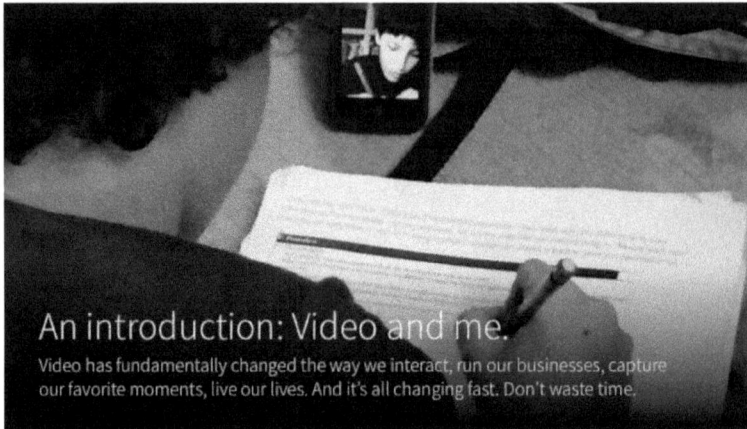

An introduction: Video and me.

Video has fundamentally changed the way we interact, run our businesses, capture our favorite moments, live our lives. And it's all changing fast. Don't waste time.

©Charles Richardson

Why I founded CakeWorks after Yahoo and several other "big" corporate jobs that I had focused on trying to make video work better, was because I realized that a lot of money and a lot of energy was being spent trying to solve a pretty easy problem, which is getting more people to watch more video and make more money.

We'll start very personally: video and me. This is my 11-year-old son. He is home from school doing homework, and on his iPhone is his best friend, Nate, who's home at his place after school on FaceTime. And they're doing their homework together. They do this pretty often. All their friends are constantly on messaging, chat, and FaceTime. It's like roving 11-year-olds on video and it's constant. Video is really changing what we're doing and it's changing the way we're interacting and it's changing the way we're communicating. It's changing the way businesses are being built and run and how they're making money. And it's changing really fast.

As you've obviously seen over the past two days, there is a lot of new technology and a lot of new models, and it's time to really embrace them and understand them. That is really what our company, CakeWorks, is doing. Hopefully we'll give you some simple takeaways to get out there and really solve this.

Arguably, Steve Jobs will be known as one of the godfathers of digital video, especially smaller screen digital video. I wanted to bring this up because, truly, if you can make things simple, you can move mountains. I think that's a really important takeaway from the last two days. There's been a lot thrown at you, a lot of machines, a lot of vision, a lot of extraordinary technology, but if you can keep it simple, you can really move.

Video is cake. This is our company's hashtag. We say it all the time. Really, it's about finding the gaps, finding the partners, meeting people out there, meeting technology partners, distribution partners, content production partners—any of the places that you can look to solve the gaps that you have in your own business and your own needs and funding. It's to really embrace relationships and embrace people and get out there. And then it doesn't have to be so confusing and hard. You've got other people joining you along the way.

Number one: solving video problems. Content matters, and often it's expensive. Especially if you're coming from a TV background, video content can be very expensive to produce. You want to produce the right content for the right format, whether it's on Facebook or YouTube or whatever. You want to really know who your audience is and how they're sharing it. Clearly you would be sad if you'd spent an extraordinary amount of money on the video being watched here only to have it be in this type of environment while it's being shared.

Audiences matter. Who is your audience? Do they like video? Do they like video on their phone? Do they like it on their gaming console? How much do they watch it? When do they watch it? That quote from earlier today, that "platform is the new black…"—I could not agree more with that. Clearly it really matters where people are when they are watching.

Monetization. We all have to make money at this somehow or we can't fund the great content experiences that you've all seen. Is your business model subscription-based? Is it based on ads? Is it some type of payment, tipping, all these other things? Who is doing it? What's the technology that you need to make it work? What are your competitors doing? Are they undercutting you on price?

Metrics. We at CakeWorks are obsessed with metrics. I have a whole other presentation called "Know Your Numbers" that is a staple for us. Every startup from the smallest to the largest companies, the Viacoms, the Googles, the Facebooks of the world—you have to know your numbers. You have to know the size of your people, your teams, your partners, how much money you're making, how many views you're getting, how many ads are being served against those views. You have to really know the things that are putting together your whole business—the engagement, the success, the day-to-day growth of your whole initiative. You really want to know your numbers.

Rebecca Paoletti,
CEO, CakeWorks

Building a Sustainab
Video Business Doesn
Be Impossibly Compl
or Outrageously Exper

Rebecca Paoletti – Co-Founder & CEO, Cak
LDV Vision Summit
May 20, 2015

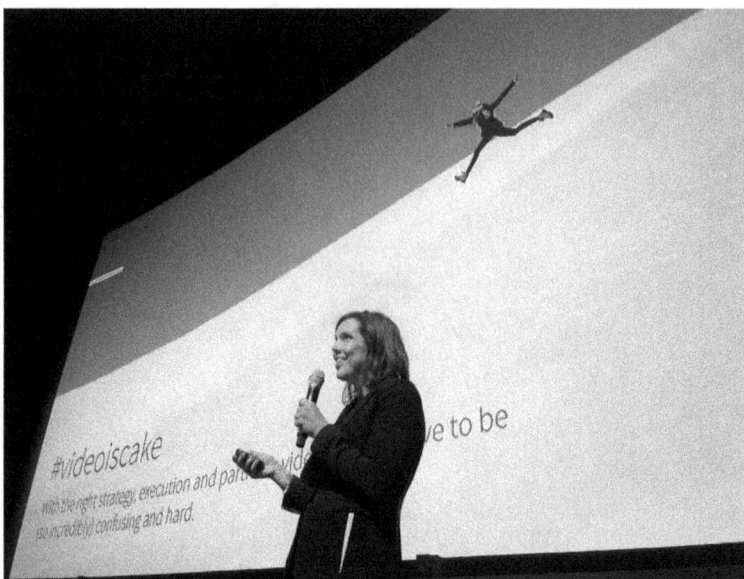

Rebecca Paoletti, CEO, CakeWorks

These are a lot of numbers. My favorite one is about Instagram just because we've spent so much time on Instagram today. The very first day that Instagram had video, the day it launched the video product, five million people uploaded a video on Instagram. The very first day. Clearly there was an opportunity, a format, a tool that worked really well.

We talked about YouTube being the Big Gorilla, so I wanted to end with this quote by Susan Wojcicki (YouTube's incredible CEO): "Work smart. Get things done. No nonsense. Move fast." That's it. Just do it.

Thank you so much.

38

WILL GLOBAL MONETIZATION AND DISTRIBUTION OF CONTENT BE DISRUPTED OR EMPOWERED?

Jamie Wilkinson, CEO & Co-Founder, VHX

HI, I'M JAMIE WILKINSON. I'M THE CO-FOUNDER AND CHIEF EXECUTIVE of VHX. We are a platform that allows people to sell video online. It's primarily film and television, but all kinds of other things. What we always say is: "If you used to bundle it up on DVD, now you can sell it online directly to your fans."

My co-founder and I have a long history with Internet video. Casey Pugh, my business partner, was one of the early engineers at Vimeo and helped create the Vimeo player. Before VHX, I created a website called Know Your

Meme, which is kind of like Wikipedia for viral videos. We've been soaking in this problem for a long time.

As anybody in this space knows, if you have videos that you want to give away for free, there's YouTube, there's plenty of websites that make this super, super easy and keep the barriers to entry very low. But if you actually want to sell your video and have that monetization problem, there really were no solutions for you.

We built a platform that sometimes we describe as "Shopify meets Netflix." The idea is that you can set up your own video store where you can sell videos, you determine the pricing, you can upload as much or as little as you want. You can sell it transactionally, you can sell a rental, and then you can also sell monthly subscriptions.

The idea is this operates from your own destination, too; that you're the one who has the relationship with the customer. We don't want to be the next Netflix. We want to be the infrastructure that powers the next Netflix. The next Netflix, as we always pitch to our extremely happy investors, is not going to necessarily be something that has thousands and thousands of titles and invests hundreds of millions of dollars in shows. It's going to be things that are built around individual people. It's going to be things that are built around individual niches, interests. That's the core of the pitch we talked about. They say that, broadly speaking, technology helps to take things that used to be expensive and difficult to do and makes them ubiquitous and available to anyone.

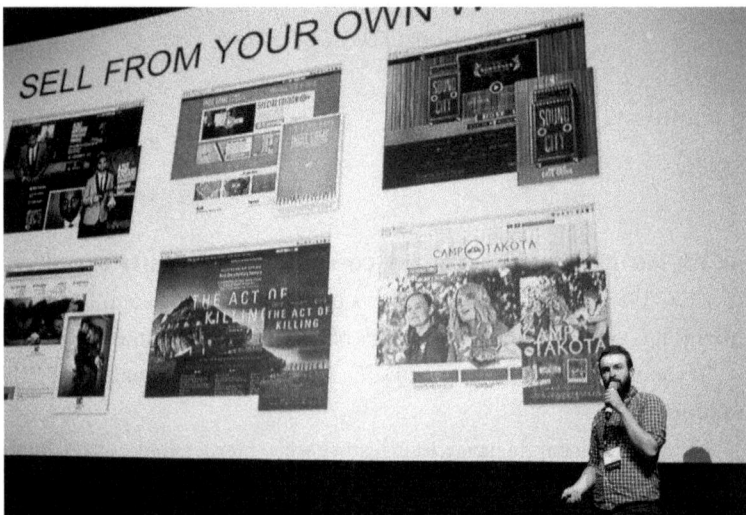

That's exactly what we're trying to do. If you're trying to sell a movie today, there's an ungodly number of middle men that you need to go through. You lose total control over the process. It's a multi-year experience. The number of filmmakers that I've met who are completely unhappy with the way that distribution works, and also a number of TV writers, would really shock you.

Jamie Wilkinson, CEO & Co-Founder, VHX

These trends are very much real and this really drives all of the innovation that you see across pretty much any sector, which is to say that there's more content being created than ever. It's more diverse than ever. People are finding it through each other rather than through advertising and through these sort of traditional means and that physical goods are going the way of the dinosaur and you just see this happening year after year.

This is the trend that we wanted to capitalize on. What we saw as the opportunity was instead of trying to build one mega store like an iTunes or an Amazon, we wanted to put the tools in the hands of the creators, the people who know the audiences the best, who know the content the best, and allow them to sell it directly to their fans. That can be 1,000 fans or that can be, in some cases, millions of fans.

We got our start working with Aziz Ansari to help him sell his comedy special. Since then, we've had the pleasure of working with a pretty wide number of other sellers. We have tens of thousands of people using the platform now. Aziz up in the corner... A wonderful video game documentary called *Indie Game: The Movie* by two first-time Canadian filmmakers was our second release and sold hundreds of thousands of dollars' worth of copies.

We've worked with Dave Grohl, we've worked with Shane Carruth, we've worked with *The Act of Killing*, which was nominated for an Oscar. Then we've also worked with famous YouTubers, like Grace Helbig, who are working on feature films, who are working on web series. Not pictured are the 5,000 other titles that are powered by VHX, including yoga videos, hunting shows, programming classes, major motion pictures, TV series, and pretty much any other kind of moving image.

The gist of our platform is providing people with super simple checkout tools, a way to pay really quickly and easily, and then a really high quality video watching experience—something that's comparable to Netflix, which is really what we consider the gold standard of a consumer experience these days.

This is a non-trivial problem. Like I was saying, every technology that comes along is making things that used to be incredibly expensive and difficult—it's taken Netflix almost 20 years to reach where they are today—and making them available to everybody. That's exactly what we're trying to do by taking our technology expertise and lending it to the content creators who know things the best, allowing them to dictate how it works, pricing it themselves, and still have the same high quality experience that they would get by working through the traditional middle men.

And it's working. We're about to celebrate our third anniversary. There's almost $6 million that have been transacted through the platform. Like I said, more than 5,000 titles. This is all available publicly. It's vhx.com/stats. There's more than a million people who own something sold through our platform and we're really excited to see what comes next.

39

FIRESIDE CHAT WITH JOANNE WILSON OF GOTHAM GAL VENTURES

Joanne Wilson, Gotham Gal Ventures
Evan Nisselson, LDV Capital

Evan Nisselson: Hello, Joanne. Thank you very much for joining us.

Joanne Wilson: Sure.

Evan Nisselson: So let's kick it off with, in your own words, two minutes about what excites you and what is the focus for your investing these days?

Joanne Wilson: There is not one thing that excites me out there and the reality is that technology is changing everything. This shared economy is certainly interesting. There's a bunch of industries that are still working on stickies, pens, and pencils that certainly still need to be disrupted. When you see one idea, you end up seeing five of the same ones within two weeks, because that's just how the world works.

Evan Nisselson: They are all just happening at the same time.

Joanne Wilson: I'm really looking for super smart entrepreneurs who are filling voids in the marketplace.

Evan Nisselson: Let's get right into that. Super smart.

Joanne Wilson: That's a big word there.

Evan Nisselson: Exactly. I think you know I have spent 18 years on the other side and now investing for the last couple of years. When you say "super smart," what are some of the personality traits that you love which relate to "super smart"? I also want to know a couple that you hate because the negative stuff also helps entrepreneurs hopefully grow.

Joanne Wilson: I am pretty honest and transparent, even to people who I turn down and they ask why. I am really specific about why it isn't for me. "Super smart" is such a broad term, but it means there's a fire in their belly and they're someone who I can work with. I believe in them. I believe they're smart, and they're scrappy, and there's a carrot dangling in front of their nose. They're going to figure it out no matter what comes their way.

Evan Nisselson: Do you have a couple examples? I'm sure you've seen so many different entrepreneurs. Give me a couple of examples. What I love is, there's often a tipping point after you see a lot of entrepreneurs and all of a sudden something clicks. It might now be the right time to invest or get involved. Maybe it's some way that somebody reached out to you or a hustle that was aggressive and direct enough but not over the top. Do you have any stories that you can tell that highlight that hustle passion?

Joanne Wilson: I'm invested in about 85 companies. I have a couple in the pipeline. I, believe it or not, am involved in all of them at some level and I am super efficient. I would say probably 35% of those, I gave their first dollar to. I think what's interesting is watching entrepreneurs who struggled to raise money at the beginning, figured out their product, and then started to get real traction. Built their business, like I always thought they would, and then finding that they're oversold on their next round. The business just keeps accelerating. Also, to hear them pitch. To watch someone, particularly women, who come in and under-promise, over-deliver, are crossing their t's, dotting their i's. And then as their businesses take off, their attitude changes in regards to the way that they pitch their business. That's really, really exciting!

Evan Nisselson: When you say that you're really involved in the companies… Tell us a little bit about the aspects where you think you add the most value, in addition to money and network. I'm sure there are many different things, but one or two specifically that is the go-to, where people say: "I've gotta talk to Joanne."

Joanne Wilson: Being an entrepreneur is a lonely job. If you don't have a co-founder or even if you do have a co-founder, many times it's really not an equal co-founder, which you brought in later to be a co-founder but they're not a 50/50 split. I think that for many, it's really just a sounding board or even a cheerleader. Today, someone emailed me and the business is finally taking off. I always knew it would take off but it's very different when you're in those trenches. I think it's just great to be able to say, "I gotta tell Joanne." And for me to say back, "That is awesome. You're doing such a great job. How can I help you get to the next point? Let's get together. Let's get on the phone and talk about the high notes, the low notes, what's the road map? Maybe I can help you think about the road map in a different way." I think most of it is really conversational.

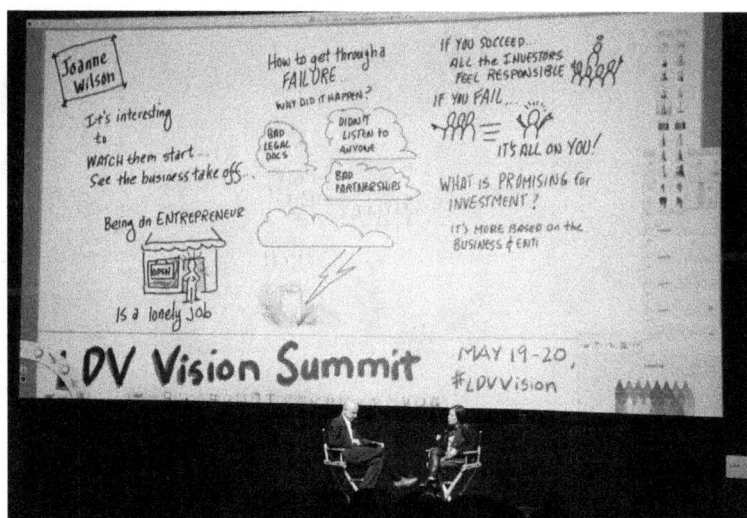

Joanne Wilson,
Gotham Gal Ventures

Evan Nisselson: That makes a lot of sense as I play a similar role, but I am sure you leverage many of the different aspects of your background. Does one come up more? Is it marketing? Is it distribution? Is it this? Is it that?

Joanne Wilson: It's everything.

Evan Nisselson: It's everything. There's not one in your mind that pops?

Joanne Wilson: No. No. I am still learning something new every day in every single business. If you look at the investments that I have made, there's really not a vertical that I'm partial to, except for women. They all have the same problems at the very beginning. They're growing a foundation. They don't know who to hire. What makes sense. One of the entrepreneurs I'm invested in, who is really starting to majorly pick up steam, she equated the whole thing to whack-a-mole. Every time something goes well, something pops up somewhere else.

Evan Nisselson: That's true.

Joanne Wilson: You thought you hired the right person and you've spent three months finding the perfect CEO. Sometimes after three weeks, you're just like, "Oh my god, this person is awful." Bad culture fit. Out the door. Start again. I think a lot of those things happen at the beginning of the businesses. Certainly, as they become more mature and they move on and they get funding from institutions and VCs, I feel great that they're in the right hands but I still stick my fingers in there.

Evan Nisselson: Yesterday we had a very interesting, very genuine, and transparent conversation with Lane Becker, who had a challenging situation with Get Satisfaction recently. Obviously not every company is a success. The majority of them aren't. When you see it's not going to work or they're asking for more money and you've decided not to invest, how do you help them through that phase in a way that it's constructive to them? How do you usually interact in those difficult times?

Joanne Wilson: I have been super lucky. I've only had three companies fail.

Evan Nisselson: Okay. Tell us about those.

Joanne Wilson: One of them was a disaster but it was one of the best things that ever happened to me because what I learned through that experience was tremendous.

Evan Nisselson: What was that?

Joanne Wilson: It really was someone that didn't listen to anyone and thought that they knew what they were doing. They had a bunch of friends who were very successful. Thought they could be part of the whole bro team and he went through a shitload of money incredibly quickly. Not listening to anything that anyone said in a board meeting and went loosey-goosey every single month to the point where it was, "Oh my god, I can't make payroll on Monday." It was a really interesting ride and there were a lot of lessons learned. A lot of red flags I saw at the beginning that I chose to ignore.

Evan Nisselson: Interesting. What were the red flags? Tell us a little bit about that side of the story so people know. I'm sure that 99% of the people in the audience will have those challenges, whether or not it kills their company—hopefully not—or it helps improve their company. I believe talking more about that reality really helps.

Joanne Wilson: Totally agree. Bad documents. Really bad documents.

Evan Nisselson: Not really keeping everything organized?

Joanne Wilson: Not organized. Terrible legal documents.

Evan Nisselson: Was that the lawyer or the entrepreneur's management, or both?

Joanne Wilson: At the end of the day, it's the entrepreneur.

Evan Nisselson: Exactly.

Joanne Wilson: At the end of the day, it's all on you.

Evan Nisselson: Okay.

Joanne Wilson: Bad legal documents. Bad partnerships that went awry that were poorly negotiated on the way out. There were so many things, but I loved the business. Big mistake.

Evan Nisselson: So did you have blinders in the beginning when you saw red flags? You loved the business and you wanted it to work, but entrepreneurs matter much more than business?

Joanne Wilson: I came in on the tail end and ended up running the whole deal with major investors—the people you want in your deal, which actually is very interesting.

Evan Nisselson: There was probably another flag there.

Joanne Wilson: Actually, now there is.

Evan Nisselson: Exactly.

Joanne Wilson: So that was interesting. I had another one, same thing. Didn't ask for help. I do have someone who went out to raise a bridge of their own. She didn't accomplish a lot of milestones and I think a lot of entrepreneurs do this. They start listening to everyone around them. One day goes by. Two days go by. Two weeks go by and they are literally going down so many paths at once because they want to make everyone happy that's involved in the businesses. I always say to entrepreneurs: if you succeed, all the investors think they are the smartest people in the room; if you fail, all on you.

Evan Nisselson: True

Joanne Wilson: The entrepreneur got into this place where she needed to raise more money and really didn't get much done in 12-18 months. I think she was really scared of failure. She went back to raise more money and I said, "I don't know if I'm going to participate in this one." I have actually participated, basically every single deal that I've been in, except the ones that failed of course. She's been hustling and she basically realized exactly what she did wrong. I was in a meeting with her where another investor is saying "blah, blah, blah, blah" and she basically is saying, "No, that is not what we're going to be doing. This is what we're focused on. This is the vision."

Evan Nisselson: Interesting.

Joanne Wilson: I was like, "Huh!"

Evan Nisselson: So did you get more positive?

Joanne Wilson: I went back and said to her, "You know what? You have been super scrappy and it's great to see. I know that I've been holding back and you keep asking and asking because other people want to know, too. I'll come back in."

Evan Nisselson: Let's talk a little bit about the blog. How do you choose stories? Since this is a visual technology summit... We just had a great panel about visual literacy, images and marketing, images in video everywhere. It's how we communicate. You tell stories about a mixture of themes on your blog—personal, business, and inspirations. How do you choose and why don't you add more pictures and video?

Joanne Wilson: I should because I have a daughter who is an artist.

Evan Nisselson: What kind of art?

Joanne Wilson: She's going through many iterations as a photographer. Now is working on technology in 3D, the new things that are happening in the art world. Actually, she worked on a piece with Ian Chang, which, if anyone went to the Frieze... When you walked in, it was that huge piece that was moving and video. It was cool. Anyway, I think visual is so easy to wrap our arms around. I am pleasantly surprised with the success of medium in regards to long form, but to watch a video, to capture a look... And I think that is why Instagram or Snapchat or any of those things have been so successful. It is so easy to communicate through a photo.

Evan Nisselson: If you were going to add more photos or a video, what would you do?

Joanne Wilson: I write what I write and then...

Evan Nisselson: Is it a spontaneous thing when you do it or do you have an outline where everything is set before? Such as, "I want to cover this in the next six months." Or do you wake up in the morning and say "Shit, I got to write this?"

Joanne Wilson: I do not wake up in the mornings and think, "Oh my god, I have no content today. What am I going to write?" I don't. My husband wakes up every morning, god knows like five in the morning, and he writes every day. I could not do that.

Evan Nisselson: I can't either.

Joanne Wilson: I can't do it. But what I do is, over the course of my week, I'll make notes to myself. I will be thinking of something, my brain starts going, and I make a note to myself that I should blog about this.

Evan Nisselson: Things move to the top of the list and then things get pushed down.

Joanne Wilson: Right. Really over the course of the weekend, usually on Sunday, I'll sit down and I'll write six or seven blog posts.

Evan Nisselson: How do you choose the images?

Joanne Wilson: After I write them, then I go and...

Evan Nisselson: So it's after. It doesn't start with an image. Why not?

Joanne Wilson: It just doesn't. It starts with an idea. Then I go back onto the Internet and I think about the kind of pictures that I want to use that will describe the post that I just wrote, because I think it's real important to have a visual.

Evan Nisselson: I can't wait to see one with all pictures and no text. Just in case.

Joanne Wilson: Food. Food is visual. Today I posted a bunch of photos. I thought of all the photos I saw at the Frieze and here is what I liked. Really simple stuff.

Evan Nisselson: In relationship to visual content, you have some companies... One is presenting later today and one is competing today in the startup competition, and one is in the medical space.

Joanne Wilson: Parts.

Evan Nisselson: Parts. I was going to figure out how you wanted to describe it. So tell us about those two companies and then others that have commerce-related aspects and how important images are to them, in their minds and in your mind.

Joanne Wilson: CaptureProof and Partpic are purely visual businesses and they are here. Both are female entrepreneurs and both dynamic. I loved what they were doing. There was no doubt in my mind that they were going to succeed at what they started. It was going to take a while to get there, but that's always the road. Partpic to me was like a total "duh." The Shazam for parts. It can be anything from something at Home Depot to medical to cars. You can go on and on. Visual versus going through the Yellow Pages. At some of these old hardware stores it's impossible to find the right screw. I mean, it's utterly ridiculous.

Evan Nisselson: It's a huge sector. I think it incorporates everything in life, but it's visual.

Joanne Wilson: It's visual.

Evan Nisselson: Are you thinking, *Hey this is interesting—I've got two new companies with great female entrepreneurs in the visual technology sector?* Are there others out there that are more interesting because it relates to what you've just invested in? Or is it really just the entrepreneurs and the market?

Joanne Wilson: It's really the entrepreneurs, first. Always, first. I always think of a house. If you buy a house, you can never move the house. But you can fix the house. You can actually rip the house down to its foundation and rebuild it.

Evan Nisselson: Probably not the right thing to do with the entrepreneur.

Joanne Wilson: Right, but you've seen them do it. You've seen plenty who have completely said, "This is not working."

Evan Nisselson: If they say it's okay, from the investor side...

Joanne Wilson: That's what the investors are supposed to do.

Evan Nisselson: Exactly.

Joanne Wilson: You're the entrepreneur. You're the idea person and the executor.

Evan Nisselson: In the other companies that you have in your portfolio... Are there other companies that really do leverage images and video and have you seen any trends that they want to increase that?

Joanne Wilson: Yeah, certainly.

Evan Nisselson: Give us a couple of examples.

Joanne Wilson: Food52 is a perfect example. They started out really as a shared economy, a recipe site from the crowd. Every week you have a recipe that wins the prize and then they get into the cookbook at the end of the year. They went into commerce and then they went into videos around how to cut an onion. How to do certain things—because we all know if there's video, then people stay longer. People watch them. They have to be very clever. I'm about to invest in a company that is all video.

Evan Nisselson: Great.

Joanne Wilson: All they do is video. I mean, it totally makes sense to me, what she's doing. The company Mouth is doing the same thing. They're an aggregator of indie food products and they're showing photos of the makers of the food, but there's also videos that are coming so you can see what's going on. In regards to clothing, I think it's super important. You can see how it's worn, how it's striped, what makes sense. Visual is so important. Someone told me yesterday on Instagram, the second largest group of pictures is food, which makes sense.

Evan Nisselson: Do you photograph or capture videos of the family? Or when you are traveling around in life or is that your daughter who is the artist?

Joanne Wilson: No, I do. Not as much.

Evan Nisselson: Posting to Instagram? Do you have as many followers—250,000—as our photographer this morning?

Joanne Wilson: No. I don't. I don't Instagram as much as I did earlier.

Evan Nisselson: You did earlier. What happened?

Joanne Wilson: You know, there are only so many hours in the day. Things just sort of fell to the wayside.

Evan Nisselson: We are going to open it up to questions from the audience in a second as well. We still have a bunch more time but I want to make it as interactive as possible. Obviously, Joanne, if you have questions for me, it can be a two-sided street. I will make it easy and keep on asking them. Whatever you want. Let's go back to the entrepreneurs because obviously you're investing a lot. You're active. You continue to be active. You have a focus on investing in female entrepreneurs.

Joanne Wilson: Yes. Men have to be even better.

Evan Nisselson: Great. What's the best way to get you to a "yes"? Other than the passion and the other things, there are probably steps. One of my favorite blog posts from Mark Suster, even though quite long, is still great. It's about how investors invest in lines, not dots.

Joanne Wilson: Yeah. One of my favorite lines from Mark ever.

Evan Nisselson: I love it. I use it all the time and quote him. What are some of those key "dots" that are signals for you that really get you excited? I know that it's sometimes hard to say.

Joanne Wilson: It is hard to say. You know, it's interesting.

Evan Nisselson: Is it easier to say which ones you don't like?

Joanne Wilson: Listen, I answer every email that comes in my box. I really look at everything that comes in my box. I feel like it's really hard being an entrepreneur and to cold call someone randomly is great. I do not answer LinkedIn emails. I'm out there. You can find me. I'm not answering you on LinkedIn. Sometimes I read these decks and I think, *Damn, that's really interesting*. Or, *Not for me*. Again, it's a lot of gut.

Evan Nisselson: So tell us a little bit about that gut. I know it's hard to describe. What's the "not for me"? You look at it and it could be anything?

Joanne Wilson: I hate the build-a-better-mousetrap.

Evan Nisselson: Okay.

Joanne Wilson: I hate it, unless it's a mousetrap that hasn't been changed in 35 years.

Evan Nisselson: Okay.

Joanne Wilson: That's one. I'm not one of these super-duper technology nerds. It's just not me and I'm not going to provide value to a company like that. If you're doing security, it's probably not for me. I really don't like the kid space.

Evan Nisselson: Why is that?

Joanne Wilson: Because my kids are adults. I'm done with it. I get it but it's not for me. I've stopped.

Evan Nisselson: It makes sense. I like it. It's very practical. I'm trying to dig in to understand some of these things.

Joanne Wilson: It's what I like.

Evan Nisselson: The goal is to help filter some of the stuff that is coming to you. But I'm helping you filter some so that the other ones are not going to send those to you.

Joanne Wilson: I don't like wearables, at all. I think it's a blip on the screen into something bigger that's going to happen. Maybe it's five years in the future.

Evan Nisselson: It's a blip that might happen or it's not going to happen.

Joanne Wilson: It's happening. There's a lot of these wearable that are all personal. I don't like wearables so I won't touch that.

Evan Nisselson: Okay.

Joanne Wilson: If you are a VC? You probably should invest in some wearables, if that's part of your thesis, because that's what you're paid to do. That's what your LPs want to see you do. You learn for the next one. As an angel, I'm not interested in the first round if I don't believe it's going to make it. I'm interested in the second round.

Evan Nisselson: Okay. Good.

Joanne Wilson: I don't really love the health and wellness area, either.

Evan Nisselson: Why?

Joanne Wilson: Because there's 4,000 things a day that are coming. What is going to end up sticking on the wall? There are so many it's ridiculous. Trust me, I'm just one of those people that, if I could get up in the morning and look next to my bed and be like, "I really feel like sweating today. Maybe I'll take the blue pill."

Evan Nisselson: That's great.

Joanne Wilson: It's just one of those things where when someone walks into your office, you just know. My assistant always laughs. She's listening in the other room. She says, "I can tell the minute you get excited."

Evan Nisselson: That makes sense.

Joanne Wilson: Yeah.

Evan Nisselson: She's usually sitting in the room or she's next door?

Joanne Wilson: We have a very open office.

Evan Nisselson: So there's a vocal sound when you get excited.

Joanne Wilson: Oh, yeah.

Evan Nisselson: In your discussion.

Joanne Wilson: I don't know. Again, I answer cold emails. I invest in people out of nowhere. I'm like, this makes great sense to me and I really think you're great.

Evan Nisselson: You're investing very early. Sometimes first money in. Do they have to have a prototype? Do they have to have a deck?

Joanne Wilson: Oh, yeah.

Evan Nisselson: They have to have all that stuff?

Joanne Wilson: No ideas.

Evan Nisselson: Okay. So no ideas, the business has to be working.

Joanne Wilson: It's got to be working.

Evan Nisselson: Okay, so that's another filter. There's a prototype or something that's somehow validated or at least a signal to validation.

Joanne Wilson: Yes.

Evan Nisselson: Okay. What is another signal that gets you excited? Is it something that deals with your practical level on a daily basis these days, or things that you interact with, or is it also, "Hey, I am fascinated with that business space. It seems like it's got to be disrupted." Is it one more than the other?

Joanne Wilson: Probably more on the latter than in the earlier.

Evan Nisselson: Because some people say, "If I won't use it on a daily basis, in a consumer world, I don't want to invest."

Joanne Wilson: No. No. No. It's more about where I think we're going down the pike. I think that is one of the reasons why I don't invest in better mousetraps. At least, I don't think really good investors invest in better mousetraps because, for instance, perfect example... You go somewhere and you're talking with a bunch of people who have nothing to do with the Internet. They are not in this whole technology space, which is like seeing the future. People say, "Wow that is so amazing, this product." And it's like, "Yes, and that's why we invested in it five years ago."

Evan Nisselson: Thinking about going down the pike, I mentioned in my keynote this morning about the future of cameras. I've been photographing for 24 years and I can't wait to be able to capture a Satellite Selfie. I click a button, we look up, and it will take a picture and send it to us. That might be much further, but it's going to happen in our lifetime. But when you say "down the pike," when you say, "Okay, it's going to happen in 10 years"... Is there a feeling of *This is going to happen in 10 years?*

Joanne Wilson: Here's a perfect example. I've been working on this project. The question is: Do we put parking spots in this project? Because if we don't put them in, you only get to have so much space. If we put them in, we get more space. That was a big conversation. And why? Because I am not so sure... Sooner than we think, less and less people are owning cars. I don't know if you read about that city in South Korea, it was totally cool. No cars. Fascinating. Next generation is much more bikes and different kinds of transportation.

Evan Nisselson: Smaller cars.

Joanne Wilson: Smaller cars. Uber and cars that are driverless.

Evan Nisselson: They need parking spaces. But not as many.

Joanne Wilson: They do need parking spaces but it's going to be different kinds of parking spaces because it will all be done through technology. You won't have a human being driving them up and down. They are going to be able to park on their own. So I end up meeting this guy—he's the second generation of family that owns all of these parking garages in New York. And I said to him, "What are you going to do with all those parking garages? What do you see as the future? Because I will tell you, there are very few spaces in New York for events where you can have 500 people over the course of a day." He said, "It's very funny you're saying that. We are actually turning some of our parking garages into that." I thought, *Smart. Very smart, second-generation man.*

Evan Nisselson: I just had a discussion last night about how I was at a meeting for a big photo agency that happened to be in the top level of a private garage in Paris where collector cars were stored. It was a great event space. We have six minutes left. I have a ton more questions but we're all in this thing together. So anyone who has questions…?

Audience member: It's been golden, everything that you've shared today. I think you invested in Maker's Row several years ago.

Joanne Wilson: Yes, I did.

Audience member (cont'd): Do you see any opportunities coming in your daily emails that is made in Bangladesh?

Joanne Wilson: No. I don't. I don't. It's interesting, though, because I actually came out of the retail/wholesale business in my earlier life. And I was involved in the retail/wholesale business when we started making a lot of goods in China. I was still making a bunch of products in Long Island City, which allowed me very quick turnover. People were doing tons of stuff in China. Again, technology was different so you had to have 1,200, 2,400, 3,600 minimums to make it worth your while before it ended up at your door, plus taxes, selling it, mark up, and what have you. Because of what is happening now in manufacturing, technology, and the strength of the dollar, it is just as worthwhile producing certain items in the United States which might not have been 10 years ago. We are seeing a shift in manufacturing. I've seen a lot of manufacturing in Vietnam. I went to this event last night my husband spoke at about Bitcoin. As we become more of a global community and you have something like Bitcoin, where it is easier to process payment in Bangladesh without having to worry about going through banks and what have you, we are going to see a lot more change in regards to production.

Evan Nisselson: What is your typical investment size for initial investment?

Joanne Wilson: Usually $50,000 and under. And I have to own 1% of the company.

Evan Nisselson: 1%?

Joanne Wilson: 1%. That's my deal. Then I get a side letter that allows me to continue to keep my pro rata share of 1% as long as I want until I have enough money in that company until I think, *You know what? I'm at a risk.*

Audience member: In these times, when you can do a video conference any-time with anyone, anywhere, how sensitive are you to investing in someone locally or do you pressure your entrepreneurs to move closer to where the action is for them?

Joanne Wilson: That's a good question. I know a lot of the people in San Francisco who say you need to move here. You know what? That's bullshit. I actually do like New York, but I have investments in San Francisco. I have investments in Los Angeles. And I have investments in Europe. I have a great investment in Berlin. A couple of them. I met this woman, we have Skyped several times. I have spent a lot of time with them and her. The company is called Cloak. Actually, two of the women in my portfolio that will be on stage here... One's in Atlanta and one's in San Francisco. I would say that they would both tell you that it's irrelevant.

Audience member: Hello. Thank you for coming. It's great that you have such a focus on women. The tech world does itself a disservice and misses out on a lot of talent because it's oftentimes not very friendly towards female talent. What are the best and worst things that a company can do to attract great female talent?

Joanne Wilson: You know, that's a great question. I was going to mention this earlier when we were talking as it was going through my head. The good news is that we are all having this conversation about women. This is not something that is behind the curtain anymore. Even this week, someone in San Francisco is getting excited about doing something, about making a stink, about women entrepreneurs and how they should be more highlighted. I will tell you that they're out there. They're at Meetups. They're at events. I think New York is much more friendly toward women and women entrepreneurs... But when I get a deck and I open it and there's four smiling men there, I delete.

Joanne Wilson: If you can't figure out at the very beginning how to create diversity in your company and create a product and a business that has a gender balance in the back end of your business, that says something.

Evan Nisselson: Totally agree. I think I mentioned that at our monthly entrepreneur dinners are 30 people with 50% female/male. There are now over 500 members in our LDV community and it's exactly half men and women executives. We're working toward having equal balance of speakers here as well.

Joanne Wilson: That is great.

Evan Nisselson: We are about 35-40% female speakers but I will not be happy until it's equal gender balance.

Joanne Wilson: The issue is that women will say, "I don't know. I'm really busy."

Evan Nisselson: That's a whole other conversation that I would love to have with you. But on a closing note, this has been fantastic. We could probably talk for a lot longer. We are both very direct. What is the personality trait of entrepreneurs that you love and hate in one word answers? What do you love?

Joanne Wilson: I love "scrappy."

Evan Nisselson: Scrappy is great. What do you hate?

Joanne Wilson: I hate "arrogance."

Evan Nisselson: I love "passion" and hate "selfish." On that note, a round of applause for Joanne. Thank you very much for coming.

©*VizWorld*

40

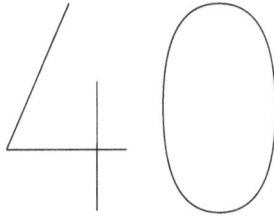

FIRESIDE CHAT WITH ANDY WEISSMAN OF UNION SQUARE VENTURES

Andy Weissman, Partner, Union Square Ventures
Evan Nisselson, LDV Capital

Evan Nisselson: What did you think of the diversity of those companies you were judging in the Startup Competition?

Andy Weissman: I actually think it was a perfect indication of what you're trying to do with these events. You take seven different slices of what "visual technologies" is or can be and so they were great. They were all great. For me at least, it was diversity around what a vertical—the breadth of what a vertical topic—can be. At the same time, the creativity of the entrepreneurial mind that is trying to take seven different slices of businesses around a common technology of, say, a camera.

Evan Nisselson: Not even just a camera. It's: *What's the output of a camera?*

Andy Weissman: Absolutely. We're imagining the output but also reimagining how you can build a community around it. And what are the business opportunities that come from that community? It's amazing.

Evan Nisselson: You're more creative than I. I'd love to hear... I'm trying to figure out what to call that sector. I've done "imaging and video." That doesn't work; it's a fact, that's what it is. I've done "visual tech," which seems like it's better but it still doesn't convey it.

Andy Weissman: It's geeky.

Evan Nisselson: It is. So what is the opportunity, the branding?

Andy Weissman: The selfie economy.

Evan Nisselson: Funny. You missed this morning, but the selfie is old school, right? The pure selfie. There's the "jelfie," which is the jumping selfie.

Andy Weissman: Really?

Evan Nisselson: There's the "delfie," which is the drone selfie.

Andy Weissman: Right.

Evan Nisselson: And everybody has these selfie sticks, which I hate.

Andy Weissman: Why?

Evan Nisselson: You carry this thing around, you hold it out in front of you which always changes the atmosphere of the moment. I've got the solution... the Satellite Selfie.

Andy Weissman: Right.

Evan Nisselson: Hit a button, you go like *this*. It'll snap a picture at you.

Andy Weissman: Yes, I saw you ask Joanne [Wilson] about that.

Evan Nisselson: I think it's great.

Andy Weissman: You know what? I actually think... I believe that the selfie is easily mocked but vastly misunderstood.

Evan Nisselson: Why do you say that?

Andy Weissman: If you think about where we are with imaging right now and imaging technology, and you think about where we are with networks to share images, we've completely flipped and inverted the model of the production, distribution, and consumption. We flipped it to where people are actually in control of their distributions, and the selfie is the actual ultimate expression, not only of self but about that change in power and change in control. I actually think it's something fundamental that's going on.

Evan Nisselson: I agree. I think you're absolutely right, but the actual feeling that I'd like is to get rid of the camera here. There's this opportunity, not for all use cases as Hans Peter [Brøndmo] said very well this morning. We talked about the different goals for every photograph and how one needs a different tool depending on the moment but the concept of having something to carry and to get in the way... My arm always has to be here.

Andy Weissman: Yes, it does, but part of the selfie is the physical expression of control. It's not you taking a photo, it's me doing it on my own terms and then doing with it what you will. Of course there's narcissism, but it's human nature. It's the way human beings are. But to me it's narcissism plus empowerment. It's represented as something more fundamental going on that the technology is enabling, regardless of what the actual selfie is. Also, there is an insatiable appetite.

Evan Nisselson: For?

Andy Weissman: For people to take selfies. It's taking control. It's control of my existence, it's control of my images, right? My partner Albert gave this TED talk called "The Great Inversion." He talked about what the Internet and digital technologies have done, how they have actually inverted class, almost every single industry, and the way we think about value and power and control in every industry. I think selfies are actually a representation of that, right? Anyone is a photographer. I can be a photographer and take a picture of myself and share it with the world and you can't stop me from being as stupid as I want to be. They're the ultimate inversion. There's no photographer, there's no editor, there's no distribution control. It's all those things.

Evan Nisselson: It's done.

Andy Weissman: Just like that. Done, exactly.

Evan Nisselson: You wrote a great blog post recently after a conversation we had, which relates to this. During the conversation we had in your office, reflecting upon something I had forgotten... Those AOL days. We've both been doing this for about the same amount of time in different places. Your blog post was about the evolution of "What is a photograph?"

Andy Weissman: Right.

Evan Nisselson: The Def Leppard quote is? There's a great Def Leppard quote.

Andy Weissman: "Photograph—all I've got is a photograph. But it's not enough."

Evan Nisselson: Tell me your view on this. Evolve on that blog post about what is a photograph, where are we going, and how does that play to what we were just talking about with selfies? A selfie is just another type of photograph. You were talking on that blog more about animations and types of images that tell different stories.

Andy Weissman: Here's one of the things I was thinking about listening to all these presentations today. There are, if you think about the LDV Vision Summit, there are two buckets. One is: there are technologies. A camera that can take a picture around a corner, or virtual reality. There are technologies that are advancing the field of visual capture and redisplay, and that's one bucket. The other bucket is the cultural element to it. They're related, but the cultural element, to me, that's interesting, because I'm not really a technologist. I'm a consumer of technology, an investor in technology. The cultural element, to me that's interesting, is that all the definitions we think we have are now being reinvented. And in being reinvented, creativity emerges. Historically, a photograph was an archival thing. The representation of that was the shoebox in your closet with the photos. They were archival. They were capturing history for the future. We have apps. The largest photo-sharing app in the world right now... The photos disappear. There is no archival element to them. So what is that image then if you've inverted? Culturally, what is it? What is the image where it can be animated? What is the image where there can be text over the image, where you tell a secret to someone? You've got reinventions because the nature of it—or the metaphor of it, which was what I was writing about—has changed so significantly that everything is up for grabs.

Evan Nisselson: I definitely agree with you. It's a great, insightful then-and-now. I don't use SnapChat—I get it, but I don't use it. Do you think those that use it a lot also take different types of photos or have different goals? Obviously certain ones are memories and certain ones are SnapChat disappearing photos. I think historically, it was 100% for memories.

Andy Weissman: Exactly.

Evan Nisselson: Now, it's... What percent do you think it is now? Obviously it'll change with the demographic, but do you send SnapChats to anybody?

Andy Weissman: I may, a little.

Evan Nisselson: My friends' kids send them to me.

Andy Weissman: Yes, exactly.

Evan Nisselson: Random ones.

Andy Weissman: That's what's interesting about it, is that there clearly is something generational associated with an application or a service where the images that historically have been archival are disappearing. There clearly is something generational. The answer to your question is, "I don't fucking know," but it clearly feels generationally-related. Again, moving for something that is archival and historical to something that is temporal, something that is much more emotional, something that exists in a certain moment and may not exist... By the way, that's just the latest rev of it. There are lots of other revs of it that we've seen here that we've invested in and that you've invested in. There are so many different slices of it, but to me the interesting thing is, it all feels like it's being reinvented. Some people will invent technology and some people will invent new metaphors that take advantage of the fact that there is the most powerful camera or video capture device ever created in the pocket of every human being. When it's every human being, especially when you think about low-cost devices, Android and all that kind of stuff, that's kind of just the beginning.

Evan Nisselson: I agree. There's a fascinating use case that I thought about which resonated recently with me, which was... We are talking about all these memories and what the majority of people are doing capturing memories. However, the other night, I looked at photos of my folks and my sister—I mentioned earlier, their wedding album and bar and bat mitzvah albums—and there were pictures coming out that I had never seen. Most of them I have in my memory because I have a photographic memory from all my work in the photography industry. I decided that I wanted to see some of those analog photos more often. I took a digital photo of the prints and it's now on my phone with the other 10,000 photos. It's no longer in that shoebox.

Andy Weissman: Right.

Evan Nisselson: It's in the digital box, which I have access to all the time. I think there's a huge opportunity in all this content which is a digital memory, when we want to see it at different times or any time.

Andy Weissman: Yes, and by the way no one has solved that.

Evan Nisselson: I know.

Andy Weissman: People have tried.

Evan Nisselson: Even just earlier, Matt Meeker, who came onstage from Bark-Box. We were talking before he came on stage about the early days when I helped him box the early BarkBoxes to send out. And I said, 'Do you have that photo when we had all the boxes?" He's like, "No." So both of us are swiping through our phone camera roles, even though both Union Square and I have invested in Clarifai, and he said, "Wait, are you using Clarifai yet for that?"

Andy Weissman: Here's what's interesting to me. There are a couple things. Here's one thing that's interesting. BarkBox is a company that sells a box full of goods for your dog. The fastest-growing segment of their business are their Instagram accounts.

Evan Nisselson: Yes.

Andy Weissman: Changing metaphors, a business that doesn't feel like it's a visual business but it is.

Evan Nisselson: You're absolutely right.

Andy Weissman: Right. Clarifai is interesting for a couple reasons and we're investors so I'll talk my own book. It feels like a lot of the services to solve your 10,000-photos-on-your-camera have been digital representations of the shoebox, file systems, nested file organization. They were just different shoeboxes, right? Clarifai takes a completely different approach. It basically says, "You don't actually ever need to organize it. It will organize itself." That's very provocative, if it works. The cognitive load on the human being is reduced in the same way that Google changed the search game by not doing directory services, by having an open search box.

Evan Nisselson: That's right and their focus is business, but it's content everywhere… But that thing of: "I don't want to actually organize myself."

Andy Weissman: I want it to be organized when I need it to be organized, right?

Evan Nisselson: Right.

Andy Weissman: I always thought that the Google metaphor was: "I don't need to know where something is, I just need to know how to search for it."

Evan Nisselson: That's right.

Andy Weissman: That 90% of the organizational cognitive overhead is now gone, leaving the key human part—just search.

Evan Nisselson: They've succeeded wildly and that makes me very happy. In Gmail, I no longer have all of those folders that I tried to organize in a smart way, which never worked. It actually made more work for me and I still could never find what I was looking for.

Andy Weissman: Exactly, right.

Evan Nisselson: Give us just a little bit in two or three sentences: the Union Square Ventures focus and your focus. Give us the parameters just for those that might not know in the audience.

Andy Weissman: We are a thesis-driven firm and so our focus is only our investment thesis. Our investment thesis is that we invest in large networks of engaged users that are differentiated through user experience and defensible through network effects. That is the 140-character representation of our investment thesis. That's all we do and that's only what we do. There is no specialization per se except we are looking for network-based businesses wherever they may exist. Network-based businesses, they can be networks of users, can be networks of data, can be networks of images, of media types. There are lots of different kinds that we had never even contemplated. The way that we operate is to look at investment opportunities through the lens of that thesis. That thesis has 10 or 11 different words that have shifting meanings. By definition, they are kind of indeterminate, except that they are networks. They are things that are interconnected and the reason that they are networks is that we believe—well, I don't know actually if we believe it, I think it's actually true—that the Internet itself is a network. The architectural structure of the Internet is a network with implications around that, around resiliency and things like… But also the only thing that we can identify that will always be a barrier to entry for a business is its network effects.

Evan Nisselson: That's right.

Andy Weissman: As investors, we think those are good investments to make because when you get them, you have a company that can be transformational. It's a company whose value increases the more people use it.

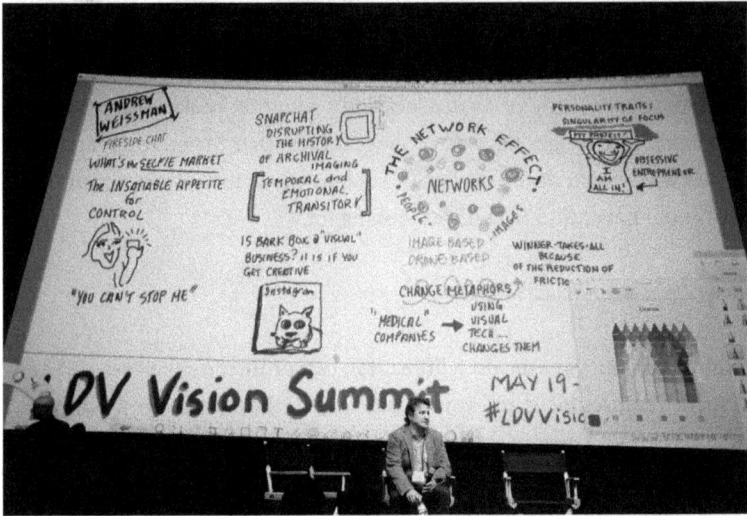

Andy Weissman is a partner at Union Square Ventures. He answers a question from the audience while the other investors join us on stage for the next investor panel.

Evan Nisselson: There are a lot of network-effects businesses that actually have images or video as a part of it.

Andy Weissman: Absolutely.

Evan Nisselson: Historically, you had a Tumblr exit, Etsy went public... And both revolve around different types of content. Some people don't think of it as a content company, but wouldn't you agree that a majority of the things sold on it has to have good to great content?

Andy Weissman: It's handmade goods where a person who makes the goods takes a picture of the goods.

Evan Nisselson: What other ones?

Andy Weissman: Lots more. I mean, there's Twitter, of course. Twitter is an image-based network. There's a host of other things. There's Clarifai, which we mentioned. We are investors in something called DroneBase, which is a network of drones that can be deployed for visual applications, whether you are a real estate agent or some other business person that needs some aerial photography. That's not a visual technology but it's a network of capabilities. It's a network of your satellite cameras that you can deploy. You are investors

418

in VHX, right, which is DIY video and YouNow, which is a live video platform? These are all portfolio companies, yes? There is Figure 1, which is a media sharing mobile app for medical professionals. They hate "Instagram for doctors"—they hate it but it kind of captures it in a way. Again, changing metaphors, think about medical professionals. The language of the medical professional is visual. The work style of medical professionals is collaboration. Both of those can be captured in a device that's with them on their job and that can connect to every other medical professional around the world. That's kind of a pretty profound metaphor.

Evan Nisselson: It's huge, absolutely.

Andy Weissman: If you can pull it off, that's a visual media network.

Evan Nisselson: There's two more medical companies that are going to have to present this afternoon. One is CaptureProof, which Joanne [Wilson] is in and I've known Meghan [Conroy] for a while, and the other one is fascinating. I know one of the co-founders. It's called Zebra Imaging. There are actually network effects in probably both of those relating to the doctors and relating to others.

Andy Weissman: When we bet on network effects, we *are* making a bet because network effects can be nebulous. We're making a bet that an application can exhibit, derive, and gain defensibility from network effects. You kind of only know it after the fact, though.

Evan Nisselson: You have a directional hunch or traction.

Andy Weissman: Yes, you got that hunch but when you know that it happens, it's kind of... You look at a service and you're like, "Well, why is that growing?" Network effects is the kind of big unknown answer for that.Evan Nisselson: I can't remember exactly but my perception is that the network effect focus was an evolution to something about 10 or 15 years ago, earlier days in Union Square. Is it something that you think is going to last forever? How long-term does that remain the thesis? Obviously, you can always evolve that but do you have an internal agreement of *We are going to do this for 10 years, unless there's a reason not to?* Or, *We are going to do it for 20?* Is there that longer-term vision? Obviously, a fund has a lifespan.

Andy Weissman: That's a good question because I think that the thing about our thesis that keeps us engaged is that... The thesis is the thesis but what is a network? What is a user?

Evan Nisselson: What is a network?

Andy Weissman: There are a lot of different networks.

Evan Nisselson: Give me a couple.

Andy Weissman: Well, you can have a network of data.

Evan Nisselson: That's great. I love that. I haven't thought about that. So what's a network of data?

Andy Weissman: A network of data could be an aggregation of data where there is some group or some service that contributes data and you have a bunch of parties that contribute data and then I'll get to derive some value from that.

Evan Nisselson: But still, you said "group." Do you mean people or could it actually mean sites?

Andy Weissman: No, it doesn't have to be people.

Evan Nisselson: That's what I'm getting at. Most people think of network effects as *people* network effects.

Andy Weissman: No, it doesn't have to be that. A collection of images can be a network, you know? Clarifai is an image technology that gets better the more the network grows.

Evan Nisselson: Perfect example.

Andy Weissman: The second part of the thesis is a unique user experience and that generally means kind of native to the Internet, even more so native to mobile. Those are being invented and changed in many different ways. One of the things we are thinking about lately is the idea that you can build... Well, there are two things. There are an enormous amount of businesses that are operating today on Instagram and it was a service that wasn't designed for that. What are the implications of that?

Evan Nisselson: That's huge. That's fascinating.

Andy Weissman: We are starting to see a lot of businesses that work via SMS. They're trying to get out of the trap of the app store. What does a unique user experience mean? Does it mean an app, does it mean a website, does it mean text in an SMS box? All of these words have enough flexibility that I think we can run with it for a while. They are changing, too.

Evan Nisselson: Agree.

Andy Weissman: When this thesis was developed, there was no freaking iPhone. We are in year eight of re-understanding the thesis because of mobile and then, what are the implications of that? There's a notion that in a mobile-first or a mobile-only world, network effects are much weaker. Part of the network effect thesis is that it generally leads to winner-take-all or winner-take-most markets. In a mobile world, is that still true? In a mobile world, where you can go between applications with almost no friction… Then, when you have a notification layer on top of that, there really is no friction. I don't know. So that keeps it interesting.

Evan Nisselson: Absolutely. We've got plenty of time left but if anybody has questions, raise your hands at any time and we'll get microphones to you. We're just going to keep chatting away. You see, it's kind of interesting there. We talk about network effects… There is a visual network effect.

Andy Weissman: That actually is the network right there.

Evan Nisselson: Amazing, huh?

Andy Weissman: There's a picture of a network.

Evan Nisselson: We are a network connected to that. We can get a little meta, but that would be another conversation as well. We talk about entrepreneurs and there are a bunch of entrepreneurs in the audience, people who might want to be entrepreneurs, there's computer vision scientists, there's professors that are thinking about maybe one day… Every investor, I think, likes and dislikes certain personalities of entrepreneurs. I'd love to hear your perspective on the personality traits that you really get excited about in entrepreneurs and ones that you don't like.

Andy Weissman: For me, the thing that is most interesting or exciting or the thing that I like the most is the obsessive nature of the entrepreneur. The singularity. They just want to talk about their company and they only want to talk about their company and they only want to talk about the challenges. That's the thing that I find is incredible, that we get to kind of participate in that.

Evan Nisselson: Let's dig into that a little bit more because I agree—but I think many people filter that in different ways. There are people who talk about their company all of the time but say nothing.

Andy Weissman: I'll give you examples. Really the only thing an investor does is pattern recognition. You have a data set that kind of keeps getting better and you're like, "I've seen that work and I've seen that doesn't work." The thing that I have seen in a number of companies that have gone on to be successful and transformational and observing them from the early stages is you have a founder or it's usually a founding team and they're always talking about some issue that's come up and they can't get their head around it. They'll even go off on a sidebar. You're sitting in a restaurant having dinner and they'll pull out their phone and they're doing their little side talk, right? They can't get out of it. They are living it all the time and they can't get out of the constant replaying analysis, problem-solving, understanding of their business. It's that kind of nature and even when you take them out of the setting of their business, it's still the same. It's like: you put an entrepreneur in front of a whiteboard—how long before they get up and start sketching?

Evan Nisselson: I totally agree and I like it. I'll just riff on it a little bit because I love the passion or the unbelievable drive of them but sometimes it gets to a point of which they might not listen to others—investors, customers, or peers. They might be driven, they might talk all the time about it, but if there's... like I said earlier with Joanne [Wilson], the personal trait that I don't like is selfish because it's not going to be a team effort.

Andy Weissman: Yes, but to me, the paradox in my answer to the question of the trait that I dislike the most is that same obsessiveness. In my mind, I actually think you need it to succeed. In my mind, it sometimes is the thing that may cause you to go off the rails.

Evan Nisselson: I totally agree. Some examples of that?

Andy Weissman: Well, you know it's...

Evan Nisselson: I'm going to ask tough questions.

Andy Weissman: These are good questions. Just the ability to get out of your head, the ability to get outside of your body, the ability to get outside of your business and look at it from a different angle, which is very hard to do because most entrepreneurs are all in. Being all in allows them to build a business when all around them, people are saying "no" to them. If you've ever raised money for a company, you just get the shit kicked out of you all the time, even if you're really good at it. It's incredibly humiliating. You're sharing your idea, your creativity, your intellectual property and being told "no." How do you get up and do that again and again? I've done it. It's hard. At my last company, we tried to raise money for Union Square Ventures many times and couldn't get them to. So you have to be obsessive.

Evan Nisselson: Tell us about that. Do you ever have flashbacks to when you were sitting there at the table, pitching Union Square Ventures?

Andy Weissman: Yes, but my flashbacks are like, *What the fuck am I doing here?* To me, there are a couple lessons. This goes back to the obsessiveness. At some point, you've got to focus on the short-term tactic. At some point, you have to take the long game. And how do you balance that and how do you balance your obsessiveness between *we have to do this today* and *what's the long game?*

Evan Nisselson: That's right.

Andy Weissman: Because it's the thing I like and dislike the most, I don't have a view on if it's a good or bad thing. It's just a thing that I've noticed and I get at some level. I understand. Our job as investors, maybe, can be just to kind of pull someone out and say, "Look at that exit sign. What if you walked up to there and you looked at that exit sign?" Maybe you see something different, maybe you won't.

Evan Nisselson: I think that is very interesting. We are going to have a question from the audience while we bring up our other panelists simultaneously because you're on this next panel as well.

Andy Weissman: Great, awesome.

Audience member: As an investor, what percentage of the time are you okay for the CEO of one of your companies to say "no" to you regarding something? Is it 40% or 60%? How often do you like to hear that "no," because ideally, they know their business better than you do?

Andy Weissman: Here's what I believe: if I'm asking a question that requires a "yes" or "no" answer, I'm asking the wrong question. I don't think the role of an investor is to be an arbiter of binary decisions for a CEO. As intimate as I am with the company or we are with companies, I don't think we can fully understand because we are not as intimate as the CEO is. I don't think the most efficient or effective role is one that requires a "yes" or "no" answer. I think the most efficient or effective role is to suggest a couple ways of thinking about a problem, suggest a couple patterns that we've seen in dealing with that problem, and then, it's your business. You're going to solve it your way.

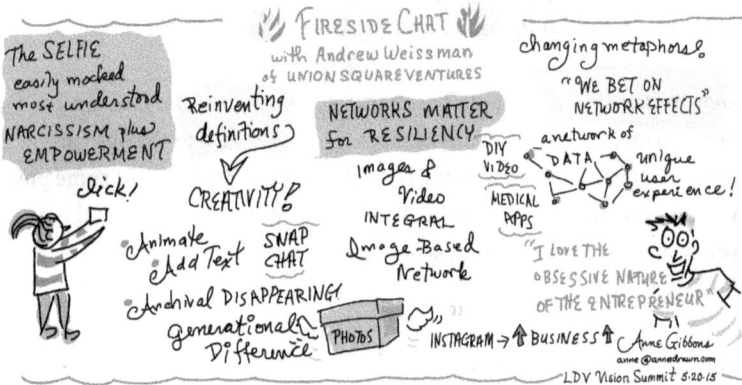

©Anne Gibbons

41

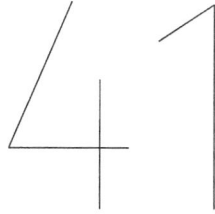

PANEL: TRENDS IN VISUAL COMMUNICATION AND TECHNOLOGY INVESTING

MODERATOR

Evan Nisselson, LDV Capital

PANELISTS

Andrew Cleland, Managing Partner, Comcast Ventures
Anu Duggal, Founding Partner, Female Founders Fund
Andy Weissman, Partner, Union Square Ventures

Evan Nisselson: Thank you for joining our panel on "Trends of Investing in Visual Technologies." Can you please give us two minutes of what you love, what's the focus of your fund, and then we're going to jump into questions and continue chatting? It's very free flow.

Anu Duggal: Sure. Thanks so much and thanks for having me.

Evan Nisselson: Our pleasure.

Anu Duggal: I am the founding partner of F Cubed, Female Founders Fund, which is a seed-stage fund focused on investing in companies founded by women in technology, primarily based in New York. We write seed-stage checks, so anywhere from $75,000 to $150,000. The funds thesis is that when you invest in really great female founders, you can get a great venture backable return.

Andrew Cleland: I'm a managing director for Comcast Ventures. We invest on behalf of Comcast, NBC, and Universal. There's really two things to understand about our fund. We are a strategic fund but we behave exactly like a traditional VC fund and there's two things that underpin that. The first is that we get compensated like a traditional VC and that's only important to relay to you for entrepreneurs to understand that we're on the same side as the entrepreneur. We're not trying to invest into the company to advantage Comcast's interests necessarily. We do well when the entrepreneur does well. We just want our portfolio companies to do as well as possible. The other thing to understand about us is that we decision-make within the partner group. As far as entrepreneurs working with us, we look and feel exactly like a traditional VC. But the plus point is, and what I think is exciting about the platform is, we can bring the assets of Comcast and NBC and Universal to bear on behalf of our portfolio companies. We get privileged access and in certain verticals—advertising is an obvious one, video is an obvious one—which I think is one of the reasons that I'm here. We can add value over and above the check.

Evan Nisselson: A little bit of perspective because I'm not sure if you guys were here yesterday. We have 350 people attending and everything is on video. This Summit doesn't end when it is over because we will be leveraging many forms of content marketing to share your wisdom from this panel. So just keep that in the back of your mind as well.

Andy Weissman: He's warning you, by the way. That was the point of that.

Evan Nisselson: Well, I did it after I gave him the flask. You see how it's all about marketing, Andy. I'll let Andy take a break for a second because he's been talking a little bit. Andrew Cleland, talk a little bit about some of the video companies that you're in. And what are the biggest challenges of those specifically and what excites you about those opportunities? Obviously Comcast is there, so that's the key piece, but personally it's a personal drive to choose those companies.

Andrew Cleland: Yes. I'll talk about the group a little bit and I'll talk about one of my companies just to highlight an example. We really think about it in relatively traditional buckets of infrastructure and content and monetization. Infrastructure, it's obviously core to our DNA. We move bits around the web as an operating company so we have a number of investments in things like IP distribution companies. We have invested into semi-conductor companies, which is quite rare these days for a range of funds. We've invested into image capture, little chips that sit behind the cameras in your smartphones for those of us in the room, presumably all of us that have smartphones. We step up the infrastructure lab. If we think about content, we invest into pure play content companies. We've invested into Fullscreen and Tastemade, a couple of better known MCNs. We've invested into a company called SnagFilms. We've invested into companies which are increasing—maybe they didn't start out as focusing on images as their core USP, but Flipboard, for example, which is becoming increasingly a visual medium.

And then in terms of monetization, we're in VHX with Andy in Union Square. And the company that perhaps I'll spend a little bit more time on is a company called SundaySky, which is a company that's one of my portfolio companies, which is really about personalized video. It breaks apart video streams and it reassembles them in intelligent ways to speak personally to the characteristics of the viewer and the impression at the time. It can be used in a couple of ways. It can be used by enterprises to do things like explain particularly complicated propositions, so insurance companies, finance companies, medical companies that need to explain really quite complicated things to its users, need personalized video that is appropriate to the particular situation of the viewer, and there's an advertising side as well. If you imagine browsing around through a session on the web, you're interested in a particular set of products, it's possible to present video to you later in that session or maybe even later in the week that reflects back some of the interests that you've demonstrated in certain products around the web.

Evan Nisselson: Great. Anu, I know we talked, as we were trying to get a mixture of great people on this panel with... A corporate VC, traditional, and early stage [investor]. You have a couple visual-based companies in your portfolio. To what extent are they leveraging images and video and how much do they think that's a part of their business?

Left to Right:

Evan Nisselson,
LDV Capital

Andy Weissman,
Partner, Union
Square Ventures

Anu Duggal, Founding
Partner, Female
Founders Fund

Andrew Cleland,
Managing Partner,
Comcast Ventures

Anu Duggal: Sure. So if I was to look through the portfolio, I would say... Actually, the first investment that we made, which I made personally, is probably the one that I think leverages images the most. That's a company called Loverly, which is in the wedding space. I think the wedding space traditionally has been quite antiquated. There hasn't been a lot of disruption. I think when you look at things from a bride's perspective, it's really about imagination and it's about visualizing what you want that day to look like. Obviously that's definitely translated into very big advertising spends. I think what Kelly, the founder, what her goal and vision with Loverly is, is to really try to create a platform which in some ways has a network. She's aggregating content across the top wedding blogs and leveraging that to build a community of brides where you have the advertising element of it but then you also have commerce on top of it. I think that that's a very clear case of a company where images are just so crucial to the customer's experience that it's hard to imagine that it didn't exist before.

Evan Nisselson: One of the slides I referenced yesterday was the teen audience and where they are spending their time. We know the names of those companies—Facebook, Snapchat, Pinterest, and other visual places—which I think wouldn't exist without images or video, at least not to the state that they're in. One question I haven't asked any of the investors—do you hear at the board level or in strategic conversations: *Maybe we should start leveraging images and video more?*

Andrew Cleland: I'm talking with a seed company at the moment where the idea of the company is really to innovate. You know, 15 years ago, it would have been a magazine and today, it's trying to figure out exactly the right blend of images, video, and text to be the most compelling proposition it can be. That's maybe a good example of the kinds of conversations that people are going through. There's some subtleties to how you implement in terms of auto play, in terms of the way that you use audio, in terms of the stills that you use, in terms of the way that you present images through the body of the text and the way that you present the text relative to these images that can dramatically change the way that people actually choose to consume the media. Video is obviously much more involving. It requires a little bit more of a time and an emotional commitment from the viewer, so you need to tailor the content as well to the particular use case. If it's things like news, where you're trying to ingest as much information in as short amount of time as possible and you're time constrained, video maybe isn't the most compelling format. If it's a little bit more discursive or it is content that we think can live for a much longer lifecycle, then it's much more interesting to use video. I think the other thing I think that's interesting is, is there's a hook for video-established brands, whether they're people brands or *brand* brands or thought leader authorities. Really interesting complement to video to pull people into that experience because you need a little bit of a hook to pull people into committing to longer time. There's 90 second views, for example.

Anu Duggal: I would say going back to your earlier comment about Bark-Box—even though when I look across the portfolio, there are definitely quite a few companies that don't involve images or video as a core part of their business, one thing that I think all of them have really tried to do is figure out what is the distribution or marketing tool that their customer base is most attracted to. In the early days, you'll test out Pinterest, you'll test out a bunch of different channels, and I think figuring out what is the type of content... Is it video content, is it images on Instagram or on Pinterest? I think that, to me, is similar to BarkBox. It's a way of thinking through how to add that layer and make it relevant. And what does your customer base most associate with your brand?

Evan Nisselson: That makes a lot of sense, but on a timing basis, it seems like that's a recent thing. What initiated this trend? Was it Instagram's sale or was it Pinterest? Does anybody recall a signal that everybody says, "Well, we have to get involved. What's our plan for this?" or "It's really, actually happening."?

Andy Weissman: I actually think it's a network effect. You want to go where people are.

Evan Nisselson: But when did it get big enough? That's the hard question, like you said earlier. You don't know a network until it is.

Andy Weissman: Yeah. It's recent, right? The last couple years.

Evan Nisselson: Initially, the reference of BarkBox was great because... I didn't get to say it earlier where, as you know, I'm very good friends with him and I helped him in the early days and I wanted to invest but I'm so thesis focused. I asked, "I bet you're going to leverage images and all types of visual content in a BarkBox community?" Matt said, "Evan, I can't. You're probably right but we don't have the plans right now, I can't commit to it." He's focused, but I knew it. "Can't we just do it?" "You can, Evan, but I can't say now that we're going to do that." And that was the right answer from him. Maybe I should have broken my rule.

Andy Weissman: There was no way to know that there would be dogs who have Instagram accounts that have a million followers, right?

Evan Nisselson: Agree.

Andy Weissman: There's no way you can predict it.

Evan Nisselson: It was just going to happen.

Andy Weissman: Yeah.

Evan Nisselson: It's easy to say now but that's what I said... That was the exact conversation.

Andy Weissman: It goes back to your question. All these things are blending, right?

Evan Nisselson: Right.

Andy Weissman: You as an investor can break your rules and you could come up with a rational thesis because these things are blending so much. There are a couple of companies that I have right now that their most effective marketing channel by far is YouTube. You know how they do it? They find someone on YouTube who makes videos that is somewhat related to what they do and they send them an email and say, "Hey, will you review our product? If you review it, we don't care what you say. Put a link in the comments." That's it. I actually shouldn't even tell this secret because they're doing so well, right?

Evan Nisselson: That's interesting. That's a perfect example of things that are happening and that's why I wanted to have us discuss this. What are other companies doing? What should companies do? What will get that to the next spike? Do you have any other examples, either of you?

Anu Duggal: Well, one company that we haven't yet announced, which I think is really interesting, is focused on YouTube and the how-to space. YouTube has just kind of skyrocketed in terms of styling and fashion. This company is basically creating a mobile video network with Tier 1 and Tier 2 influencers who are creating very short, 30-second to 1-minute videos on how to put on mascara, various styling and fashion tips. It's interesting because, again, you have the network effect. You don't necessarily need to start with the most well-known influencers, but they bring their own audience. It's not a traditional advertising model in that they've already signed up with brands... How do you cuff a sleeve properly with Jenna Lyons from J. Crew? Talking to J. Crew, video is actually their, by far, their most effective advertising tool. They just don't have enough platforms to use video as a tool. To me, that company hasn't launched yet and I can't name it, but it's a really interesting way of kind of leveraging YouTube's existence but also acknowledging that 60% of YouTube videos are still viewed on web and not on mobile. That's an interesting company, which I think is kind of taking advantage of it.

Evan Nisselson: That's great.

Andrew Cleland: I was just going to make a slightly different point. I think we're seeing so much activity and will continue to see activity. I think it is a single inflection point. I don't think it'll be a pendulum swing back again because it's not that long since mobile networks appeared. It's the interplay between mobile and the load on the networks that compound video and images. It's not that long since mobile networks have been able to support this kind of content. It needs a year or 18 months for entrepreneurs to figure out what to do with this new capability. Things like auto play in the Facebook stream, things like Instagram, things like Snapchat—it just takes a little bit of time for the ecosystem to figure out. Even Gmail's innovations around inbox and threading photos into the stream, it's taken some time and I now think all of that stuff is coming to market. It's clear that the future of web is going to be far more visual.

Evan Nisselson: You're giving me a flashback to a 10k file taking 10 minutes to download on a 9600 baud modem.

Andy Weissman: They don't even know what you're talking about.

Evan Nisselson: I know they don't. They're young. Anybody have questions? We've got about 12 minutes left so feel free to raise your hand if you've got questions. What questions do you guys have for each other? I have one while you guys start thinking about questions. The most important thing to me, and I think all of us, is it's all about the people. It's all about the entrepreneurs. It's all about you guys [pointing to the audience] and us. What do you love and hate about your job?

Anu Duggal: I can start. I'll start with what I love. I would say the part that I love about my job is obviously meeting entrepreneurs and seeing what, over time, new business models are emerging, new industries that are being disrupted. As an investor, I think you have the advantage, unlike an operator where you're very singly focused, of getting to see a little bit of everything. I think what I really dislike about my job is that you spend a significant portion of time saying "no." And having been an entrepreneur before, in some ways it goes against your makeup, which is to try to help and build things and help people. I would say that's something that I'm still trying to get better at.

Evan Nisselson: I have similar challenges all of the time.

Andrew Cleland: I'm terrified of being bored in my job. I have been in jobs where I have been bored and it's a horrible place to be. This job is fascinating. You're always intellectually challenged. You're very rarely the smartest person in the room. It's an incredible privilege to have a job where people are explaining their deepest passions to you and providing you with insight all day. There's incredible intellectual variety. That's the highlight of my job. There really aren't many lowlights. I do love what I do. I think if I was to point toward anything, it's a slow cycle in terms of seeing the fruits of your work. You only start to see in five or six years' time really how good you are. It would be nice to have a shorter cycle of feedback, I would say.

Evan Nisselson: We have all been entrepreneurs on this panel so this is a great operator/investor group, which I believe is the ideal mixture. The biggest thing I struggle with as an entrepreneur daily was either you are moving the ball forward quickly, slowly, or not at all, or possibly going backwards. You knew what was happening daily. There were metrics that you really knew. And doing this, I had gotten used to it for 18 years and I love it but it is a whole different game as an investor. You have a perception that you're more in control as an entrepreneur, and you might be, but we're never really in control.

Andrew Cleland, Managing Partner, Comcast Ventures sharing his wisdom.

Andy Weissman: Related to that, the thing that I think is pretty interesting about venture is that it's a near perfect system for allocating very risky capital. Your question was "What do you love about it?" One of the things that I love about it is that we are in the business of constructing portfolios that hopefully are diversified in a manner that we define, but we're not in the business of having to be right all the time. We're actually probably in the business of having to be right once every five years. That's a pretty interestingly-constructed environment, which I like. The thing I dislike the most is exactly that, which is that you're going to, if you're good at it, you're probably going to be wrong 70, 80, 90% of the time. The relationship between us and our companies is vastly asymmetrical, you know?

Evan Nisselson: Right. That's why I ask this question. I like talking about the dual perspective. What's the perspective of the entrepreneur from the investors and what's the investor's perspective of the entrepreneurs? What do you like and dislike?

Andy Weissman: The success of your company. I invested in you and I invested in 23 other companies in constructing a portfolio. The success of your company is kind of irrelevant to me, to my portfolio. The success of your company is actually all you have. I find one way to resolve that is just to be up front about that. Be up front that that's the relationship.

Evan Nisselson: Which is why I like bringing it up here because it is a real thing. They're all relevant but it actually is not life or death, I think, is the situation. It's a percentage game where you're building a portfolio and every-body working together to try to figure out how to make them all successful as a goal, ideally.

Andy Weissman: Yeah but...

Evan Nisselson: The numbers are numbers. They're not going to be, right?

Andy Weissman: They're not going to be.

Evan Nisselson: Right. You're absolutely right. Questions in the audience? Yes, go ahead over there. Stand up.

Audience member: Andrew, I have a question for you.

Evan Nisselson: Andy or Andrew?

Audience member: Andy Weissman. You mentioned earlier about investing in network businesses and the good thing about network businesses is that they are very defensible and you have great barriers to enter it. Fantastic. At the same time, usually it's winner takes-it-all markets and that really increases the risk. Do you think that is a high risk strategy and how do you mitigate that risk?

Andy Weissman: Yes, it is a high risk strategy and we're not in the business of mitigating. We accept that risk because that's the risk that drives returns that are meaningful to our investors to allow them to invest in a liquid security where they won't see return for 10 years. That's the dynamic that we want. I have a question in my mind, and we talk about this in USV, whether things are still winner-take-all or winner-take-most, but that's separate. That's a dynamic. Binary outcomes drive the returns that make our investors happy. And if our investors are happy, we are happy. If our investors aren't, we aren't. That's a feature, not a bug, maybe. Now, the more interesting question to me is that that's not necessarily a mode that is appropriate. That's not necessarily a mode that every entrepreneur should aspire to. There are lots of flavors of running a business. There are lots of flavors of being an entrepreneur. There are lots of flavors of financing your company and there are lots of flavors of exiting and getting liquid from the investment you make in your company. We are just one flavor, but we're a flavor that requires outsized return or it doesn't work.

Audience member (cont'd): Thank you.

Audience member: This is for mainly Andrew Cleland. You mentioned that companies are only now utilizing the bandwidth of mobile networks and with the gigabyte networks coming I saw demonstrated from Google Fiber and... They're just playing 10 videos at the same time, which is not really what people are going to be doing. Have you started to see what people are going to be actually doing with this besides higher resolution or the boring stuff?

Andrew Cleland: There certainly is higher resolution. It may be boring, but 4K is going to stress these networks. I don't think it's that well known by the general public how stressed these networks are by the rise of video. It's really pretty dramatic and the network operators are having to invest large sums of money to keep pace and to make sure that people get the experiences that they need. 4K is going to be interesting. Augmented reality is going to be interesting. VR is going to be interesting. These are all very bandwidth-intensive applications that are going to stress the network. I'd say even more than that, and just to take our US hats off for a second, think about the infrastructure internationally and the fact that plenty of people in the world can't access some of the most basic picture-oriented services. That is also a challenge.

Evan Nisselson: Any other questions in the audience? Like I said earlier, if you don't raise your hand, we don't know who you are. [Pause] Virtual reality, augmented reality. What do we think about it? Is it really here this time? Do you believe virtual reality is really going to meet the hype within the next two years—for anybody who wants to talk about it—or is it really going to take another 10 to 20 years? Anybody want to answer?

Andrew Cleland: Yeah, I'll give you my perspective.

Evan Nisselson: What do you think?

Andrew Cleland: Yeah, so I'll give you my personal point of view.

Evan Nisselson: Right.

Andrew Cleland: My group is quite split around this point.

Evan Nisselson: I think a lot of people are split on their views.

Andrew Cleland: Yeah, so our group spent time with Oculus, our group has spent time with Magic Leap—two of the more interesting companies in this space. It feels magical when you experience these services. My personal point of view is that it's going to take a long time, longer than people expect. It's going to be fundamentally world changing, but it's going to take a long time and I think investing is partly about understanding when to invest. I think these are very important technologies. I think in 30 years' time, they're going to change the way our world works in many, many ways.

Evan Nisselson: Andrew, I agree. I've got to jump in because the definition of a long time for you...

Andrew Cleland: Yeah.

Evan Nisselson: ... an entrepreneur. My sister, me, and everybody else in this room is totally different. It's the same kind of question of *what's a lot of money*? "A lot of money" is very different to every single person in this room. There's enough money to eat and there's playing money. They're different. What's a long time to you? It's going to take a long time to get to the tipping point that we just talked about, about images really being marketing and trending. When does that happen for VR, roughly? Within 10 years, 20 years? I mean like time frame? Are we 5 years, 10 years, 30? When is it going to happen, which then relates to when you'd be interested, obviously? What's in your gut?

Andrew Cleland: Really, I don't have an answer for that, but let me give you just a stab in the dark. I think the technology will arrive ahead of scaled usage because I think the content will take time.

Evan Nisselson: Agreed.

Andrew Cleland: I think creating that corpus of content is really interesting. It's going to take six, seven, eight years before we're really using these services on a daily, weekly basis.

Evan Nisselson: Six to eight years?

Andrew Cleland: Mm-hmm..

Evan Nisselson: Okay. What do you think?

Andy Weissman: I have no idea.

Evan Nisselson: Anu, any idea?

Anu Duggal: I have no idea.

Evan Nisselson: Yeah, so I'll continue the discussion because I asked the question. Basically, I'm trying to figure this out because I'm looking at a lot of these companies. I'm putting on the headsets and I think it's fascinating. However, I get dizzy sometimes and what are the use cases other than a couple of the obvious ones, which are the leading edge of all technologies, anyway? I think it's going to be infrequently used in the beginning. I think in terms of images and VR and the content, it's going to be unbelievably world-changing, but I think it's going to take much longer. I don't think a majority of people will use it for 10 to 15 years.

Andy Weissman: You know what's an interesting question is that... I think it's hard. The pace of change, it's so ridiculous. It really is hard to predict anything. One thing though that you can do is I'd look at... We have an incredibly healthy, enriched ecosystem for entrepreneurs to exist as entrepreneurs and it didn't exist 5 years ago or 10 years ago and it sure as hell didn't exist when I started in the business. Entrepreneurs are really frickin' smart. I just watch what they do. That dude is building that camera. I wouldn't have thought of that camera with like six things and so I'm like, they'll fill gaps in ways that I think we don't see because we're up at a higher level. That, to me, is maybe a little more indicative.

Evan Nisselson: Right. Sure. I think you're absolutely right. I don't like putting set times and I've been bleeding edge a little too often with my article in 2003 that camera phones will replace point and shoot cameras and other ones. I was right but it took a long time. I feel that I am correct about satellite selfies.

Please give one word answers—what is the personality trait that you love and hate in entrepreneurs?

Anu Duggal: Hustle.

Evan Nisselson: Hustle—that is what you love?

Anu Duggal: I love it because I think that you need it as an entrepreneur. You have to be resilient and to keep going with a lot of rejection, but sometimes it can be annoying.

Evan Nisselson: You love and hate the same word?

Anu Duggal: Yeah.

Evan Nisselson: We've got a trend going on here. Andy, you started this trend. Andrew?

Andrew Cleland: Transparency and that means intellectual honesty to me. I find that entrepreneurs tend to divide into two camps. One is "here is my business, here are my challenges, let's figure through this together" and kind of "fake it 'til you make it." I prefer working with the former.

Evan Nisselson: Great. Andy, you already did yours. Do you want to go again or are you sticking to it? Good man. You can start on the next one. What is the one-sentence advice to entrepreneurs building a business?

Andy Weissman: Oh God—one, just one?

Evan Nisselson: One sentence. I just like sound bytes. Does anybody want to go? Andy's thinking.

Anu Duggal: I can go.

Evan Nisselson: Good, go.

Anu Duggal: My one piece of advice would be to surround yourself, in whatever capacity you can—advisors, mentors, investors hopefully—with people who are knowledgeable about the industry that you're entering. The majority of entrepreneurs I meet don't necessarily come from the industry and I think that's a good thing. If you don't and you're willing to disrupt, then try to at least get market knowledge by the community around you.

Evan Nisselson: Great.

Andrew Cleland: Set 3 month, 6 month, 12 month goals. Make sure they're consistent. You don't have to stick to them religiously. You can alter them on the field of battle as circumstances change, but quite often you'll sit down with an entrepreneur and they'll sketch out what they're trying to do now and where they think they'll be in a year and there isn't the coherency.

Evan Nisselson: Okay, great. Great advice. Andy?

Andy Weissman: Give a lot of thought to whether when you hear "no," whether that's a positive indicator or a negative indicator. Be really honest in thinking about it because it often... In different circumstances, it's both.

Evan Nisselson: That's great. I want to riff on that in mine. In my mind, "no" never means "no." It means "not now." Until you evolve to either validation from customers, significant others, or parents, in anything in life, "no" never means "no." It's just not a "yes" yet. The "yes" could take a long time, but that's my advice. Round of applause. Thank you very much, fantastic. I'm awed with this panel and we're moving on. Thank you very much.

©VizWorld

42

COMPUTER VISION KNOWS
WHAT COLORS ARE BEST

Asmau Ahmed, CEO, Founder, Engineer, Plum Perfect

MY NAME IS ASMAU AHMED. I AM THE FOUNDER AND ENGINEER OF PLUM Perfect. We read your consumer photos to recommend items perfect for you. Imagine for a second a world where you take an image of anything, and no matter where you are, within seconds you can get personalized recommendations from retailers or brands based on the content of that image.

For the ladies in the room, imagine typing into your Google search engine: "Find that perfect foundation for my skin tone" and instantly getting it. For the gentlemen in the room: "Find a tie to match a blue checkered shirt or a blue blazer that I have in my closet" and immediately get that. If you're decorating your home: "Find that painting that would go perfectly well with my living room" and again, instantly getting it. With Plum Perfect, there's no typing necessary. You just take a photo.

We really see this as the future of shopping. We think that in the very near future, everybody is going to have this behavior of taking a photo and using that as a way to get personalized, individual recommendations to shop. We actually started with a beauty platform where users can take a selfie and we then extract and read her hair, her skin, her eye, and her lip color. Within 30 seconds, we scour the web and find the perfect lipstick for her or the perfect foundation for her. She can also buy directly from our app. The same applies in fashion. You can take a photo of your shirt and within seconds we can find that perfect tie or blazer and pair of shoes that you can click to buy from the app or go into the store to pick up. In the case of home, taking a photo of a couch. We read the photo, extract the colors, and again within seconds find the perfect rug or the perfect painting to go with it.

I shared earlier that we started with beauty. Anyone know who Michelle Phan is here? Okay, great. A couple of people do. We learned very early on that adopters of our technology were girls 13 to 25 years old. It was second nature to them to just take a selfie, find the perfect lip gloss, and click to buy. A couple of weeks after we launched our app, Michelle Phan found us, somehow, and did a short video about us.

MICHELLE "Dang. I wish I had this when I
PHAN   first started wearing makeup."

Video of Michelle Phan: "The first app I want to recommend is for those who are looking for an easy way to apply makeup without having to think about it. If you're looking for the right colors for your complexion but you don't know where to start, check out this app called Plum Perfect. All it really takes is a simple selfie, and I'll show you how it works. After taking your selfie, Plum Perfect analyzes your lips, hair, eyes, and skin color. Dang. I wish I had this when I first started wearing makeup."

Asmau Ahmed, CEO,
Founder, Engineer,
PlumPerfect

THE
S

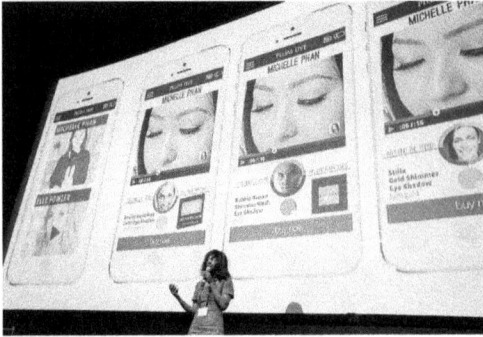

I woke up the next morning after this video was done and our servers were completely down, completely overloaded. I couldn't figure out what had happened. Prior to this, just to give you guys some context, we had been on ABC News, we had been in every single magazine you can possibly think of. She drove 10 times as much traffic as everybody combined. That was great for us.

Fun aside, there's real science and technology behind this. I have a chemical engineering background where I spent a lot of time in the color space in the manufacturing industry that has translated to what we're doing now at Plum Perfect. Seven years of color research. We have three patents pending. We've analyzed 16.7 million colors to date. Finally, one of our goals, which we've attained, was to get people from taking the photo to getting their perfect color recommendations within 30 seconds. We really feel like this is just the beginning.

Someone mentioned on the previous panel the use of 30-second YouTube videos and live tutorials to drive commerce. Absolutely right. One of the new features that we're launching is live and trending tutorials. We don't have to create any new content; instead, we leverage the massive amount of YouTube and Instagram videos that exist. Imagine three women who look completely different. They have different color profiles and are watching the exact same video. The experience is very unique to them. They would get recommendations that are unique to their own color from the video. Imagine this in fashion as well or in home decor.

In-app advertising—today, retailers really have a blind spot. They don't know what you look like and are unable to match their products to you. What Plum Perfect can do, and does, is match each product to consumers for which they are a match and instantly message them.

In this example, Brooks Brothers can send you a personal notification that says, "I have a tie that would match that shirt that you have in your closet."

Some data from our technology: over 15% as of this morning click-to-buy from our application. That's compared to an average of 2% to 4%. We ran a program with a national retailer, the largest retailer in the US. You know who they are. Three out of four people that got a personalized recommendation from taking a photo actually bought the product that they were recommended. All great statistics in the right direction.

Just a few things about the technology and how it works, again using beauty as an example. We take a selfie, we analyze the ambient lighting, and do a contrast enhancement. Then we

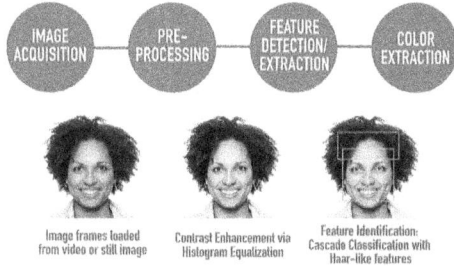

| IMAGE ACQUISITION | PRE-PROCESSING | FEATURE DETECTION/ EXTRACTION | COLOR EXTRACTION |

Image frames loaded from video or still image

Contrast Enhancement via Histogram Equalization

Feature Identification: Cascade Classification with Haar-like features

do a feature detection. Once we extract the color... The key thing here to know when we extract the colors from the different features is that we categorize the colors based on how the human brain perceives color. The human brain perceives color based on three dimensions: temperature, value, and chroma. In this case, we were able to tell that she has medium-toned skin, cool and soft undertones, etc. How we make the recommendations is really a combination of two things: theoretical rules and data. Rules—if she's going for a glam look, we want high-contrast level based on these dimensions. Data—when I use our app, it automatically builds a cluster of women that have similar colors as I do and looks at their behavior and considers that in making recommendations for me.

It's pretty interesting and this is our team. Thank you.

©Anne Gibbons

43

THE WAY YOU MOVE IS THE WAY YOU EXPERIENCE THE WORLD

Kegan Schouwenburg, CEO & Co-Founder, SOLS

FOR THOSE OF YOU WHO DON'T KNOW ME OR MY COMPANY, I'M THE founder and CEO of SOLS. We're a company focused on building innovative technology that introduces the future of movement and the notion of mass customization in footwear. Our first product is a 3D printed orthotic, which we're shipping now through our network of over 650 doctors and to thousands of their patients. Over the summer we launched a three month pop-up shop on the Bowery. This was an incredibly exciting time as it allowed us to have a one-on-one interaction with our consumer. This initiative created an urgency for us to develop scalable platforms for mass customization that make the world of a truly customized future with 3D printing possible.

Kegan Schouwenburg, CEO & Co-Founder, SOLS

Before we kick into the more technical aspects, I thought I would show you a video giving you an overarching view of our platform and our technology. When I started SOLS, the key differentiating factor between our company and others in the 3D printing and mass customized space is looking at the technology from a systematic perspective. It isn't about *How do we take a Kinect sensor?* or *How do we take off-the-shelf scanning technology?* or *How do I get somebody to go to Best Buy and put something on an iPad?* It is about *How do I take technology that is in everybody's phone, how do I take an iPhone, how do I take an Android, how do I use that and enable the customer with that technology to customize products that they need to reflect the uniqueness of themselves?*

When I first started with that idea and I went out to raise our first round of financing, the investors, they looked at me and they thought I was crazy. They said "Hey, there's a million different off-the-shelf opportunities out there. You're going into the medical space. You could easily just ship all these doctors a scanner. You could let them use their existing scanning technology." But for me, for this world to make sense, for mass customization to be possible, it had to happen on that existing device. That's what we've done here at SOLS.

Thanks everyone. Going back to what I was saying earlier, what we, at SOLS, do is something that I refer to as "product as process." We develop systems and those systems enable scalable mass customization. What that looks like from a technical perspective is that we have an iPhone application, we enable our doctors with our app, and soon we'll enable consumers with the app as well, by simply taking three photos of the foot. They take a photo of the inside, the arch view; of the back of the ankle; and of the plantar surface of the foot. We then use those photos... We extract a series of data points, we plug them into a script, we use that script to generate an STL file, and that STL file goes down to production at our factory in Texas. It's then shipped back up to us and then ultimately it becomes the bespoke product that the consumer receives at home.

Obviously when you look at these things, you're like "Okay, we're building an automated, scalable system for mass customization." Obviously there are a series of challenges behind that that I don't think I could've imagined that we would've run into when starting the company. It's very easy to say "Okay, we're going to take iPhones and we're going to use iPhones to make customized orthotics." It's a whole other thing to look at some of the real-world challenges that happen and that occur when you are actually trying to take technology, you're trying to make it invisible, and you're ultimately using it to create a product that, to the consumer, is just meant to make their lives better. Most of our consumers don't even know that SOLS are 3D printed. To them, it's an orthotic that they got from their podiatrist that is hopefully much better than anything else that they would have gotten otherwise.

It all starts with something that we call Sally. Sally is basically leveraging semantic segmentation to analyze these photos. We're breaking the images apart. We're saying "Okay, how can we identify the foot in the series of photos?" We have a team of trainers on site that are effectively processing images and allowing us to get smarter and smarter and smarter with each pair of SOLS that we ship. The cool thing about this is that our algorithms are constantly improving. It means as we ship more and more product into the world, our product becomes better, it becomes more refined and accurate, there's more data behind it, and ultimately SOLS—the product we ship to consumers—improve.

Obviously there are some real-world problems that we've run into. We've encountered blurry photos, we've encountered reflectivity, we've see that humans—patients and doctors—struggle when taking photos. And, even with the help in the iPad app that we give them, where we're using the accelerometer to help them hold the iPad level, to help them take photos in the appropriate lighting conditions, soon we'll actually be doing on-the-fly detection of photo quality to help them improve it further. There are so many challenges. Even something as slight as somebody tipping their foot in slightly the wrong direction or having their ankle collapse in the wrong way... We're looking at fleshy-bodied objects in photos and we're using that to drive a physical, hard-bodied form. The multitude of challenges that come with that are extremely complex.

SALLY FINDS FEET IN
PHOTOS THROUGH
SEMANTIC
SEGMENTATION
ACHIEVING 93%
HUMAN MACHINE
OVERLAP

@GET_SOLS

Because of that, what happens is we have the next stage of our process. We start with these photos, we extract the data points, we say "Okay, this is what we think the foot looks like" and then we say "Okay, based upon known factors, based upon this probabilistic model that we've made"—we take a statistical approach to that and we say—"How do we then edit these data points? How do we take what we know quantitatively about the individual and apply that and combine it with what we know qualitatively about them?" Qualitative fit is something that's really, really interesting. I don't think that I ever could've imagined that it would've been so challenging and it would've had such an impact.

We think about customization. We want to apply this technical approach to it and we want to say "Okay, well I have my body. I scanned my body. I can make something that's perfect for me." Yet what we find is that people like different things. Some people like their SOLS to fit looser, some people like them to be tighter, some people wear shoes that are two sizes too small, other people wear them that are two sizes too big. Other ones wear them with really, really pointy toe boxes or want to be able to wiggle their toes in a way that maybe a runner or an athlete has an entirely different perspective on. We've worked hard to apply both the quantitative and the qualitative aspects to this step in the process.

It's as much about knowing what somebody's biomechanics is like as it is about knowing how much they weigh, what they're going to wear the SOLS for, what kind of shoes that they're going to wear them in, what kind of footwear they've previously worn, and taking all of that and combining it into the algorithm to get the fit right. Does that mean we get it right every time? No.

Kegan Schouwenburg,
CEO & Co-Founder,
SOLS

Right now, I think we're averaging... Let's call it 92% first-time-right for custom fit which, in my understanding, is pretty great. I would love, obviously, to be at a higher point than that. The fact that we're shipping 90% of SOLS that are great based upon a series of six photos that somebody's taken out the gate I think is pretty exciting.

Finally, it's about looking at all of what we analyze daily and saying it's not just about the digital input and taking that and putting it into our model, but it's about actually closing that loop and seeing how can we take a real-time approach to product development? How can we then take the data that we get back from the customers, analyze that data, and say: "Okay, so we shipped a pair of SOLS. Is it good?" Yes, check. "Okay, we know this worked for this demographic, for this age group, for this weight class." Is it not right? "Okay, how do we need to change it to make it right?" Are all feet the same? Do we need to adjust our algorithm to account for more and more variability? As we grow as a company and as we ship more SOLS and we encounter more feet types and, frankly, our database of feet gets bigger, we begin to make it more and more complex in this way and we begin to constantly feed that data and feed that input back into our model so that it can learn, so that it can grow.

A great example of this is... I think last week or the week before, we pushed out a fix that was accounting for 80% of the remakes that we were encountering, which dealt with how we defined the toe box shape. It was as simple as looking at the model and saying "Okay, mapping back this section of complaints that we were getting to this portion of the model, pushing that out, and then seeing the immediate impact." The great thing about 3D printing, the great thing about digital manufacturing, is the fact that you truly are developing products in real time. This isn't about looking at what the market wants let's say a year from now, predicting what the market wants and then making a product that's going to serve that. It's about saying *What does the market want tomorrow? What are we shipping today? What are our doctors saying tomorrow?* And then feeding that back so our product is constantly getting better, so the moat between us and our competitors is constantly getting wider. Ultimately, we want to continue to position ourselves as the market leader in mass customization and 3D printing.

What you see here is effectively what that model looks like. We're training the algorithms. We're using machine learning to apply that data that we're receiving back from our customers to the model so that it improves constantly. We're using that to generate 3D files. The way that we've designed that is pretty cool in that we're enabling our biomechanical engineers to work in their language, which is math, and we're combining that with our software engineers so that we're actually plugging those equations into code and enabling the biomechanics people to design and code without actually having to know how to write the scripts. It enables us to work much, much faster than we have been able to in the past and really put out a real-time product creating an entirely new language and an entirely new framework around it.

Finally, obviously, pushing those products out into the world, capturing that human feedback, trying to close that loop between the moment where that product ends up on the user's foot where we're capturing that data, where we're capturing that feedback from them, and doing an analysis on that. Right now, a lot of that's happening manually, but we do hope that that will happen through machine learning in the future. From there pushing that back, starting again, going back to training, and we repeat this cycle almost on a daily basis.

Finally, what does that lead to? What are we doing? We're building products that adapt to us. We're building a foundation, we're building software, we're building a platform that enables products that adapt to us. Today we make SOLS, but I could plug any kind of product I wanted to in our platform. It could be braces, it could be casts, it could be helmets, it could be shoulder pads. If I wanted to change the company from an orthotics company to a footwear company to a helmet company to a bicycle seat company tomorrow, I could do that.

I think for that, that is what enables the future that we want to be possible. That's what enables mass customization to be possible. When you combine that with advanced data and you combine that with this idea of iterative real time product development, you effectively have an entirely new kind of mass customization. You introduce a new way to launch a product into the world, and you move from a world where we have marketing-driven companies to a world where we have data-driven product companies. I think we can all imagine what that looks like, where we are all truly enabled to be ourselves, to be individuals, and to have things which are uniquely designed for us. I think, I hope, that world starts with SOLS and I'm excited to be pioneering some of the technology behind that.

Thank you.

44

DECOUPLING MEDICAL CARE FROM LOCATION AND TIME

Meghan Conroy, CEO & Founder, CaptureProof

WE HAVE BEEN FORTUNATE TO WITNESS A REVOLUTION—A CONNECTION and communication revolution. In 2005, Vatican City looked like *this* and five short years later, like *this*.

2005   2010

Each of those white lights signifies someone capturing the event in a photo. We are living in the "selfie" generation. Every two minutes, we take more pictures than all of humanity did in the 1800s. Facebook gets 350 million photos a day and more than 100 hours of video uploaded to YouTube each minute. CaptureProof is bringing this revolution to healthcare, improving patient outcomes across the globe.

Photos and videos give us a way to communicate more clearly, like never before. While social media encourages public posting and sharing, healthcare has strong safeguards for the security and privacy of patient data. These regulations have heavy fines—up to $50,000 for every piece of data lost. And more than half of the fines by HHS are due to unencrypted data lost on hardware. And there is no hardware easier to lose than the mobile phone. A provider losing her phone with eight photos on it could be slapped with a $400,000 fine. CaptureProof removes this risk.

What's at RISK?

- $50,000 per piece of PHI
- 1 year in jail
- Loss of license

Medical images start as a small data set, like above, but very quickly turn into the data set below. Providers have hundreds of patients who will create thousands of photos.

CaptureProof is the HIPAA-compliant solution for secure medical photo and video organization and sharing. CaptureProof allows clear communication for the entire care team. A resident can ask their attending's advice from the patient's bedside or a mother can show the pediatric neurologist the exact movement her baby is doing.

This communication leads to more accurate triage and more accurate monitoring, leading to the same health care outcomes—or even better—at a fraction of the cost.

CaptureProof is telemedicine for providers who don't run on time. Even with the increasingly-used video visit... This requires both patient and provider to be on time and have enough bandwidth for the connection. Not with CaptureProof. Photos, videos, and chat are shared at the provider and patient's convenience, from the clinic or the comfort of their home. CaptureProof allows you to do what you learned as a five-year-old and have it be very powerful: show and tell symptoms versus telling alone. What do we do? CaptureProof is the prescription pad for photos and videos. Providers have the ability to request a very specific set of information from their patients.

When the patient receives this media Rx, it is overlaid at the top of the screen in the active camera so the provider's directions are clear and easily viewable. Once you've taken the first photo, that initial image is overlaid in the active screen so that you can use it as a guide to have subsequent, consistent images captured. Once a photo or video is captured, it is immediately shared with the provider, who can view the media right away or when their busy schedule allows.

CaptureProof is more than just secure medical communication. It enables providers to monitor the patient over time. Using CaptureProof's "Compare" feature, in three easy clicks providers can view uniform images side by side and see if the patient is getting better or not. CaptureProof takes it a step further with video as well, allowing for two videos to be played back side by side. CaptureProof worked with Kaiser Permanente in order to help them with their total knee replacement follow-up care. The use of CaptureProof for incision site management and range of motion recovery reduced in-person visits by 75%, saving $7,500 per patient.

At Cornell, Dr. Claire Henchcliffe is using CaptureProof to automate the Parkinson's Scale from home. Patients have instructional videos shared with them by the provider and they record themselves doing these exercises at home. The providers can use these videos to monitor the patients and make medication modifications remotely, reducing the need for patients with mobility issues to come into an appointment.

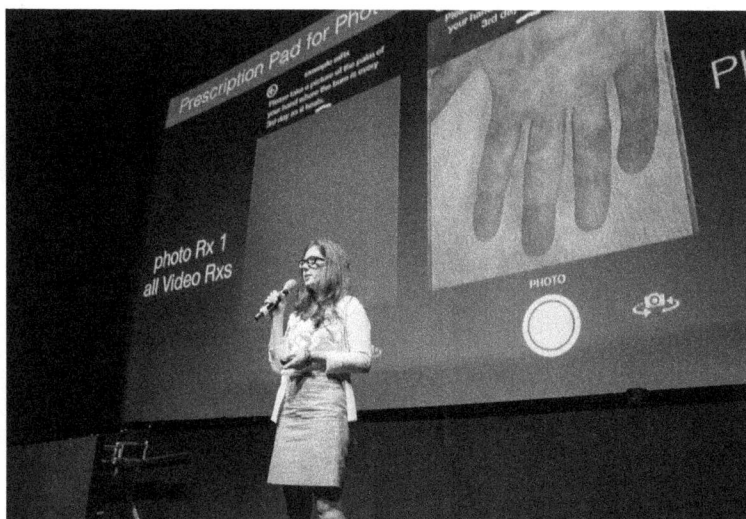

Meghan Conroy, CEO & Founder, CaptureProof

CaptureProof is focused on optimizing healthcare and our study results have demonstrated that we are achieving exactly that. CaptureProof is reducing in-person visits at Kaiser by 80% but even more interestingly, saving providers 66% of their time per patient. CaptureProof is not in the business... People are asking: "Well, don't doctors want to see the patient?" Many surgeries are covered by bundled payment which means one payment is made for the next 90 days of care. Whether the provider sees the patient once or 20 times in that period, the payment is the same. Therefore, with CaptureProof, the provider can more closely monitor a patient and bring them in when it is medically necessary to do so, saving the patient the time and hassle of coming in to the appointment and opening up the provider's schedule for other appointments.

CaptureProof is the whole-hospital solution. We are currently being used in 32 sub-specialties across the hospital, saving time and money and optimizing the way care teams communicate and practice.

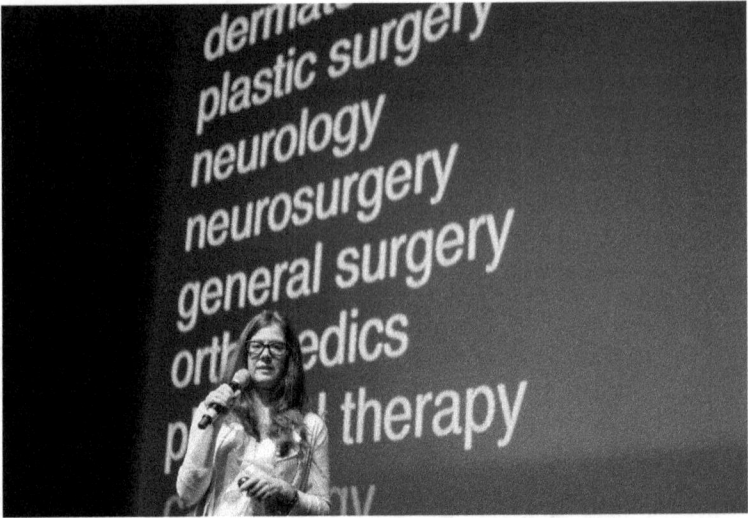

Meghan Conroy, CEO & Founder, CaptureProof

For neurology and seizures, CaptureProof has reduced urgent patients by 78% and cut post-op ER visits in half. If you are a patient at any of the above institutions, ask about using CaptureProof.

Currently, 80% of doctors across the globe have smartphones in their pocket and 40% of them actually believe that mobile technologies can reduce the number of office visits. It's only a matter of time until smartphones become a tool in the provider's belt, and CaptureProof will be the photo and video solution optimizing their practice.

The most interesting statistic up there is actually 88% of doctors want to monitor their patients at home. The time is definitely now, and CaptureProof is definitely the product that's allowing doctors to have a virtual visit at any moment from anywhere.

45

DESPITE MODERN IMAGING TECHNIQUES, THE NUMBER OF MISSED MAJOR DIAGNOSES HAS NOT ESSENTIALLY CHANGED

Elad Benjamin, CEO, Zebra Medical Vision

MY NAME IS ELAD. I'M WITH ZEBRA MEDICAL VISION, AND I'M GOING TO talk to you in the next few minutes about what we're doing at Zebra to revolutionize the world of medical imaging and radiology. Most of us do not encounter that world on a day-to-day basis, so it's not always clear as to how much radiology impacts us on a day-to-day basis. So I wanted to give you a sense of what radiology does and what its impact is. This is a bit hardcore for the end of the day, but it hopefully will give you some inspiration.

Lung cancer is the most common cancer in the world. About two million people are diagnosed every year. Hundreds of thousands die from it. If it's detected early before it spreads out of the lung, you have a 54% survival rate. If it's detected a little bit later, you have a 4% survival rate.

Huge, huge, huge implications for many hundreds of thousands of people every year. Given that it's so important to detect lung cancer early, how many lung cancers do you think are actually detected early enough to increase the survival rate? The answer, unfortunately, is only 15%. Imagine how many lives that impacts every year and how many families of those people it impacts every year.

Osteoporosis is a disease. There's a fracture somewhere in the world every three seconds. In the US alone, there's about 50 million people that are affected by osteoporosis. It costs the healthcare system about $17 billion to deal with this disease that one out of every three women and one out of every five men over 55 has. Osteoporosis is an easily treatable disease through a combination of medication, diet, lifestyle changes, awareness, and, if treated early, can significantly improve the lives of millions of people. In the US alone, it can decrease about one million fractures every year... And of those one million, about a quarter of those elderly end up in nursing homes.

Tuberculosis kills a million and a half people every year somewhere around the world. Tuberculosis is easily treatable and easily identifiable on a chest x-ray, which hints at the power of radiology.

These three diseases, along with most of the major diseases that kill or affect hundreds of millions of people globally, are all easily detectable at a relatively early stage by medical imaging, by radiology. And yet, this doesn't happen on a regular basis and hundreds of thousands of people every year are not diagnosed when they should be.

Why does this happen? What's holding us back? In general, there's a lot of macroeconomics here, but the basic thing that's happening is that demand is far outpacing supply. Over the last couple of decades, hundreds of millions of people are joining the middle class, imaging technologies are improving, and a radiologist today sits down in the morning in front of two high resolution monitors to look at medical imaging studies coming in—CTs, MRIs, ultrasounds, x-rays—and through the course of a shift one day, they will look at about 5,000 images over a few hours, 8 to 10 to 12 hours, depending on the length of the shift. They have to find a pathology that's maybe a few pixels by a few pixels. They're only human. The amount of mistakes has been steadily growing over the last couple of decades. I don't mean to scare you too much, but one out of three people who have walked out of the hospital with a diagnosis from a radiologist has walked out with the wrong diagnosis, either a false positive or a false negative, and that's kind of scary when you think about it.

What is Zebra doing? We believe that there is a key. We're already holding a key to solving a lot of the problems. Around the world there are hundreds and hundreds of petabytes of medical imaging studies stored in archives of hospitals and healthcare institutions, and these are just sitting there. Once a study is performed, it's stored for legal purposes for 7 to 15 years, depending on the type of study, and that's it. It's never referred to again. There are very few screening programs around the world that try and catch things early.

So what Zebra has set out to do was to take this vast amount of hundreds of millions of imaging studies from around the world and use them for early detection. We've set out to teach computers to read medical images and to do that using the largest data set that has ever been curated and released into private hands.

Essentially, we have spent the last few years curating millions and millions of medical imaging studies from a variety of hospitals, patients, and diseases. What we are now doing is using machine learning tools to identify diseases and to build high end algorithms, next generation algorithms, which can provide early detection of some of the major disease that afflict us globally. Think of it ultimately as a virtual black box. A medical imaging study comes in and runs against a host of algorithms. One algorithm knows how to identify breast cancer in mammography, another knows how to do lung cancer, another early signs of Alzheimer's, etc. So an imaging study comes in and a diagnosis goes out the other end.

That black box, as I said, is going to be built from hundreds of different algorithms. Each knows how to identify a particular pathology or a particular disease. We can't do it alone, so what we've done actually is we released our research platform, and that's a platform that allows computer vision researchers globally to access Zebra, to build; create algorithms; and test, run, and validate them against the medical imaging data that we have on our platform. We provide the commercialization services and tools back into hospitals and healthcare organizations and then share subsequent revenue with our research partners. On the one hand, we provide data and a platform and a research environment to create innovation, and on the other hand, we help take that innovation to market and implement it in a clinical setting, improving patient care.

Where are we? The company's about a year old. We are on the verge of releasing our very first algorithm. It's an osteoporosis algorithm. I hinted first at the prevalence of this disease and how much it hurts people and economies and healthcare institutions. Our algorithm is a fully-automated algorithm, meaning we come into a healthcare institution, we run through their entire existing database of CT studies, and we are able to extract from their existing studies all the people that are at high risk of an osteoporotic fracture. That allows the healthcare institution to then understand the risk profile of their patient population. From that risk profile, they can insert those people into early fracture prevention programs, which will hopefully avoid some of the fractures.

Elad Benjamin, CEO, Zebra Medical Vision

Fracture prevention programs exist. For example, in the US, these have proven to reduce 40% to 50% of fractures. So this algorithm alone, if we fast forward a few years and it's implemented across the healthcare system across the US, could save up to about $10 billion in healthcare costs, save a million fractures a year, and help about 250,000 people a year avoid getting into a nursing home because of a fractured hip and disability.

That's just the first. We have other algorithms in the works. I've hinted at some. Detecting early signs of breast cancer, lung cancer, chronic emphysema, and other diseases which severely debilitate and hurt hundreds of millions of people globally. So what we're doing now is both developing those algorithms and beginning to work with a global community of researchers to join us in this effort to create the next generation of imaging algorithms to help humanity.

Thank you very much.

LDV VISI

Partners

LDV CAPITAL

Dropb

 CAKEWORKS

 olapi

CORNELL TECH

sprout by

INTERNATIONAL CENTER OF PHOTOGRAPHY

MFA PHOTO VIDEO
SVA NYC

ORRICK

N SUMMIT

#LDVvision

Media Partners

JWPLAYER

Vidcaster

QUALCOMM®

aws activate

Kaptur.co
The first magazine about the photo:tech world

VizWorld
News and community for the visual thinker

www.ingramcontent.com/pod-product-compliance
Lightning Source LLC
Chambersburg PA
CBHW071532200326

41519CB00021BB/6453